Solutions Manual
for
Introduction to Genetic Analysis
Tenth Edition

DAVID SCOTT, ELAINE A. SIA, MIRJANA BROCKETT, WILLIAM D. FIXSEN, AND DIANE K. LAVITT

W. H. FREEMAN AND COMPANY
NEW YORK

ISBN: 1-4292-3255-2
ISBN-13: 978-1-4292-3255-5

Printed in the United States of America

Fourth printing

Contents

1

The Genetics Revolution in the Life Sciences

PROBLEMS

In each chapter, a set of problems tests the reader's comprehension of the concepts in the chapter and their relation to concepts in previous chapters. Each problem set begins with some problems based on the figures in the chapter, which embody important concepts. These are followed by problems of a more general nature.

WORKING WITH THE FIGURES

1. In considering Figure 1-2, if you were to extend the diagram, what would the next two stages of "magnification" beyond DNA be?

   Answer: The next stage of the diagram would be an in-depth look into the DNA molecule, as two long molecular strands of nucleotides wound around each other in a double helix and the basic structure of those monomers. Focus on specific nucleotides and their organic molecule parts: a deoxyribose sugar, a phosphate group, and a nitrogenous base (adenine, thymine, guanine, and cytosine). Second, would be atomic composition of those organic molecules (C, H, O, P, and N).

2. In considering Figure 1-3,

   a. what do the small blue spheres represent?
   b. what do the brown slabs represent?
   c. do you agree with the analogy that DNA is structured like a ladder?

   Answer:
   a. Blue ribbon represents sugar phosphate backbone (deoxyribose and a phosphate group), while the blue spheres signify atoms.
   b. Brown slabs show complementary bases (A, T, G, and C)
   c. Yes, it is a helical structure.

3. In Figure 1-4, can you tell if the number of hydrogen bonds between adenine and thymine is the same as that between cytosine and guanine? Do you think that a DNA molecule with a high content of A + T would be more stable than one with high content of G + C?

Answer: There are two hydrogen bonds between adenine and thymine; three between guanine and cytosine. No, the molecule with a high content of G-C would be more stable.

4. From Figure 1-6, can you predict how many chromosomes there would be in a muntjac sperm? How many purple chromosomes would there be in a sperm cell?

Answer: There would be only three chromosomes in a sperm cell of this species. Since each homologous chromosome pair is stained with a different color, there would be only one purple chromosome in a sperm cell.

5. In examining Figure 1-7, state one major difference between the chromosomal "landscapes" of yeast and *Drosophila*.

Answer: Yeast chromosome landscape shows fewer introns and less space between the coding genes.

6. In Figure 1-8, is it true that the direction of transcription is from right to left as written for all the genes shown in these chromosomal segments?

Answer: No, there is one gene that would be read from left to right, since RNA polymerase can assemble polynucleotides only in the 5′ -3′ direction.

7. In Figure 1-9, estimate what length of DNA is shown in the right-hand part of the figure.

Answer: The right-hand part of this figure shows a section of a 30 nm fiber, which is composed of nucleosomes (10 nm fibers) DNA alone is a 2 nm fiber. If stretched out, a DNA molecule of each chromosome would be about 4 cm long, thousands of times the diameter of a cell nucleus.

8. From Figure 1-12, what is the main difference in the locales of transcription and translation?

Answer: In a eukaryotic cell the nucleus provides a separate location for transcription, while translation continues in the cell cytoplasm.

9. In Figure 1-14, what do the colors blue and gold represent?

Answer: Blue represents original DNA (chromatid) in a cell before replication, while gold represents new DNA (sister chromatids) after the semi-conservative replication of the chromosome.

10. From Figure 1-17, locate the chromosomal positions of three genes involved in tumor production in the human body.

Answer: There are many genes involved in tumor production in humans, such as a gene for: neurofibromatosis on chromosome 2, familial colon cancer on chromosome 2, for malignant melanoma on chromosome 9, and retinoblastoma on chromosome 13.

11. In Figure 1-18, calculate the approximate number of nucleotide differences between humans and dogs in the cytochrome *c* gene. Repeat for humans and moths. Considering that the gene is several hundred nucleotides long, do these numbers seem large or small to you? Explain.

Answer: Cytochrome *c* appears somewhat different when compared between humans and dogs, since they diverged with approximately 14 nucleotide substitutions since the common ancestor. Humans and moths differ even more, in about 32 nucleotide substitutions, yet the difference is not as large as expected based on the broad biological differences between insects and mammals. These could tell us that the cytochrome *c* gene has been highly conserved due to its significance in metabolism of aerobic organisms.

12. In Figure 1-21, why are colored ladders of bands shown in all three electrophoretic gels? If the molecular labels used in all cases were radioactive, do you think the black bands in the bottom part of the figure would all be radioactive?

Answer: In this figure we see three different types of electrophoresis in different colors (Southern blot of DNA fragments, Northern blot of RNA and Western blot of a protein product). Such a mixture of macromolecules could be hybridized with a radioactive probe, and the bands in the lower part of the figure would indicate radioactivity.

BASIC QUESTIONS

13. In this chapter, the statement is made that most of the major questions of biology

have been answered through genetics. What are the main questions of biology, and do you agree with the above statement? (State your reasons.)

Answer: Biological sciences inquire about life and its properties. Many themes connect concepts and study life's properties at the different levels of biological hierarchy. One main theme is the continuity of life, which is based on heredity. Genetics studies this theme in a great detail. Another theme is of course, evolution of life, where again genetics plays a major role in understanding life history and unity, as well as diversity of life.

14. It has been said that the DNA → RNA → protein discovery was the "Rosetta stone" of biology. Do you agree?

Answer: Yes, this is the main aspect of the information processing in a cell: from DNA to RNA and protein; from genotype to phenotype. Although this has been a "central dogma of molecular biology" for decades, we know today that it has its exceptions, such as the reverse transcription (RNA viruses) or the small RNA and their role in gene regulation.
Understanding of this essential concept of life gives us an insight into another, such as evolution, and the nature of mutations as a basic source of variability upon which evolutionary processes might act.

15. Who do you think had the greatest impact on biology, Charles Darwin or the research partners James Watson and Francis Crick?

Answer: Charles Darwin made an enormous impact on Biological sciences and society, and his works are studied in many different areas, continuing to make an impact today. A hundred years later, James Watson and Francis Crick discovered the double helix, a molecule of life's heritable information. This was perhaps the most significant milestone in genetics and beyond. If Darwin had any information about genes, even more about the properties of life's blueprints, his theory would have an important mechanism. It was the scientists of the first half of the twentieth century who made a connection between the works of Darwin and Mendel in the "great synthesis" and those in the second half of the twentieth century who made a connection of all these milestones in the field of molecular evolution. Today, in the twenty-first century both themes grew into studies of genomes and phylogenies and at the even higher level into integrative and systems biology. It is hard to say whose contribution is greater, but we must see their presence in the entire realm of the biological sciences.

16. How has genetics affected (a) agriculture, (b) medicine, (c) evolution, and (d) modern biological research?

Answer:

a. Genetics has affected agriculture for thousands of years, yet since the early twentieth century this impact has been essential. Knowledge of the genetic basis of traits and the experimental crossing allowed the growth in all of the fields of agriculture. Besides artificial selection and breeding strategies, recombinant DNA technology lead to genetic engineering and amazing results in this filed.

b. One of the fastest growing areas of genetics is the area involved with human health and medicine. Genetics plays an essential role in studies of many diseases, such as numerous hereditary diseases, cancer, diabetes, etc. Many genetic disciplines are constantly involved in such studies and practices to understand and diagnose human diseases. In addition, genetics plays an essential role in reproductive biology. Finally, genetics might be used to cure diseases, whether through gene therapy, stem cell treatments, or pharmacogenomics.

c. Evolution could be defined as a change in genetic makeup of a population over time or, at a more broad level in Darwin's words, as a descent with modification. In the light of modern genetics, we could see how changes in genomes support the concept that all of the living species descend from ancestral species. Evolution is supported by an extensive amount of evidence, above all genetic evidence, which continues to enrich our understanding of life's unity and diversity. For example, phylogenies constructed on genetic studies of species show evolutionary relationships, enabling scientists to construct the tree of life. Besides such studies based on genetics, studies in population, quantitative and developmental genetics, molecular genetics, and bioinformatics bring new insights on evolution as a unifying theory of all biology.

d. Genetics is the essential discipline in modern biological research and it is present in almost every field of study. DNA cloning and polymerase chain reaction techniques changed the way modern biology operates. At the same time, DNA technology allows us to find genes of interest and study their function. Reproductive cloning of mammals, genetic engineering, forensics, stem cell research, diagnosis of human diseases, and gene therapy are new areas in modern biology founded on genetics.

17. Assume for the sake of this question that the human body contains a trillion cells (a low estimate). We know that a human genome contains about 1 meter of DNA. If all the DNA in your body were laid end to end, do you think it could stretch to the Moon and back? Justify your answer with a calculation. (**Note:** The average distance to the Moon is 385,000 kilometers.)

Answer: Yes, if we could take DNA molecules from all of the nuclei in an individual human and lay them straight, one after another, a total length of such nucleic acid polymer would be equal: number of cells in a human body (trillion or 1,000,000,000,000) × length of each cell's DNA (1 m or 0.001 km) = 1,000,000,000 km, enough to reach the Moon and return. The key to such enormous lengths is the chromosome packaging.

2

Single Gene Inheritance

WORKING WITH THE FIGURES

(The first 14 questions require inspection of text figures.)

1. In the left-hand part of Figure 2-4, the red arrows show selfing as pollination within single flowers of one F_1 plant. Would the same F_2 results be produced by cross-pollinating two different F_1 plants?

    Answer: No, the results would be different. While self pollination produces 3 : 1 ratio of yellow versus gene phenotype, cross pollination would result in 1 : 1 ratio, in the F_2. This is because F_1 yellow are heterozygous, while green are homozygous genotypes.

2. In the right-hand part of Figure 2-4, in the plant showing an 11 : 11 ratio, do you think it would be possible to find a pod with all yellow peas? All green? Explain.

    Answer: Yes, it is possible to find a pod with only yellow peas or heterozygous for the seed color gene, if all the flowers had dominant allele in a given fruit/pod. This could be also one example of rare changes at a physiological level.

3. In Table 2-1, state the recessive phenotype in each of the seven cases.

    Answer: wrinkled seeds; green seeds; white petals; pinched pods; yellow pods; terminal flowers; short stems

4. Considering Figure 2-8, is the sequence "pairing → replication → segregation → segregation" a good shorthand description of meiosis?

    Answer: No, it should say either: "pairing, recombination, segregation, segregation" or: "replication, pairing, segregation, segregation."

5. Point to all cases of bivalents, dyads, and tetrads in Figure 2-11.

Answer: Replicate sister chromosomes or dyads are at any chromatid after the replication (S phase). A pair of synapsed dyads is called a bivalent and it would represent two dyads together (sister chromatids on the right), while the four chromatids that make up a bivalent are called a tetrad and they would be the entire square (with same or different alleles on the bivalents).

6. In Figure 2-12, assume (as in corn plants) that A encodes an allele that produces starch in pollen and allele a does not. Iodine solution stains starch black. How would you demonstrate Mendel's first law directly with such a system?

Answer: One would use this iodine dye to color the starch producing corn pollen. Since pollen is a plant gametophyte generation (haploid) it will be produced by meiosis. Mendel's first law predicts segregation of alleles into gametes, therefore we would expect 1 : 1 ratio of starch producing (A) versus non-starch producing (a) pollen grains, from a heterozygous (A/a) parent/male flower. It would be easy to color the pollen and count the observed ratio.

7. In the text figure on page 43, assume the left-hand individual is selfed. What pattern of radioactive bands would you see in a Southern analysis of the progeny?

Answer: If an individual is selfed, the restriction fragments should be identical to the parents fragments. In this case, a heterozygous parent to the left had three bands (two from a mutant allele "a" and one from dominant allele "A").

8. Considering Figure 2-15, if you had a homozygous double mutant $m3/m3$ $m5/m5$, would you expect it to be mutant in phenotype? (**Note:** This line would have two mutant sites in the same coding sequence.)

Answer: Yes, this double mutant $m3/m3$ and $m5/m5$ would be a null mutation, because $m3$ mutation changes the exon sequence.

9. In which of the stages of the *Drosophila* life cycle (represented in the box on page 52) does meiosis take place?

Answer: Meiosis happens in adult ovaries and testes, therefore before fertilization. After fertilization, fruit flies would lay their eggs (with now diploid embryos). That would be Stage 1 on the figure.

10. If you assume Figure 2-17 also applies to mice and you irradiate male sperm with X rays (known to inactivate genes), what phenotype would you look for in progeny in order to find cases of individuals with an inactivated *SRY* gene?

Answer: If we inactivate the *SRY* gene in mammals with radiation, the offspring should all be phenotypically females, yet on the chromosome level there would be both XX and XY (in this case sterile, female looking males).

11. In Figure 2-19, how does the 3 : 1 ratio in the bottom-left-hand grid differ from the 3 : 1 ratios obtained by Mendel?

Answer: It differs because in Mendel's experiments, we learned about autosomal genes, while in this case we have a sex linked gene for eye color. 3 : 1 ratio means that all females have red eyes ($X^{+/-}$), while half the males have red (X^+/Y) and half white (X^W/Y).
Careful sex determination when counting F_2 offspring would point out to a sex linked trait.

12. In Figure 2-21, assume that the pedigree is for mice, in which any chosen cross can be made. If you bred IV-1 with IV-3, what is the probability that the first baby will show the recessive phenotype?

Answer: The answer would be:
$2/3 \times 2/3 \times 1/4 = 1/9$ or 0.11

Probability that IV 1 and IV 3 mice are heterozygous is 2/3. This is because both of their parents are known heterozygotes (*A/a*) and since they are dominant phenotype they could only be A/A or A/a. Now, probability that two heterozygotes have a recessive homozygote offspring is 1/4.

13. Which part of the pedigree in Figure 2-23 in your opinion best demonstrates Mendel's first law?

Answer: Any part of this pedigree demonstrates the law, showing segregation of alleles into gametes. The middle part of generation II marriage shows a typical test cross (expected 1:1). Neither ratio in the pedigree could be confirmed because of a small sample size in any given family, but allele segregation is obvious.

14. Could the pedigree in Figure 2-31 be explained as an autosomal dominant disorder? Explain.

Answer: Yes, it could in some cases, but in this case we have clues that the pedigree is for a sex linked dominant trait. First, if fathers have a gene, daughters will receive it only, and second, if mother has a gene, both sons and daughters would receive it.

BASIC PROBLEMS

15. Make up a sentence including the words *chromosome, genes,* and *genome.*

Answer: The human genome contains an estimated 20,000–25,000 genes located on 23 different chromosomes.

16. Peas (*Pisum sativum*) are diploid and $2n = 14$. In *Neurospora,* the haploid fungus, $n = 7$. If it were possible to fractionate genomic DNA from both species by using pulsed field electrophoresis, how many distinct DNA bands would be visible in each species?

Answer: PFGE separates DNA molecules by size. When DNA is carefully isolated from *Neurospora* (which has seven different chromosomes) seven bands should be produced using this technique. Similarly, the pea has seven different chromosomes and will produce seven bands (homologous chromosomes will co-migrate as a single band).

17. The broad bean (*Vicia faba*) is diploid and $2n = 18$. Each haploid chromosome set contains approximately 4 m of DNA. The average size of each chromosome during metaphase of mitosis is 13 μm. What is the average packing ratio of DNA at metaphase? (Packing ratio = length of chromosome/length of DNA molecule therein.) How is this packing achieved?

Answer: There is a total of 4 m of DNA and nine chromosomes per haploid set. On average, each is $4/9$ m long. At metaphase, their average length is 13 μm, so the average packing ratio is 13×10^{-6} m : 4.4×10^{-1} m or roughly 1 : 34,000! This remarkable achievement is accomplished through the interaction of the DNA with proteins. At its most basic, eukaryotic DNA is associated with histones in units called nucleosomes and during mitosis, coils into a solenoid. As loops, it associates with and winds into a central core of nonhistone protein called the scaffold.

18. If we call the amount of DNA per genome "*x*," name a situation or situations in diploid organisms in which the amount of DNA per cell is:
    a. *x*   **b.** $2x$ **c.** $4x$

Answer: Because the DNA levels vary four-fold, the range covers cells that are haploid (gametes) to cells that are dividing (after DNA has replicated but prior to cell division). The following cells would fit the DNA measurements:

x^+ haploid cells
$2x$ diploid cells in G_1 or cells after meiosis I but prior to meiosis II
$4x$ diploid cells after S but prior to cell division

19. Name the key function of mitosis.

Answer: The key function of mitosis is to generate two daughter cells genetically identical to the original parent cell.

20. Name two key functions of meiosis.

Answer: Two key functions of meiosis are to halve the DNA content and to reshuffle the genetic content of the organism to generate genetic diversity among the progeny.

21. Can you design a different nuclear-division system that would achieve the same outcome as that of meiosis?

Answer: It's pretty hard to beat several billions of years of evolution, but it might be simpler if DNA did not replicate prior to meiosis. The same events responsible for halving the DNA and producing genetic diversity could be achieved in a single cell division if homologous chromosomes paired, recombined, randomly aligned during metaphase, and separated during anaphase, etc. However, you would lose the chance to check and repair DNA that replication allows.

22. In a possible future scenario, male fertility drops to zero, but, luckily, scientists develop a way for women to produce babies by virgin birth. Meiocytes are converted directly (without undergoing meiosis) into zygotes, which implant in the usual way. What would be the short- and long-term effects in such a society?

Answer: In large part, this question is asking, why sex? Parthenogenesis (the ability to reproduce without fertilization—in essence, cloning) is not common among multicellular organisms. Parthenogenesis occurs in some species of lizards and fishes, and several kinds of insects, but it is the only means of reproduction in only a few of these species. In plants, about 400 species can reproduce asexually by a process called apomixis. These plants produce seeds without fertilization. However, the majority of plants and animals reproduce sexually. Sexual reproduction produces a wide variety of different offspring by forming new combinations of traits inherited from both the father and the mother. Despite the numerical advantages of asexual reproduction, most multicellular species that have adopted it as their only method of reproducing have become extinct. However, there is no agreed upon explanation of why the loss of sexual reproduction usually leads to early extinction or conversely, why sexual reproduction is associated with evolutionary success.

On the other hand, the immediate effects of such a scenario are obvious. All offspring will be genetically identical to their mothers, and males would be extinct within one generation.

23.   In what ways does the second division of meiosis differ from mitosis?

Answer: As cells divide mitotically, each chromosome consists of identical sister chromatids that are separated to form genetically identical daughter cells. Although the second division of meiosis appears to be a similar process, the "sister" chromatids are likely to be different. Recombination during earlier meiotic stages has swapped regions of DNA between sister and nonsister chromosomes such that the two daughter cells of this division typically are not genetically identical.

24.   Make up mnemonics for remembering the five stages of prophase I of meiosis and the four stages of mitosis.

Answer: The four stages of mitosis are: prophase, metaphase, anaphase, and telophase. The first letters, PMAT, can be remembered by a mnemonic such as: Playful Mice Analyze Twice.

The five stages of prophase I are: leptotene, zygotene, pachytene, diplotene, and diakinesis. The first letters, LZPDD, can be remembered by a mnemonic such as: Large Zoos Provide Dangerous Distractions.

25.   In an attempt to simplify meiosis for the benefit of students, mad scientists develop a way of preventing premeiotic S phase and making do with having just one division, including pairing, crossing over, and segregation. Would this system work, and would the products of such a system differ from those of the present system?

Answer: Yes, it could work but certain DNA repair mechanisms (such as postreplication recombination repair) could not be invoked prior to cell division. There would be just two cells as products of this meiosis, rather than four.

26.   Theodor Boveri said, "The nucleus doesn't divide; it is divided." What was he getting at?

Answer: The nucleus contains the genome and separates it from the cytoplasm. However, during cell division, the nuclear envelope dissociates (breaks down). It is the job of the microtubule-based spindle to actually separate the chromosomes (divide the genetic material) around which nuclei reform during telophase. In this sense, it can be viewed as a passive structure that is divided by the cell's cytoskeleton.

27.   Francis Galton, a geneticist of the pre-Mendelian era, devised the principle that half of our genetic makeup is derived from each parent, one-quarter from each grandparent, one-eighth from each great-grandparent, and so forth. Was he right? Explain.

Answer: Yes, half of our genetic makeup is derived from each parent, each parent's genetic makeup is derived half from each of their parents, etc.

28.   If children obtain half their genes from one parent and half from the other parent, why aren't siblings identical?

Answer: Because the "half" inherited is very random, the chances of receiving exactly the same half is vanishingly small. Ignoring recombination and focusing just on which chromosomes are inherited from one parent (for example, the one they inherited from their father or the one from their mother?), there are $2^{23} =$ 8,388,608 possible combinations!

29.   State where cells divide mitotically and where they divide meiotically in a fern, a moss, a flowering plant, a pine tree, a mushroom, a frog, a butterfly, and a snail.

Answer:

	Mitosis	Meiosis
fern	sporophyte gametophyte	(sporangium)
moss	sporophyte gametophyte	sporophyte (antheridium and archegonium)
plant	sporophyte gametophyte	sporophyte (anther and ovule)
pine tree	sporophyte gametophyte	sporophyte (pine cone)
mushroom	sporophyte gametophyte	sporophyte (ascus or basidium)
frog	somatic cells	gonads
butterfly	somatic cells	gonads

snail	somatic cells	gonads

30. Human cells normally have 46 chromosomes. For each of the following stages, state the number of nuclear DNA molecules present in a human cell:
   a. metaphase of mitosis.
   b. metaphase I of meiosis.
   c. telophase of mitosis.
   d. telophase I of meiosis.
   e. telophase II of meiosis.

Answer: This problem is tricky because the answers depend on how a cell is defined. In general, geneticists consider the transition from one cell to two cells to occur with the onset of anaphase in both mitosis and meiosis, even though cytoplasmic division occurs at a later stage.
   a. 46 chromosomes, each with two chromatids = 92 chromatids
   b. 46 chromosomes, each with two chromatids = 92 chromatids
   c. 46 physically separate chromosomes in each of two about-to-be-formed cells
   d. 23 chromosomes in each of two about-to-be-formed cells, each with two chromatids = 46 chromatids
   e. 23 chromosomes in each of two about-to-be-formed cells

31. Four of the following events are part of both meiosis and mitosis, but only one is meiotic. Which one? (1) chromatid formation, (2) spindle formation, (3) chromosome condensation, (4) chromosome movement to poles, (5) synapsis.

Answer: (5) chromosome pairing (synapsis)

32. In corn, the allele f' causes floury endosperm and the allele f'' causes flinty endosperm. In the cross f'/f' [female symbol] $\times$ f''/f'' [male symbol], all the progeny endosperms are floury, but in the reciprocal cross, all the progeny endosperms are flinty. What is a possible explanation? (Check the legend for Figure 2-7.)

Answer: First, examine the crosses and the resulting genotypes of the endosperm:

Female	Male	Polar nuclei	Sperm	Endosperm
f'/f'	f''/f''	f' and f'	f''/f''	$f'/f'/f''$ (floury)
f''/f''	f'/f'	f'' and f''	f'/f'	$f''/f''/f'$ (flinty)

As can be seen, the phenotype of the endosperm correlates to the predominant allele present.

33.   What is Mendel's first law?

Answer: Mendel's first law states that alleles segregate into gametes during meiosis. This discovery came from his monohybrid experimental crosses.

34.   If you had a fruit fly (*Drosophila melanogaster*) that was of phenotype *A*, what test would you make to determine if the fly's genotype was *A/A* or *A/a*?

Answer: Do a test-cross (cross to *a/a*). If the fly was *A/A*, all the progeny will be phenotypically A; if the fly was *A/a*, half the progeny will be A, and half will be *a*.

35.   In examining a large sample of yeast colonies on a petri dish, a geneticist finds an abnormal-looking colony that is very small. This small colony was crossed with wild type, and products of meiosis (ascospores) were spread on a plate to produce colonies. In total, there were 188 wild-type (normal-size) colonies and 180 small ones.
   a.   What can be deduced from these results regarding the inheritance of the small-colony phenotype? (Invent genetic symbols.)
   b.   What would an ascus from this cross look like?

Answer:
   a.   A diploid meiocyte that is heterozygous for one gene (for example, s^+/s where s is the allele that confers the small colony phenotype) will, after replication and segregation, give two meiotic products of genotype s^+ and two of s. If the random spores of many meiocytes are analyzed, you would expect to find about 50 percent normally sized colonies and 50 percent small colonies if the abnormal phenotype is the result of a mutation in a single gene. Thus, the actual results of 188 normally sized and 180 small-sized colonies support the hypothesis that the phenotype is the result of a mutation in a single gene.
   b.   The following represents an ascus with four spores. The important detail is that two of the spores are s and two are s^+.

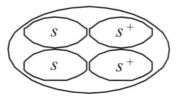

36. Two black guinea pigs were mated and over several years produced 29 black and 9 white offspring. Explain these results, giving the genotypes of parents and progeny.

Answer: The progeny ratio is approximately 3:1, indicating classic heterozygous-by-heterozygous mating. Since black (B) is dominant to white (b):
Parents: $B/b \times B/b$
Progeny: 3 black:1 white (1 B/B : 2 B/b : 1 b/b)

This ratio indicates that black parents were probably heterozygous and that black is dominant over white.

37. In a fungus with four ascospores, a mutant allele *lys-5* causes the ascospores bearing that allele to be white, whereas the wild-type allele *lys-5+* results in black ascospores. (Ascospores are the spores that constitute the four products of meiosis.) Draw an ascus from each of the following crosses:
   a. *lys-5* × *lys-5+*
   b. *lys-5* × *lys-5*
   c. *lys-5+* × <u>lys</u>-5+

Answer:
   a. You expect two *lys-5*$^+$ (black) spores and two *lys-5* (white) spores.

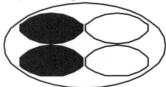

   b. You expect all *lys-5* (white) spores.

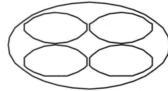

   c. You expect all *lys-5*$^+$ (black) spores.

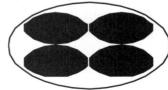

38. For a certain gene in a diploid organism, eight units of protein product are needed for normal function. Each wild-type allele produces five units.
   a. If a mutation creates a null allele, do you think this allele will be recessive or mutant?

   b. What assumptions need to be made to answer part a?

   Answer:
   a. You do not expect the mutation to be recessive. This would be an example of a haploinsufficient gene since one copy of the wild-type allele does produce enough protein product for normal function.
   b. An important assumption would be that having five of eight units of protein product would result in an observable phenotype. It also assumes that the regulation of the single wild-type allele is not affected. Finally, if the mutant allele was leaky rather than null, there might be sufficient protein function when heterozygous with a wild-type allele.

39. A *Neurospora* colony at the edge of a plate seemed to be sparse (low density) in comparison with the other colonies on the plate. This colony was thought to be a possible mutant, and so it was removed and crossed with a wild type of the opposite mating type. From this cross, 100 ascospore progeny were obtained. None of the colonies from these ascospores was sparse, all appearing to be normal. What is the simplest explanation of this result? How would you test your explanation? (**Note:** *Neurospora* is haploid.)

   Answer: The simplest explanation is that the abnormal phenotype was not due to an genetic change. Perhaps the environment (edge of plate) was less favorable for growth. Since *Neurospora* is haploid and forms ascospores, isolating individual asci from a cross of the possible "mutant" to wild type and individually growing the spores should yield 50 percent wild-type and 50 percent "mutant" colonies. If all spores yield wild-type colonies, the low density phenotype was not heritable.

40. From a large-scale screen of many plants of *Collinsia grandiflora,* a plant with three cotyledons was discovered (normally, there are two cotyledons). This plant was crossed with a normal pure-breeding wild-type plant, and 600 seeds from this cross were planted. There were 298 plants with two cotyledons and 302 with three cotyledons. What can be deduced about the inheritance of three cotyledons? Invent gene symbols as part of your explanation.

   Answer: Since half of the F_1 progeny are mutant, it suggests that the mutation that results in three cotyledons is dominant, and the original mutant was heterozygous. Assuming C = the mutant allele and c = the wild-type allele, the cross becomes:
   P $C/c \times c/c$
   F_1 C/c three cotyledons
   c/c two cotyledons

41. In the plant *Arabidopsis thaliana,* a geneticist is interested in the development

of trichomes (small projections). A large screen turns up two mutant plants (A and B) that have no trichomes, and these mutants seem to be potentially useful in studying trichome development. (If they were determined by single-gene mutations, then finding the normal and abnormal functions of these genes would be instructive.) Each plant is crossed with wild type; in both cases, the next generation (F_1) had normal trichomes. When F_1 plants were selfed, the resulting F_2's were as follows:

     F_2 from mutant A: 602 normal; 198 no trichomes

     F_2 from mutant B: 267 normal; 93 no trichomes

a. What do these results show? Include proposed genotypes of all plants in your answer.

b. Under your explanation to part a, is it possible to confidently predict the F_1 from crossing the original mutant A with the original mutant B?

Answer:

a. The data for both crosses suggest that both A and B mutant plants are homozygous for recessive alleles. Both F_2 crosses give 3:1 ratios of normal to mutant progeny. For example, let A = normal and a = mutant, then

    P $A / A \times a/a$

    F_1 A/a

    F_2 1 A/A phenotype: normal

          2 A/a phenotype: normal

          1 $a\,a$ phenotype: mutant (no trichomes).

b. No. You do not know if the a and b mutations are in the same or different genes. If they are in the same gene then the F_1 will all be mutant. If they are in different genes, then the F_1 will all be wild type.

42. You have three dice: one red (R), one green (G), and one blue (B). When all three dice are rolled at the same time, calculate the probability of the following outcomes:

a. 6 (R), 6 (G), 6 (B)

b. 6 (R), 5 (G), 6 (B)

c. 6 (R), 5 (G), 4 (B)

d. No sixes at all

e. A different number on all dice

Answer: Each die has six sides, so the probability of any one side (number) is $1/6$. To get specific red, green, and blue numbers involves "and" statements that are independent. So each independent probability is multiplied together.

a. $(1/6)(1/6)(1/6) = (1/6)^3 = 1/216$

b. $(1/6)(1/6)(1/6) = (1/6)^3 = 1/216$

c. $(1/6)(1/6)(1/6) = (1/6)^3 = 1/216$

d. To not roll any sixes is the same as getting anything but sixes:

$(1 – {}^1/_6)(1 – {}^1/_6)(1 – {}^1/_6) = ({}^5/_6)^3 = {}^{125}/_{216.}$

e. The easiest way to approach this problem is to consider each die separately. The first die thrown can be any number. Therefore, the probability for it is 1.

The second die can be any number except the number obtained on the first die. Therefore, the probability of not duplicating the first die is $1 – p$(first die duplicated) $= 1 – {}^1/_6 = {}^5/_6$.

The third die can be any number except the numbers obtained on the first two dice. Therefore, the probability is $1 – p$(first two dice duplicated) $= 1 – {}^2/_6 = {}^2/_3$.

Finally, the probability of all different dice is $(1)({}^5/_6)({}^2/_3) = {}^{10}/_{18} = {}^5/_9$.

43. In the pedigree below, the black symbols represent individuals with a very rare blood disease.

If you had no other information to go on, would you think it more likely that the disease was dominant or recessive? Give your reasons.

Answer: You are told that the disease being followed in this pedigree is very rare. If the allele that results in this disease is recessive, then the father would have to be homozygous and the mother would have to be heterozygous for this allele. On the other hand, if the trait is dominant, then all that is necessary to explain the pedigree is that the father is heterozygous for the allele that causes the disease. This is the better choice as it is more likely, given the rarity of the disease.

44. a. The ability to taste the chemical phenylthiocarbamide is an autosomal dominant phenotype, and the inability to taste it is recessive. If a taster woman with a nontaster father marries a taster man who in a previous marriage had a nontaster daughter, what is the probability that their first child will be:
(1) A nontaster girl
(2) A taster girl
(3) A taster boy
b. What is the probability that their first two children will be tasters of either sex?

Answer:
a. By considering the pedigree (see below), you will discover that the cross in question is $T/t \times T/t$. Therefore, the probability of being a taster is $^3/_4$, and the probability of being a nontaster is $^1/_4$.

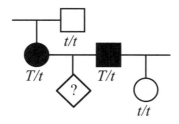

Also, the probability of having a boy equals the probability of having a girl equals $1/2$.

(1) p(nontaster girl) = p(nontaster) $\times$ p(girl) = $1/4 \times 1/2 = 1/8$

(2) p(taster girl) = p(taster) $\times$ p(girl) = $3/4 \times 1/2 = 3/8$

(3) p(taster boy) = p(taster) $\times$ p(boy) = $3/4 \times 1/2 = 3/8$

b. p(taster for first two children) = p(taster for first child) $\times$ p(taster for second child) = $3/4 \times 3/4 = 9/16$

45. John and Martha are contemplating having children, but John's brother has galactosemia (an autosomal recessive disease) and Martha's great-grandmother also had galactosemia. Martha has a sister who has three children, none of whom have galactosemia. What is the probability that John and Martha's first child will have galactosemia?

Unpacking the Problem

1. Can the problem be restated as a pedigree? If so, write one.

    Answer: Yes. The pedigree is given below.

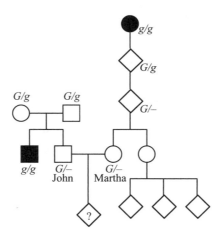

2. Can parts of the problem be restated by using Punnett squares?

    Answer: In order to state this problem as a Punnett square, you must first know the genotypes of John and Martha. The genotypes can be determined only through considering the pedigree. Even with the pedigree, however, the genotypes can be stated only as *G/–* for both John and Martha.

The probability that John is carrying the allele for galactosemia is $2/3$, rather than the $1/2$ that you might guess. To understand this, recall that John's parents must be heterozygous in order to have a child with the recessive disorder while still being normal themselves (the assumption of normalcy is based on the information given in the problem). John's parents were both *G/g*. A Punnett square for their mating would be:

		Father	
		G	*g*
Mother	*G*	*G/G*	*G/g*
	g	*G/g*	*g/g*

The cross is:

P		*G/g* × *G/g*
F$_1$	*g/g*	John's brother
	G/–	John (either *G/G* or *G/g*)

The expected ratio of the F$_1$ is 1 *G/G* : 2 *G/g* : 1 *g/g*. Because John does not have galactosemia (an assumption based on the information given in the problem), he can be either *G/G* or *G/g*, which occurs at a ratio of 1:2. Therefore, his probability of carrying the *g* allele is $2/3$.

The probability that Martha is carrying the *g* allele is based on the following chain of logic. Her great-grandmother had galactosemia, which means that she had to pass the allele to Martha's grandparent. Because the problem states nothing with regard to the grandparent's phenotype, it must be assumed that the grandparent was normal, or *G/g*. The probability that the grandparent passed it to Martha's parent is $1/2$. Next, the probability that Martha's parent passed the allele to Martha is also $1/2$, assuming that the parent actually has it. Therefore, the probability that Martha's parent has the allele and passed it to Martha is $1/2 \times 1/2$, or $1/4$.

In summary:

John	$p(G/G) = 1/3$
	$p(G/g) = 2/3$
Martha	$p(G/G) = 3/4$
	$p(G/g) = 1/4$

This information does not fit easily into a Punnett square.

3.   Can parts of the problem be restated by using branch diagrams?

Answer: While the above information could be put into a branch diagram, it does not easily fit into one and overcomplicates the problem, just as a Punnett square would.

4.   In the pedigree, identify a mating that illustrates Mendel's first law.

Answer: The marriage between John's parents illustrates Mendel's first law.

5.   Define all the scientific terms in the problem, and look up any other terms about which you are uncertain.

Answer: The scientific words in this problem are *galactosemia*, *autosomal*, and *recessive*.

Galactosemia is a metabolic disorder characterized by the absence of the enzyme galactose-1-phosphate uridyl transferase, which results in an accumulation of galactose. In the vast majority of cases, galactosemia results in an enlarged liver, jaundice, vomiting, anorexia, lethargy, and very early death if galactose is not omitted from the diet (initially, the child obtains galactose from milk).

Autosomal refers to genes that are on the autosomes.

Recessive means that in order for an allele to be expressed, it must be the only form of the gene present in the organism.

6.   What assumptions need to be made in answering this problem?

Answer: The major assumption is that if nothing is stated about a person's phenotype, the person is of normal phenotype. Another assumption that may be of value, but is not actually needed, is that all people marrying into these two families are normal and do not carry the allele for galactosemia.

7.   Which unmentioned family members must be considered? Why?

Answer: The people not mentioned in the problem, but who must be considered, are John's parents and Martha's grandparent and parent descended from her affected great-grandmother.

8.   What statistical rules might be relevant, and in what situations can they be applied? Do such situations exist in this problem?

Answer: The major statistical rule needed to solve the problem is the product rule (the "and" rule). It is used to calculate the cumulative probabilities described in part 2 of this unpacked solution (e.g., What is the probability that Martha's parent inherited the galactosemia allele AND passed that allele onto Martha AND Martha will pass that allele on to her child?).

9.   What are two generalities about autosomal recessive diseases in human populations?

Answer: Autosomal recessive disorders are assumed to be rare and to occur equally frequently in males and females. They are also assumed to be expressed if the person is homozygous for the recessive genotype.

10.   What is the relevance of the rareness of the phenotype under study in pedigree analysis generally, and what can be inferred in this problem?

Answer: Rareness leads to the assumption that people who marry into a family that is being studied do not carry the allele, which was assumed in entry (6) above.

11.   In this family, whose genotypes are certain and whose are uncertain?

Answer: The only certain genotypes in the pedigree are John's parents, John's brother, and Martha's great-grandmother and grandmother. All other individuals have uncertain genotypes.

12.   In what way is John's side of the pedigree different from Martha's side? How does this difference affect your calculations?

Answer: John's family can be treated simply as a heterozygous-by-heterozygous cross, with John having a $2/3$ probability of being a carrier, while it is unknown if either of Martha's parents carry the allele. Therefore Martha's chance of being a carrier must be calculated as a series of probabilities.

13.   Is there any irrelevant information in the problem as stated?

Answer: The information regarding Martha's sister and her children turns out to be irrelevant to the problem.

14. In what way is solving this kind of problem similar to solving problems that you have already successfully solved? In what way is it different?

Answer: The problem contains a number of assumptions that have not been necessary in problem solving until now.

15. Can you make up a short story based on the human dilemma in this problem?

Answer: Many scenarios are possible in response to this question.

Now try to solve the problem. If you are unable to do so, try to identify the obstacle and write a sentence or two describing your difficulty. Then go back to the expansion questions and see if any of them relate to your difficulty.

Solution to the Problem

Answer: p(child has galactosemia) = p(John is G/g) × p(Martha is G/g) × p(both parents passed g to the child) = $(2/3)(1/4)(1/4) = 2/48 = 1/24$

46. Holstein cattle are normally black and white. A superb black-and-white bull, Charlie, was purchased by a farmer for $100,000. All the progeny sired by Charlie were normal in appearance. However, certain pairs of his progeny, when interbred, produced red-and-white progeny at a frequency of about 25 percent. Charlie was soon removed from the stud lists of the Holstein breeders. Use symbols to explain precisely why.

Answer: Charlie, his mate, or both, obviously were not homozygous for one of the alleles (pure-breeding), because his F_2 progeny were of two phenotypes. Let A = black and white, and a = red and white. If both parents were heterozygous, then red and white would have been expected in the F_1 generation. Red and white were not observed in the F_1 generation, so only one of the parents was heterozygous. The cross is:

P A/a × A/A
F_1 1 A/a : 1 A/A

Two F_1 heterozygotes (A/a) when crossed would give 1 A/A (black and white) : 2 A/a (black and white) : 1 a/a (red and white). If the red and white F_2 progeny were from more than one mate of Charlie's, then the farmer acted correctly.

However, if the F$_2$ progeny came only from one mate, the farmer may have acted too quickly.

47. Suppose that a husband and wife are both heterozygous for a recessive allele for albinism. If they have dizygotic (two-egg) twins, what is the probability that both the twins will have the same phenotype for pigmentation?

Answer: Because the parents are heterozygous, both are *A/a*. Both twins could be albino or both twins could be normal (and = multiply, or = add). The probability of being normal (*A/−*) is $^3/_4$, and the probability of being albino (*a/a*) is $^1/_4$.

$$p(\text{both normal}) + p(\text{both albino})$$
$$p(\text{first normal}) \times p(\text{second normal}) + p(\text{first albino}) \times p(\text{second albino})$$
$$(^3/_4)(^3/_4) + (^1/_4)(^1/_4) = {}^9/_{16} + {}^1/_{16} = {}^5/_8$$

48. The plant blue-eyed Mary grows on Vancouver Island and on the lower mainland of British Columbia. The populations are dimorphic for purple blotches on the leaves—some plants have blotches and others don't. Near Nanaimo, one plant in nature had blotched leaves. This plant, which had not yet flowered, was dug up and taken to a laboratory, where it was allowed to self. Seeds were collected and grown into progeny. One randomly selected (but typical) leaf from each of the progeny is shown in the accompanying illustration.

a. Formulate a concise genetic hypothesis to explain these results. Explain all symbols and show all genotypic classes (and the genotype of the original plant).

b. How would you test your hypothesis? Be specific.

Answer: The plants are approximately 3 blotched : 1 unblotched. This suggests that blotched is dominant to unblotched and that the original plant which was selfed was a heterozygote.

a.   Let A = blotched, a = unblotched.

P A/a (blotched) × A/a (blotched)

F_1 1 A/A : 2 A/a : 1 a/a

3 $A/-$ (blotched) : 1 a/a (unblotched)

b.   All unblotched plants should be pure-breeding in a testcross with an unblotched plant (a/a), and one-third of the blotched plants should be pure-breeding.

49.   Can it ever be proved that an animal is not a carrier of a recessive allele (that is, not a heterozygote for a given gene)? Explain.

Answer: In theory, it cannot be proved that an animal is not a carrier for a recessive allele. However, in an $A/-$ × a/a cross, the more dominant-phenotype progeny produced, the less likely it is that the parent is A/a. In such a cross, half the progeny would be a/a and half would be A/a. With n dominant phenotype progeny, the probability that the parent is A/a is $(1/2)^n$. (DNA sequencing can be used to prove heterozygosity, but without sequence level information, the level of certainty is limited by sample size.)

50.   In nature, the plant *Plectritis congesta* is dimorphic for fruit shape; that is, individual plants bear either wingless or winged fruits, as shown in the illustration. Plants were collected from nature before flowering and were crossed or selfed with the following results:

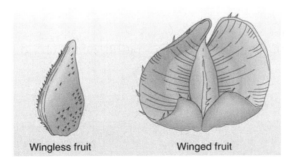

Wingless fruit   Winged fruit

	Number of progeny	
Pollination	Winged	Wingless
Winged (selfed)	91	1*
Winged (selfed)	90	30
Wingless (selfed)	4*	80

Winged × wingless	161	0
Winged × wingless	29	31
Winged × wingless	46	0
Winged × winged	44	0
Winged × winged	24	0

*Phenotype probably has a nongenetic explanation.

Interpret these results, and derive the mode of inheritance of these fruit-shaped phenotypes. Use symbols. What do you think is the nongenetic explanation for the phenotypes marked by asterisks in the table?

Answer: The results suggest that winged (*A/–*) is dominant to wingless (*a/a*) (cross 2 gives a 3 : 1 ratio). If that is correct, the crosses become:

		Number of progeny plants	
Pollination	Genotypes	Winged	Wingless
winged (selfed)	*A/A × A/A*	91	1*
winged (selfed)	*A/a × A/a*	90	30
wingless (selfed)	*a/a × a/a*	4*	80
winged × wingless	*A/A × a/a*	161	0
winged × wingless	*A/a × a/a*	29	31
winged × wingless	*A/A × a/a*	46	0
winged × winged	*A/A × A/–*	44	0
winged × winged	*A/A × A/–*	24	0

The five unusual plants are most likely due either to human error in classification or to contamination. Alternatively, they could result from environmental effects on development. For example, too little water may have prevented the seed pods from becoming winged even though they are genetically winged.

51. The accompanying pedigree is for a rare, but relatively mild, hereditary disorder of the skin.

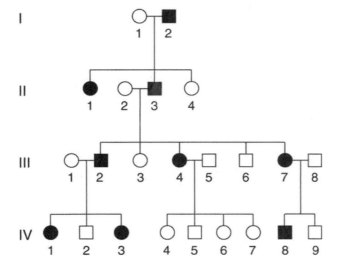

a. How is the disorder inherited? State reasons for your answer.

b. Give genotypes for as many individuals in the pedigree as possible. (Invent your own defined allele symbols.)

c. Consider the four unaffected children of parents III-4 and III-5. In all four-child progenies from parents of these genotypes, what proportion is expected to contain all unaffected children?

Answer:

a. The disorder appears to be dominant because all affected individuals have an affected parent. If the trait was recessive, then I-1, II-2, III-1, and III-8 would all have to be carriers (heterozygous for the rare allele).

b. Assuming dominance, the genotypes are:

I : *d/d, D/d*

II : *D/d, d/d, D/d, d/d*

III : *d/d, D/d, d/d, D/d, d/d, d/d, D/d, d/d*

IV : *D/d, d/d, D/d, d/d, d/d, d/d, d/d, D/d, d/d*

c. The mating is $D/d \times d/d$. The probability of an affected child (D/d) equals $1/2$, and the probability of an unaffected child (d/d) equals $1/2$. Therefore, the chance of having four unaffected children (since each is an independent event) is: $1/2 \times 1/2 \times 1/2 \times 1/2 = 1/16$.

52. Four human pedigrees are shown in the accompanying illustration. The black symbols represent an abnormal phenotype inherited in a simple Mendelian manner.

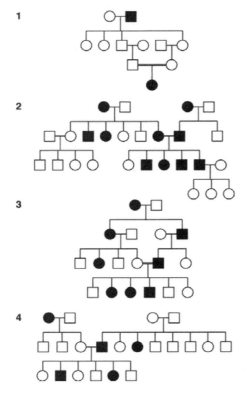

a. For each pedigree, state whether the abnormal condition is dominant or recessive. Try to state the logic behind your answer.

b. For each pedigree, describe the genotypes of as many persons as possible.

Answer:

a. *Pedigree 1*: The best answer is recessive because two unaffected individuals had affected progeny. Also, the disorder skips generations and appears in a mating between two related individuals.

Pedigree 2: The best answer is dominant because two affected parents have an unaffected child. Also, it appears in each generation, roughly half the progeny are affected, and all affected individuals have an affected parent.

Pedigree 3: The best answer is dominant, for many of the reasons stated for pedigree 2. Inbreeding, while present in the pedigree, does not allow an explanation of recessive because it cannot account for individuals in the second or third generations.

Pedigree 4: The best answer is recessive. Two unaffected individuals had affected progeny.

b. Genotypes of pedigree 1:
Generation I: *A/–, a/a*
Generation II: *A/a, A/a, A/a, A/–, A/–, A/a*
Generation III: *A/a, A/a*

Generation IV: *a/a*

Genotypes of pedigree 2:
Generation I: *A/a, a/a, A/a, a/a*
Generation II: *a/a, a/a, A/a, A/a, a/a, a/a, A/a, A/a, a/a*
Generation III: *a/a, a/a, a/a, a/a, a/a, A/–, A/–, A/–, A/a, a/a*
Generation IV: *a/a, a/a, a/a*

Genotypes of pedigree 3:
Generation I: *A/–, a/a*
Generation II: *A/a, a/a, a/a, A/a*
Generation III: *a/a, A/a, a/a, a/a, A/a, a/a*
Generation IV: *a/a, A/a, A/a, A/a, a/a, a/a*

Genotypes of pedigree 4:
Generation I: *a/a, A/–, A/a, A/a*
Generation II: *A/a, A/a, A/a, a/a, A/–, a/a, A/–, A/–, A/–, A/–, A/–*
Generation III: *A/a, a/a, A/a, A/a, a/a, A/a*

53. Tay-Sachs disease (infantile amaurotic idiocy) is a rare human disease in which toxic substances accumulate in nerve cells. The recessive allele responsible for the disease is inherited in a simple Mendelian manner. For unknown reasons, the allele is more common in populations of Ashkenazi Jews of eastern Europe. A woman is planning to marry her first cousin, but the couple discovers that their shared grandfather's sister died in infancy of Tay-Sachs disease.

a. Draw the relevant parts of the pedigree, and show all the genotypes as completely as possible.

b. What is the probability that the cousins' first child will have Tay-Sachs disease, assuming that all people who marry into the family are homozygous normal?

Answer:
a.   The pedigree is

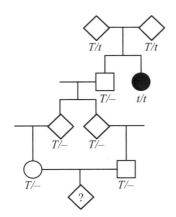

b. The probability that the child of the two first cousins will have Tay-Sachs disease is a function of three probabilities: p(the woman is T/t) $\times$ p(the man is T/t) $\times$ p(both donate t);

$$= (^2/_3)(^1/_2)(^1/_2) \times (^2/_3)(^1/_2)(^1/_2) \times ^1/_4 = ^1/_{144}$$

To understand the probabilities of the first two events, see the discussion for problem 8 part (2) of this chapter.

54. The following pedigree was obtained for a rare kidney disease.

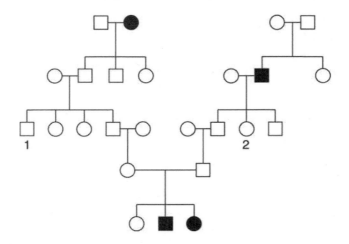

a. Deduce the inheritance of this condition, stating your reasons.
b. If persons 1 and 2 marry, what is the probability that their first child will have the kidney disease?

Answer:
a. Autosomal recessive: affected individuals inherited the trait from unaffected parents and a daughter inherited the trait from an unaffected father.
b. Both parents must be heterozygous to have a $^1/_4$ chance of having an affected child. Parent 2 is heterozygous, since her father is homozygous for the recessive allele and parent 1 has a $^1/_2$ chance of being heterozygous, since his father is heterozygous because 1's paternal grandmother was affected. Overall, $1 \times ^1/_2 \times ^1/_4 = ^1/_8$.

55. This pedigree is for Huntington disease, a late-onset disorder of the nervous system. The slashes indicate deceased family members.

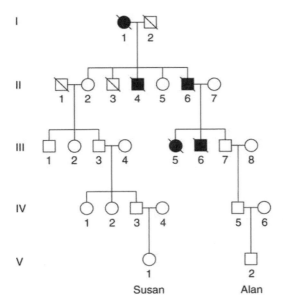

a. Is this pedigree compatible with the mode of inheritance for Huntington disease mentioned in the chapter?

b. Consider two newborn children in the two arms of the pedigree, Susan in the left arm and Alan in the right arm. Study the graph in Figure 2-24 and form an opinion on the likelihood that they will develop Huntington disease. Assume for the sake of the discussion that parents have children at age 25.

Answer:

a. Yes. It is inherited as an autosomal dominant trait.

b. Susan is highly unlikely to have Huntington's disease. Her great-grandmother (individual II-2) is 75 years old and has yet to develop it, when nearly 100 percent of people carrying the allele will have developed the disease by that age. If her great-grandmother does not have it, Susan cannot inherit it.

Alan is somewhat more likely than Susan to develop Huntington's disease. His grandfather (individual III-7) is only 50 years old, and approximately 20 percent of the people with the allele have yet to develop the disease by that age. Therefore, it can be estimated that the grandfather has a 10 percent chance of being a carrier (50 percent chance he inherited the allele from his father × 20 percent chance he has not yet developed symptoms). If Alan's grandfather eventually develops Huntington's disease, then there is a probability of 50 percent that Alan's father inherited it from him, and a probability of 50 percent that Alan received that allele from his father. Therefore, Alan has a $1/10 \times 1/2 \times 1/2 = 1/40$ current probability of developing Huntington's disease and a $1/2 \times 1/2 = 1/4$ probability if his grandfather eventually develops it.

56. Consider the accompanying pedigree of a rare autosomal recessive disease, PKU.

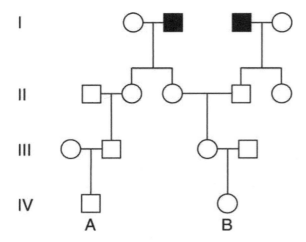

  a. List the genotypes of as many of the family members as possible.
  b. If persons A and B marry, what is the probability that their first child will have PKU?
  c. If their first child is normal, what is the probability that their second child will have PKU?
  d. If their first child has the disease, what is the probability that their second child will be unaffected?
(Assume that all people marrying into the pedigree lack the abnormal allele.)

Answer:
  a. Assuming the trait is rare, expect that all individuals marrying into the pedigree do not carry the disease-causing allele.

  $\quad\quad$ I : $P/P, p/p, p/p, P/P$
  $\quad\quad$ II : $P/P, P/p, P/p, P/p, P/p$
  $\quad\quad$ III : $P/P, P/-, P/-, P/P$
  $\quad\quad$ IV : $P/-, P/-$

  b. For their child to have PKU, both A and B must be carriers and both must donate the recessive allele.

  $\quad$ The probability that individual A has the PKU allele is derived from individual II-2. II-2 must be P/p since her father must be p/p. Therefore, the probability that II-2 passed the PKU allele to individual III-2 is $1/2$. If III-2 received the allele, the probability that he passed it to individual IV-1 (A) is $1/2$. Therefore, the probability that A is a carrier is $1/2 \times 1/2 = 1/4$.

  $\quad$ The probability that individual B has the allele goes back to the mating of II-3 and II-4, both of whom are heterozygous. Their child, III-3, has a $2/3$ probability of having received the PKU allele and a probability of $1/2$ of passing it to IV-2 (B). Therefore, the probability that B has the PKU allele is $2/3 \times 1/2 = 1/3$.

If both parents are heterozygous, they have a $1/4$ chance of both passing the p allele to their child.

p(child has PKU) $= p$(A is P/p) $\times p$(B is P/p) $\times p$(both parents donate p)

$$1/4 \quad \times \quad 1/3 \quad \times \quad 1/4 \quad = 1/48$$

c. If the first child is normal, no additional information has been gained and the probability that the second child will have PKU is the same as the probability that the first child will have PKU, or $1/48$.

d. If the first child has PKU, both parents are heterozygous. The probability of having an affected child is now $1/4$, and the probability of having an unaffected child is $3/4$.

57. A man has attached earlobes, whereas his wife has free earlobes. Their first child, a boy, has attached earlobes.
   a. If the phenotypic difference is assumed to be due to two alleles of a single gene, is it possible that the gene is X-linked?
   b. Is it possible to decide if attached earlobes are dominant or recessive?

Answer:
   a. Sons inherit their X chromosome from their mother. The mother has earlobes, the son does not. If the allele for earlobes is dominant and the allele for lack of earlobes recessive, then the mother could be heterozygous for this trait and the gene could be X-linked.
   b. It is not possible from the data given to decide which allele is dominant. If lack of earlobes is dominant, then the father would be heterozygous and the son would have a 50 percent chance of inheriting the dominant "lack-of-earlobes" allele. If lack of earlobes is recessive, then the trait could be autosomal or X-linked, but in either case, the mother would be heterozygous.

58. A rare recessive allele inherited in a Mendelian manner causes the disease cystic fibrosis. A phenotypically normal man whose father had cystic fibrosis marries a phenotypically normal woman from outside the family, and the couple consider having a child.
   a. Draw the pedigree as far as described.
   b. If the frequency in the population of heterozygotes for cystic fibrosis is 1 in 50, what is the chance that the couple's first child will have cystic fibrosis?
   c. If the first child does have cystic fibrosis, what is the probability that the second child will be normal?

Answer:

a. Let C stand for the normal allele and c stand for the allele that causes cystic fibrosis.

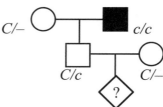

b. The man has a 100 percent probability of having the c allele. His wife, who is from the general population, has a $1/50$ chance of having the c allele. If both have the allele, then $1/4$ of their children will have cystic fibrosis. The probability that their first child will have cystic fibrosis is:

p(man has c) × p(woman has c) × p(both pass c to the child)

$$1.0 \quad \times \quad 1/50 \quad \times \quad 1/4 = 1/200 = 0.005$$

c. If the first child does have cystic fibrosis, then the woman is a carrier of the c allele. Because both parents are C/c, the chance that the second child will be normal is the probability of a normal child in a heterozygous × heterozygous mating, or $3/4$.

59. The allele c causes albinism in mice (C causes mice to be black). The cross C/c × c/c produces 10 progeny. What is the probability of all of them being black?

Answer: The cross is C/c × c/c so there is a $1/2$ chance that a progeny would be black (C/c). Because each progeny's genotype is independent of the others, the chance that all 10 progeny are black is $(1/2)^{10}$.

60. The recessive allele s causes *Drosophila* to have small wings and the s^1 allele causes normal wings. This gene is known to be X linked. If a small-winged male is crossed with a homozygous wild-type female, what ratio of normal to small-winged flies can be expected in each sex in the F_1? If F_1 flies are intercrossed, what F_2 progeny ratios are expected? What progeny ratios are predicted if F_1 females are backcrossed with their father?

Answer:

P $\qquad s^+/s^+ \quad \times \quad s/Y$

$\qquad\qquad\qquad\qquad\qquad\qquad \downarrow$

$F_1 \qquad 1/2\ s^+/s \qquad\qquad$ normal female

$\qquad\quad 1/2\ s^+/Y \qquad\qquad$ normal male

$s^+/s \quad \times \quad s^+/Y$
$\downarrow$

F2	$^1/_4\ s^+/s^+$	normal female
	$^1/_4\ s^+/s$	normal female
	$^1/_4\ s^+/Y$	normal male
	$^1/_4\ s/Y$	small wings male

P $s^+/s \quad \times \quad s/Y$
$\downarrow$

Progeny	$^1/_4\ s^+/s$	normal female
	$^1/_4\ s/s$	small wings female
	$^1/_4\ s^+/Y$	normal male
	$^1/_4\ s/Y$	small wings male

61. An X-linked dominant allele causes hypophosphatemia in humans. A man with hypophosphatemia marries a normal woman. What proportion of their sons will have hypophosphatemia?

Answer: Let H = hypophosphate and h = normal. The cross is $H/Y \times h/h$, yielding H/h (females) and h/Y (males). The answer is 0% because sons always inherit an X chromosome from their mothers and a Y chromosome from their fathers.

62. Duchenne muscular dystrophy is sex linked and usually affects only males. Victims of the disease become progressively weaker, starting early in life.

a. What is the probability that a woman whose brother has Duchenne's disease will have an affected child?

b. If your mother's brother (your uncle) had Duchenne's disease, what is the probability that you have received the allele?

c. If your father's brother had the disease, what is the probability that you have received the allele?

Answer:

a. You should draw pedigrees for this question.

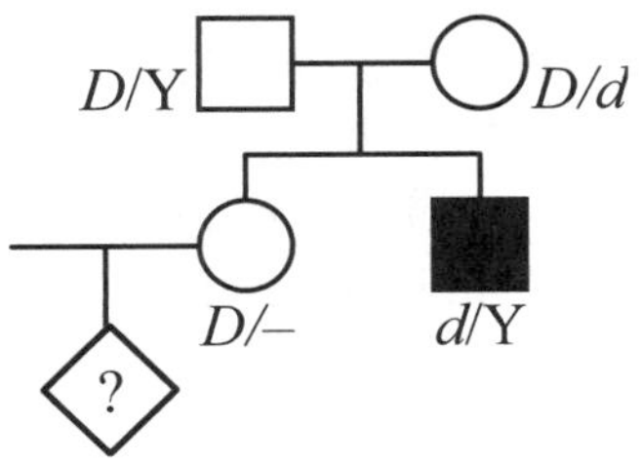

The "maternal grandmother" had to be a carrier, *D/d*. The probability that the woman inherited the *d* allele from her is $^1/_2$. The probability that she passes it to her child is $^1/_2$. The probability that the child is male is $^1/_2$. The total probability of the woman having an affected child is $^1/_2 \times ^1/_2 \times ^1/_2 = ^1/_8$.

b. The same pedigree as part (a) applies. The "maternal grandmother" had to be a carrier, *D/d*. The probability that your mother received the allele is $^1/_2$. The probability that your mother passed it to you is $^1/_2$. The total probability is $^1/_2 \times ^1/_2 = ^1/_4$.

c.

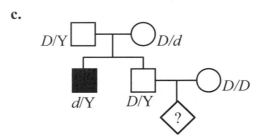

Because your father does not have the disease, you cannot inherit the allele from him. Therefore, the probability of inheriting an allele will be based on the chance that your mother is heterozygous. Since she is "unrelated" to the pedigree, assume that this is zero.

63. A recently married man and woman discover that each had an uncle with alkaptonuria, otherwise known as "black urine disease," a rare disease caused by an autosomal recessive allele of a single gene. They are about to have their first baby. What is the probability that their child will have alkaptonuria?

Answer: For the recently married man and woman to each have an uncle with alkaptonuria means that each may have one parent (the parent related to the uncle) that is heterozygous for the disease-causing allele. Specifically, this parent (and related uncle) must have had parents that were both heterozygous for alkaptonuria. Any child of parents that are both heterozygous for a recessive trait, but does not have that trait, has a $^2/_3$ chance of being heterozygous. (Remember, if both parents are heterozygous, we expect a 1 : 2 : 1 ratio of genotypes, but once we know a person is not homozygous recessive, the only possibilities left are 1 (homozygous dominant) to 2 (heterozygous) or $^2/_3$ chance of being heterozygous.) So the man and woman each have a $^2/_3 \times ^1/_2 = ^1/_3$ of being carriers (heterozygous), and the chance of their having an affected child would be $^1/_3 \times ^1/_3 \times ^1/_4 = ^1/_{36}$.

64. The accompanying pedigree concerns an inherited dental abnormality, amelogenesis imperfecta.

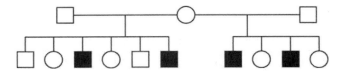

  a. What mode of inheritance best accounts for the transmission of this trait?
  b. Write the genotypes of all family members according to your hypothesis.

Answer:
a. Because none of the parents are affected, the disease must be recessive. Because the inheritance of this trait appears to be sex-specific, it is most likely X-linked. If it were autosomal, all three parents would have to be carriers, and by chance, only sons and none of the daughters inherited the trait (which is quite unlikely).

b. I   *A/Y, A/a, A/Y*
      II   *A/Y, A/–, a/Y, A/–, A/Y, a/Y, a/Y, A/–, a/Y, A/–*

65. A couple who are about to get married learn from studying their family histories that, in both their families, their unaffected grandparents had siblings with cystic fibrosis (a rare autosomal recessive disease).
  a. If the couple marries and has a child, what is the probability that the child will have cystic fibrosis?
  b. If they have four children, what is the chance that the children will have the precise Mendelian ratio of 3 : 1 for normal : cystic fibrosis?
  c. If their first child has cystic fibrosis, what is the probability that their next three children will be normal?

Answer:
a. This question is similar to question 49, but this time the discussion begins with the grandparents rather than the parents. Again, given that a sibling is affected with a recessive disease, the related unaffected brother/sister will have a $2/3$ chance of being heterozygous. In this case, that is one of the grandparents of both the man and woman about to be married. Given this, the couple will both have a $2/3 \times 1/2 \times 1/2 = 1/6$ chance of being carriers (heterozygous) and the chance of their having an affected child will be $1/6 \times 1/6 \times 1/4 = 1/144$.

b. If both parents are carriers, there is a $3/4$ chance a child will be normal and a $1/4$ chance a child will have cystic fibrosis. Each child is an independent event, but since birth order is not considered, there are four ways to have the desired outcome. The child with cystic fibrosis may be the first, second, third, or fourth so assuming the first is affected, the specified outcome would be a $1/4 \times 3/4 \times 3/4 \times 3/4$ or a $27/256$ chance. Now, taking into account the four possible birth orders and the chance that both parents are heterozygous, the chance of an exact 3:1 ratio becomes $4 \times \frac{1}{6} \times \frac{1}{6} \times \frac{27}{256} = \frac{3}{256}$.

c.   In this case, knowing the first child has cystic fibrosis lets us now deduce that the parents must both be heterozygous. Given this, there is a $\frac{3}{4}$ chance than any future child will be normal. Since each is independent, the chance that their next three are normal is simply $\frac{3}{4} \times \frac{3}{4} \times \frac{3}{4}$ or $\frac{27}{64}$.

66.   A sex-linked recessive allele c produces a red–green color blindness in humans. A normal woman whose father was color blind marries a color-blind man.

a.   What genotypes are possible for the mother of the color-blind man?

b.   What are the chances that the first child from this marriage will be a color-blind boy?

c.   Of the girls produced by these parents, what proportion can be expected to be color blind?

d.   Of all the children (sex unspecified) of these parents, what proportion can be expected to have normal color vision?

Answer:
You should draw the pedigree before beginning.

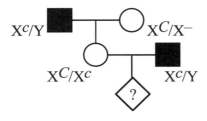

a. X^C/X^c, X^c/X^c

b. $p(\text{colorblind}) \times p(\text{male}) = (\frac{1}{2})(\frac{1}{2}) = \frac{1}{4}$

c.   The girls will be 1 normal (X^C/X^c) : 1 colorblind (X^c/X^c).

d.   The cross is $X^C/X^c \times X^c/Y$, yielding 1 normal : 1 colorblind for both sexes.

67.   Male house cats are either black or orange; females are black, orange, or calico.

a.   If these coat-color phenotypes are governed by a sex-linked gene, how can these observations be explained?

b.   Using appropriate symbols, determine the phenotypes expected in the progeny of a cross between an orange female and a black male.

c.   Half the females produced by a certain kind of mating are calico, and half are black; half the males are orange, and half are black. What colors are the parental males and females in this kind of mating?

d.   Another kind of mating produces progeny in the following proportions: one-fourth orange males, one-fourth orange females, one-fourth black males, and one-fourth calico females. What colors are the parental males and females in this kind of mating?

Answer:

a. This problem involves X-inactivation. Let B = black and b = *orange*.

Females	Males
X^B/X^B = black	X^B/Y = black
X^b/X^b = orange	X^b/Y = orange
X^B/X^b = calico	

b. P X^b/X^b (orange) × X^B/Y (black)

       F_1 X^B/X^b (calico female)

             X^b/Y (orange male)

c. Because the males are black or orange, the mother had to have been calico. Half the daughters are black, indicating that their father was black.

d. Males were orange or black, indicating that the mothers were calico. Orange females indicate that the father was orange.

68. The pedigree below concerns a certain rare disease that is incapacitating but not fatal.

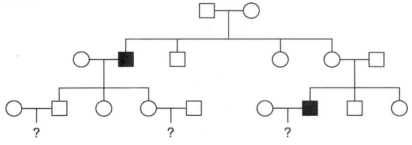

a. Determine the most likely mode of inheritance of this disease.

b. Write the genotype of each family member according to your proposed mode of inheritance.

c. If you were this family's doctor, how would you advise the three couples in the third generation about the likelihood of having an affected child?

Answer:

a. Recessive (unaffected parents have affected progeny) and X-linked (only assumption is that the grandmother, I-2, is a carrier). If autosomal, then I-1, I-2, and II-6 would all have to be carriers.

b. Generation I: X^A/Y, X^A/X^a

    Generation II: X^A/X^A, X^a/Y, $X^A Y$, X^A/X^-, X^A/X^a, X^A/Y

    Generation III: X^A/X^A, X^A/Y, X^A/X^a, X^A/X^a, X^A/Y, $X^A X^A$, X^a/Y,

                        X^A/Y, X^A/X^-

c. Because it is stated that the trait is rare, the assumption is that no one marrying into the pedigree carries the recessive allele. Therefore, the first couple has no chance of an affected child because the son received a Y chromosome from his father. The second couple has a 50 percent chance of having affected sons and no chance of having affected daughters. The third couple has no chance of having an affected child, but all of their daughters will be carriers.

69. In corn, the allele *s* causes sugary endosperm, whereas *S* causes starchy. What endosperm genotypes result from each of the following crosses?
   a. *s/s* female × *S/S* male
   b. *S/S* female × *s/s* male
   c. *S/s* female × *S/s* male

Answer: Remember, the endosperm is formed from two polar nuclei (which are genetically identical) and one sperm nucleus.

Female	Male	Polar nuclei	Sperm	Endosperm
s/s	*S/S*	*s* and *s*	*S*	*S/s/s*
S/S	*s/s*	*S* and *S*	*s*	*S/S/s*
S/s	*S/s*	$^1/_2$ *S* and *S*	$^1/_2$ *S*	$^1/_4$ *S/S/S*
				$^1/_4$ *S/S/s*
		$^1/_2$ *s* and *s*	$^1/_2$ *s*	$^1/_4$ *S/s/s*
				$^1/_4$ *s/s/s*

70. A plant geneticist has two pure lines, one with purple petals and one with blue. She hypothesizes that the phenotypic difference is due to two alleles of one gene. To test this idea, she aims to look for a 3 : 1 ratio in the F_2. She crosses the lines and finds that all the F_1 progeny are purple. The F_1 plants are selfed and 400 F_2 plants are obtained. Of these F_2 plants, 320 are purple and 80 are blue. Do these results fit her hypothesis well? If not, suggest why.

Answer: To use the chi-square test, first state the hypothesis being tested and the expected results. In this case, the hypothesis is that the phenotypic difference is due to two alleles of one gene. The expected results would be 300 purple and 100 blue (or an expected 3:1 ratio in the F_2). The formula to use is:

$$\chi^2 = \sum (\text{observed - expected})^2/\text{expected}$$

Class	Observed (*O*)	Expected (*E*)	$(O{-}E)^2$		$(O{-}E)^2/E$
purple	320	300	400	1.33	
blue	80	100	400	4.00	

Total = χ^2 = 5.33

This χ^2 value must now be looked up in a table (Table 2-2 in the companion text) which will give you the probability that these data fit the stated hypothesis. But first, you must also determine the degrees of freedom (df) or the number of independent variables in the data. Generally, the degrees of freedom can be calculated as the number of phenotypes in the problem minus one. In this example, there are two phenotypes (purple and blue) and therefore, one degree of freedom. Using the table, you will find that the data have a p value that is between 0.025 and 0.01. The p value is the probability that the deviation of the observed from that expected is due to chance alone. The relative standard commonly used in biological research for rejecting a hypothesis is $p < 0.05$. In this case, the data do not support the hypothesis.

Unpacking the Problem

71. A man's grandfather has galactosemia, a rare autosomal recessive disease caused by the inability to process galactose, leading to muscle, nerve, and kidney malfunction. The man married a woman whose sister had galactosemia. The woman is now pregnant with their first child.
    a. Draw the pedigree as described.
    b. What is the probability that this child will have galactosemia?
    c. If the first child does have galactosemia, what is the probability that a second child will have it?

Solution to the Problem

Answer:
   a. Galactosemia pedigree

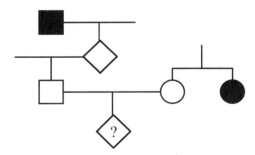

   b. Both parents must be heterozygous for this child to have a $1/4$ chance of inheriting the disease. Since the mother's sister is affected with galactosemia, their parents must have both been heterozygous. Since the mother does not have the trait, there is a $2/3$ chance that she is a carrier (heterozygous). One of the father's parents must be a carrier since his grandfather had the recessive trait. Thus, the father had a $1/2$ chance of inheriting the allele from that parent. Since these are all independent events, the child's risk is:

$$1/4 \times 2/3 \times 1/2 = 1/12$$

   c. If the child has galactosemia, both parents must be carriers and thus those probabilities become 100 percent. Now all future children have a $1/4$ chance of inheriting the disease.

CHALLENGING PROBLEMS

72. A geneticist working on peas has a single plant monohybrid *Y/y* (yellow) and, from a self of this plant, wants to produce a plant of genotype *y/y* to use as a tester. How many progeny plants need to be grown to be 95% sure of obtaining at least one in the sample?

Answer: The probability of obtaining *y/y*; *r/r* from this cross is $^1/_{16}$, and the probability of not obtaining this is $^{15}/_{16}$. Since only one plant is needed, the probability of not getting this genotype in *n* trials is $(^{15}/_{16})^n$. Because the probability of failure must be no greater than 5 percent:

$$(^{15}/_{16})^n < 0.05$$
$$n > 46.42, \text{ or } 47 \text{ plants}$$

73. A curious polymorphism in human populations has to do with the ability to curl up the sides of the tongue to make a trough ("tongue rolling"). Some people can do this trick, and others simply cannot. Hence, it is an example of a dimorphism. Its significance is a complete mystery. In one family, a boy was unable to roll his tongue but, to his great chagrin, his sister could. Furthermore, both his parents were rollers, and so were both grandfathers, one paternal uncle, and one paternal aunt. One paternal aunt, one paternal uncle, and one maternal uncle could not roll their tongues.

  a. Draw the pedigree for this family, defining your symbols clearly, and deduce the genotypes of as many individual members as possible.
  b. The pedigree that you drew is typical of the inheritance of tongue rolling and led geneticists to come up with the inheritance mechanism that no doubt you came up with. However, in a study of 33 pairs of identical twins, both members of 18 pairs could roll, neither member of 8 pairs could roll, and one of the twins in 7 pairs could roll but the other could not. Because identical twins are derived from the splitting of one fertilized egg into two embryos, the members of a pair must be genetically identical. How can the existence of the seven discordant pairs be reconciled with your genetic explanation of the pedigree?

Answer:
  a. In order to draw this pedigree, you should realize that if an individual's status is not mentioned, then there is no way to assign a genotype to that person. The parents of the boy in question had a phenotype (and genotype) that differed from his. Therefore, both parents were heterozygous and the boy, who is a non-roller, is homozygous recessive. Let *R* stand for the ability to roll the tongue and *r* stand for the inability to roll the tongue. The pedigree becomes:

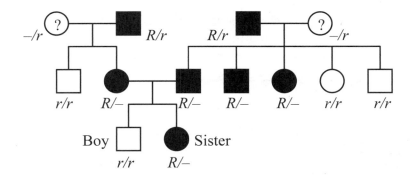

b. Assuming the twins are identical, there might be either an environmental component to the expression of that gene or developmental noise (see Chapter 1). Another possibility is that the *R* allele is not fully penetrant and that some genotypic "rollers" do not express the phenotype.

74. Red hair runs in families, as in the following pedigree. (Pedigree from W. R. Singleton and B. Ellis, *Journal of Heredity* 55, 1964, 261.)

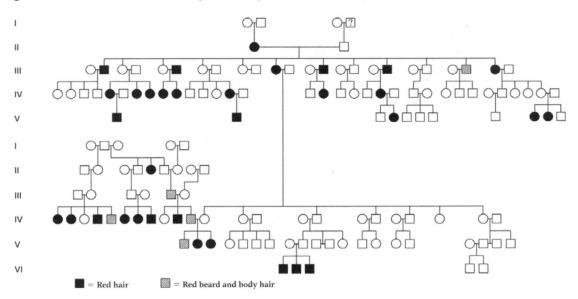

a. Does the inheritance pattern in this pedigree suggest that red hair could be caused by a dominant or a recessive allele of a gene that is inherited in a simple Mendelian manner?
b. Do you think that the red-hair allele is common or rare in the population as a whole?

Answer:
a. The inheritance pattern for red hair suggested by this pedigree is recessive since most red-haired individuals are from parents without this trait.
b. In most populations, the allele appears to be somewhat rare.

75. When many families were tested for the ability to taste the chemical phenylthiocarbamide, the matings were grouped into three types and the progeny were totaled, with the results shown below:

		Children	
Parents	Number of families	Tasters	Nontasters
Taster × taster	425	929	130
Taster × nontaster	289	483	278
Nontaster × nontaster	86	5	218

With the assumption that PTC tasting is dominant (*P*) and nontasting is recessive (*p*), how can the progeny ratios in each of the three types of mating be accounted for?

Answer: *Taster by taster cross*: Tasters can be either *T/T* or *T/t*, and the genotypic status cannot be determined until a large number of offspring are observed. A failure to obtain a 3 : 1 ratio in the marriage of two tasters would be expected because there are three types of marriages:

Mating	Genotypes	Phenotypes
T/T × *T/T*	all *T/T*	all tasters
T/T × *T/t*	1 *T/T* : 1 *T/t*	all tasters
T/t × *T/t*	1 *T/T* : 2 *T/t* : 1 *t/t*	3 tasters : 1 non-tasters

Taster by non-taster cross: There are two types of mating that resulted in the observed progeny:

Mating	Genotypes	Phenotypes
T/T × *t/t*	all *T/t*	all tasters
T/t × *t/t*	1 *T/t* : 1 *t/t*	1 tasters : 1 non-tasters

Again, the failure to obtain either a 1:0 ratio or a 1:1 ratio would be expected because of the two mating types.

Non-taster by non-taster cross: There is only one mating that is non-taster by non-taster (*t/t* × *t/t*), and 100 percent of the progeny would be expected to be non-tasters. Of 223 children, five were classified as tasters. Some could be the result of mutation (unlikely), some could be the result of misclassification (likely), some could be the result of a second gene that affects the expression of the gene in question (possible), some could be the result of developmental noise (possible), and some could be due to illegitimacy (possible).

76. A condition known as *icthyosis hystrix gravior* appeared in a boy in the early eighteenth century. His skin became very thick and formed loose spines that were sloughed off at intervals. When he grew up, this "porcupine man" married and had six sons, all of whom had this condition, and several daughters, all of whom were normal. For four generations, this condition was passed from father to son. From this evidence, what can you postulate about the location of the gene?

Answer: If the historical record is accurate, the data suggest Y linkage. Another explanation is an autosomal gene that is dominant in males and recessive in females. This has been observed for other genes in both humans and other species.

77. The wild-type (W) *Abraxas* moth has large spots on its wings, but the lacticolor (L) form of this species has very small spots. Crosses were made between strains differing in this character, with the following results:

	Parents		Progeny	
Cross	♀	♂	F_1	F_2
1	L	W	♀ W	♀ $\frac{1}{2}$ L, $\frac{1}{2}$ W
			♂ W	♂ W
2	W	L	♀ L	♀ $\frac{1}{2}$ W, $\frac{1}{2}$ L
			♂ W	♂ $\frac{1}{2}$ W, $\frac{1}{2}$ L

Provide a clear genetic explanation of the results in these two crosses, showing the genotypes of all individual moths.

Answer: The different sex-specific phenotypes found in the F_1 indicate sex-linkage—the females inherit the trait of their fathers. The first cross also indicates that the wild-type large spots are dominant over the lacticolor small spots. Let A = wild type and a = lacticolor.

Cross 1: If the male is assumed to be the hemizygous sex, then it soon becomes clear that the predictions do not match what was observed:

P   *a/a* female × *A/Y* male

F_1 *A/a* wild-type females
           a/Y lacticolor males

Therefore, assume that the female is the hemizygous sex. Let Z stand for the sex-determining chromosome in females. The cross becomes:

P   *a/Z* female × *A/A* male

F_1 *A/a*   wild-type male

	A/Z		wild-type female
F$_2$	1/4	A/Z	wild-type females
	1/2	A/–	wild-type males
	1/4	a/Z	lacticolor females

Cross 2:

P	A/Z female × a/a male		
F$_1$	a/Z		lacticolor females
	A/a		wild-type males
F$_2$	1/4	A/Z	wild-type females
	1/4	A/a	wild-type males
	1/4	a/Z	lacticolor females
	1/4	a/a	lacticolor males

78. The following pedigree shows the inheritance of a rare human disease. Is the pattern best explained as being caused by an X-linked recessive allele or by an autosomal dominant allele with expression limited to males?
(Pedigree modified from J. F. Crow, *Genetics Notes,* 6th ed. Copyright 1967 by Burgess Publishing Co., Minneapolis.)

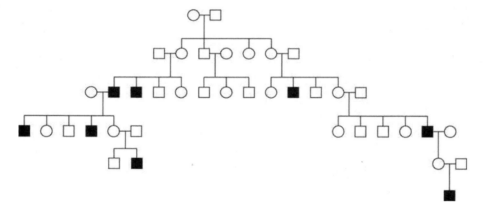

Answer: Note that only males are affected and that in all but one case, the trait can be traced through the female side. However, there is one example of an affected male having affected sons. If the trait is X-linked, this male's wife must be a carrier. Depending on how rare this trait is in the general population, this suggests that the disorder is caused by an autosomal dominant with expression limited to males.

79. A certain type of deafness in humans is inherited as an X-linked recessive trait. A man who suffers from this type of deafness married a normal woman, and they are expecting a child. They find out that they are distantly related. Part of the family tree is shown here.

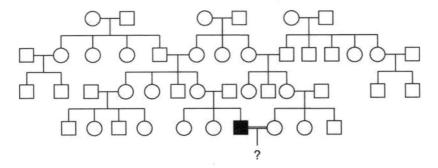

?

How would you advise the parents about the probability of their child being a deaf boy, a deaf girl, a normal boy, or a normal girl? Be sure to state any assumptions that you make.

Answer: Because the disorder is X-linked recessive, the affected male had to have received the allele, a, from the female common ancestor in the first generation. The probability that the affected man's wife also carries the a allele is the probability that she also received it from the female common ancestor. That probability is $1/8$.

The probability that the couple will have an affected boy is:
$$p(\text{father donates Y}) \times p(\text{the mother has } a) \times p(\text{mother donates } a)$$
$$1/2 \times 1/8 \times 1/2 = 1/32$$

The probability that the couple will have an affected girl is:
$$p(\text{father donates } X^a) \times p(\text{the mother has } a) \times p(\text{mother donates } a)$$
$$1/2 \times 1/8 \times 1/2 = 1/32$$

The probability of normal children is:
$$= 1 - p(\text{affected children})$$
$$= 1 - p(\text{affected male}) - p(\text{affected female})$$
$$= 1 - 1/32 - 1/32 = 30/32 = 15/16$$

Half the normal children will be boys, with a probability of $15/32$, and half the normal children will be girls, with a probability of $15/32$.

80. The accompanying pedigree shows a very unusual inheritance pattern that actually did exist. All progeny are shown, but the fathers in each mating have been omitted to draw attention to the remarkable pattern.

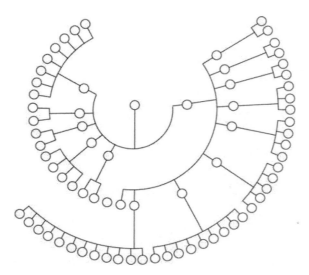

a. Concisely state exactly what is unusual about this pedigree.
b. Can the pattern be explained by Mendelian inheritance?

Answer:

a. The complete absence of male offspring is the unusual aspect of this pedigree. In addition, all progeny that mate carry the trait for lack of male offspring. If the male lethality factor were nuclear, the male parent would be expected to alter this pattern. Therefore, cytoplasmic inheritance is suggested.

b. If all females resulted from chance alone, then the probability of this result is $(1/2)^n$, where n = the number of female births. In this case n are 72. Chance is an unlikely explanation for the observations.

The observations can be explained by cytoplasmic factors by assuming that the proposed mutation in mitochondria is lethal only in males.

A modified form of Mendelian inheritance, an autosomal dominant, sex-limited lethal trait, might also explain these data, but it is an unlikely answer, due to the probability arguments above.

3

Independent Assortment of Genes

WORKING WITH THE FIGURES

1. Using Table 3-1, answer the following questions:

    a. If χ^2 is calculated to be 17 with 9 df, what is the approximate probability value?

    b. If χ^2 is 17 with 6 df, what is the probability value?

    c. What trend ("rule") do you see in the previous two calculations?

    Answer:
    a. For 9 degrees of freedom, with this value probability (p) is between 0.05 and 0.025

    b. For 6 degrees of freedom, with the same chi-square value, probability is between 0.01 and 0.005

    c. We could see that critical chi-square values differ depending on the number of categories (proportional to the "df") and with fewer categories, a statistical test yields a lower probability of the null hypothesis

2. Inspect Figure 3-8: which meiotic stage is responsible for generating Mendel's second law?

    Answer: Anaphase I, when homologous chromosomes independently assort from the equatorial plane in the metaphase I.

3. In Figure 3-9,

    a. identify the diploid nuclei.

    b. identify which part of the figure illustrates Mendel's first law.

    Answer:
    a. Diploid nuclei could be found in the diploid meiocytes, just before

meiosis.

b. Mendel's first law of the segregation is best illustrated in ascus (pl., asci) formation, where alleles from diploid meiocytes are distributed in the haploid sexual spores (ascospore).

4.   Inspect Figure 3-10: what would be the outcome in the octad if on rare occasions a nucleus from the postmeiotic mitotic division of nucleus 2 slipped past a nucleus from the postmeiotic mitotic division of nucleus 3? How could you measure the frequency of such a rare event?

Answer: In Figure 3-10, the four spore pairs (1-4) would change the number of nuclei per ascospore, so that the middle pair has one spore with no nuclei and one with 2. This would be easy to detect using micro-dissection of the large spore sample and calculating the frequency of such events.

5.   In Figure 3-11, if the input genotypes were a · B and A · b, what would be the genotypes colored blue?

Answer: Recombinant genotypes would be: AB and ab.

6.   In Figure 3-13, what are the origins of the chromosomes colored dark blue, light blue, and very light blue?

Answer: Dark blue chromosome comes from the parent with dominant homozygous genotype for an allele B, while light blue comes from the parent with a recessive homozygous genotype for the allele b. Very light blue chromosome also carries a recessive allele, but this chromosome comes from a testcross individual in F1 generation.

7.   In Figure 3-17, in which bar of the histogram would the genotype $R_1/r_1 · R_2/R_2 · r_3/r_3$ be found?

Answer: This genotype has three dominant alleles (doses) and belongs to the middle part of the histogram (with value of 20).

8.   In examining Figure 3-19, what do you think is the main reason for the difference in size of yeast and human mtDNA?

Answer: Yeast and human DNA differ in size probably because of their evolutionary distance. Yeast mitochondria have 78kb, yet most are nongenic DNA, while human mitochondria have only about 17kb, with energy-producing and other important gene sequences.

9. In Figure 3-20, what color is used to denote cytoplasm containing wild-type mitochondria?

Answer: Green is used to denote cytoplasm containing wild-type mitochondria?

10. In Figure 3-21, what would be the leaf types of progeny of the apical (top) flower?

Answer: The apical flower has variegated leaves and such gametes could be either with both chloroplast (therefore producing variegated offspring) or with only white or green, if chloroplasts segregate in the mature egg cell.

11. From the pedigree in Figure 3-25, what principle can you deduce about the inheritance of mitochondrial disease from affected fathers?

Answer: Human mitochondrial DNA is only inherited from the mothers.

BASIC PROBLEMS

12. Assume independent assortment and start with a plant that is dihybrid A/a ; B/b.

   a. What phenotypic ratio is produced from selfing it?
   b. What genotypic ratio is produced from selfing it?
   c. What phenotypic ratio is produced from testcrossing it?
   d. What genotypic ratio is produced from testcrossing it?

Answer:
   a. The expected phenotypic ratio from the self cross of A/a ; B/b is

9	$A/—$; $B/—$
3	$A/—$; b/b
3	a/a ; $B/—$
1	a/a ; b/b

   b. The expected genotypic ratio from the self cross of A/a ; B/b is

1	A/A ; B/B
2	A/A ; B/b
1	A/A ; b/b
2	A/a ; B/B
4	A/a ; B/b
2	A/a ; b/b
1	a/a ; B/B
2	a/a ; B/b
1	a/a ; b/b

c. and d. The expected phenotypic and genotypic ratios from the testcross of *A/a ; B/b* is

1	*A/a ; B/b*
1	*A/a ; b/b*
1	*a/a ; B/b*
1	*a/a ; b/b*

13. Normal mitosis takes place in a diploid cell of genotype *A/a ; B/b*. Which of the following genotypes might represent possible daughter cells?
    a. *A ; B*
    b. *a ; b*
    c. *A ; b*
    d. *a ; B*
    e. *A/A ; B/B*
    f. *A/a ; B/b*
    g. *a/a ; b/b*

    Answer: The resulting cells will have the identical genotype as the original cell: *A/a ; B/b*.

14. In a diploid organism of $2n = 10$, assume that you can label all the centromeres derived from its female parent and all the centromeres derived from its male parent. When this organism produces gametes, how many male- and female-labeled centromere combinations are possible in the gametes?

    Answer: The general formula for the number of different male/female centromeric combinations possible is $2n$, where n = number of different chromosome pairs. In this case, $2^5 = 32$.

15. It has been shown that when a thin beam of light is aimed at a nucleus, the amount of light absorbed is proportional to the cell's DNA content. Using this method, the DNA in the nuclei of several different types of cells in a corn plant were compared. The following numbers represent the relative amounts of DNA in these different types of cells:

    0.7, 1.4, 2.1, 2.8, and 4.2

    Which cells could have been used for these measurements? (**Note:** In plants, the endosperm part of the seed is often triploid, $3n$.)

    Answer: Because the DNA levels vary six-fold, the range covers cells that are haploid (spores or cells of the gametophyte stage) to cells that are triploid (the endosperm) and dividing (after DNA has replicated but prior to cell division). The following cells would fit the DNA measurements:

0.7   haploid cells
1.4   diploid cells in G1 or haploid cells after S but prior to cell division
2.1   triploid cells of the endosperm
2.8   diploid cells after S but prior to cell division
4.2   triploid cells after S but prior to cell division

16. Draw a haploid mitosis of the genotype a^+ ; b.

Answer:

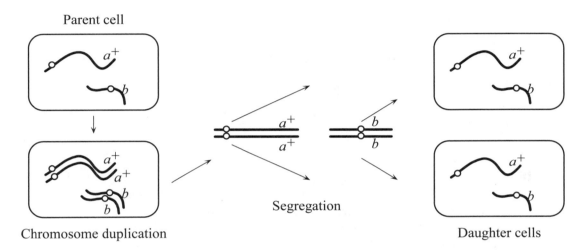

17. In moss, the genes *A* and *B* are expressed only in the gametophyte. A sporophyte of genotype *A/a* ; *B/b* is allowed to produce gametophytes.
a. What proportion of the gametophytes will be *A* ; *B*?
b. If fertilization is random, what proportion of sporophytes in the next generation will be *A/a* ; *B/b*?

Answer:
a. A sporophyte of *A/a* ; *B/b* genotype will produce gametophytes in the following proportions:
   1/4 *A* ; *B*
   1/4 *A* ; *b*
   1/4 *a* ; *B*
   1/4 *a* ; *b*
b. Random fertilization of the spores from the above gametophytes can occur 4′ 4 = 16 possible ways. Four of these combinations (*A* ; *B*′ *a* ; *b, a* ; *b*′ *A* ; *B, A* ; *b*′ *a* ; *B, a* ; *B*′ *A* ; *b*) will result in the desired *A/a* ; *B/b* sporophyte genotype. Therefore, 1/4 of the next generation should be of this genotype.

18. When a cell of genotype A/a ; B/b ; C/c having all the genes on separate chromosome pairs divides mitotically, what are the genotypes of the daughter cells?

Answer: Mitosis produces cells with the same starting genotype: A/a ; B/b ; C/c.

19. In the haploid yeast *Saccharomyces cerevisiae,* the two mating types are known as MAT*a* and MATα. You cross a purple (ad^-) strain of mating type a and a white (ad^+) strain of mating type αα. If ad^- and ad^+ are alleles of one gene, and a and α are alleles of an independently inherited gene on a separate chromosome pair, what progeny do you expect to obtain? In what proportions?

Answer:

P	ad^- ; a' ad^+ ; a
Transient diploid	ad^+/ad^- ; a/a
F$_1$	1/4 ad^+ ; a, white
	1/4 ad^- ; a, purple
	1/4 ad^+ ; a, white
	1/4 ad^- ; a, purple

20. In mice, dwarfism is caused by an X-linked recessive allele, and pink coat is caused by an autosomal dominant allele (coats are normally brownish). If a dwarf female from a pure line is crossed with a pink male from a pure line, what will be the phenotypic ratios in the F$_1$ and F$_2$ in each sex? (Invent and define your own gene symbols.)

Answer: The cross is female X^d/X^d ; p/p' male X^D/Y ; P/P where P = dominant allele for pink and d = recessive allele for dwarf.

F$_1$ 1/2 X^D/X^d : P/p (pink female)
1/2 X^d/Y ; P/p (dwarf, pink male)

F$_2$ 1/16 X^D/X^d : P/P (pink female)
1/8 X^D/X^d : P/p (pink female)
1/16 X^D/X^d : p/p (wild type female)
1/16 X^d/X^d : P/P (dwarf, pink female)
1/8 X^d/X^d : P/p (dwarf, pink female)
1/16 X^d/X^d : p/p (dwarf female)
1/16 X^D/Y ; P/P (pink male)
1/8 X^D/Y ; P/p (pink male)
1/16 X^D/Y ; p/p (wild type male)
1/16 X^d/Y ; P/P (dwarf, pink male)
1/8 X^d/Y ; P/p (dwarf, pink male)
1/16 X^d/Y ; p/p (dwarf male)

21. Suppose you discover two interesting *rare* cytological abnormalities in the karyotype of a human male. (A karyotype is the total visible chromosome complement.) There is an extra piece (satellite) on *one* of the chromosomes of pair 4, and there is an abnormal pattern of staining on one of the chromosomes of pair 7. With the assumption that all the gametes of this male are equally viable, what proportion of his children will have the same karyotype that he has?

Answer: His children will have to inherit the satellite-containing 4 (probability = 1/2), the abnormally-staining 7 (probability = 1/2), and the Y chromosome (probability = 1/2). To get all three, the probability is (1/2)(1/2)(1/2) = 1/8.

22. Suppose that meiosis occurs in the transient diploid stage of the cycle of a haploid organism of chromosome number *n*. What is the probability that an individual haploid cell resulting from the meiotic division will have a complete parental set of centromeres (that is, a set all from one parent or all from the other parent)?

Answer: The parental set of centromeres can match either parent, which means there are two ways to satisfy the problem. For any one pair, the probability of a centromere from one parent going into a specific gamete is 1/2. For *n* pairs, the probability of all the centromeres being from one parent is $(1/2)n$. Therefore, the total probability of having a haploid complement of centromeres from either parent is $2(1/2)n = (1/2)n–1$.

23. Pretend that the year is 1868. You are a skilled young lens maker working in Vienna. With your superior new lenses, you have just built a microscope that has better resolution than any others available. In your testing of this microscope, you have been observing the cells in the testes of grasshoppers and have been fascinated by the behavior of strange elongated structures that you have seen within the dividing cells. One day, in the library, you read a recent journal paper by G. Mendel on hypothetical "factors" that he claims explain the results of certain crosses in peas. In a flash of revelation, you are struck by the parallels between your grasshopper studies and Mendel's pea studies, and you resolve to write him a letter. What do you write? (Based on an idea by Ernest Kroeker.)

Answer:
Dear Monk Mendel:

I have worked with grasshoppers, however, not your garden peas. Although you are a man of the cloth, you are also a man of science, and I pray that you will not be offended when I state that I have specifically studied the reproductive organs of male grasshoppers. Indeed, I did not limit myself to studying the organs themselves; instead, I also studied the smaller units that make up the male organs and have beheld structures most amazing within them.

These structures are contained within numerous small bags within the male organs. Each bag has a number of these structures, which are long and threadlike at some times and short and compact at other times. They come together in the middle of a bag, and then they appear to divide equally. Shortly thereafter, the bag itself divides, and what looks like half of the threadlike structures goes into each new bag. Could it be, Sir, that these threadlike structures are the very same as your factors? I know, of course, that garden peas do not have male organs in the same way that grasshoppers do, but it seems to me that you found it necessary to emasculate the garden peas in order to do some crosses, so I do not think it too far-fetched to postulate a similarity between grasshoppers and garden peas in this respect.

Pray, Sir, do not laugh at me and dismiss my thoughts on this subject even though I have neither your excellent training nor your astounding wisdom in the Sciences. I remain your humble servant to eternity!

24. From a presumed testcross $A/a \times a/a$, in which A represents red and a represents white, use the χ^2 test to find out which of the following possible results would fit the expectations:

   a. 120 red, 100 white
   b. 5000 red, 5400 white
   c. 500 red, 540 white
   d. 50 red, 54 white

Answer: The hypothesis is that the organism being tested is a heterozygote and that the A/a and a/a progeny are of equal viability. The expected values would be that phenotypes occur with equal frequency. There are two genotypes in each case, so there is one degree of freedom.

$$c^2 = \sum (\text{observed-expected})^2/\text{expected}$$

   a. $c^2 = [(120-110)^2 + (100-110)^2]/110$
            $= 1.818; p > 0.10$, nonsignificant; hypothesis cannot be rejected

   b. $c^2 = [(5000-5200)^2 + (5400-5200)^2]/5200$
            $= 15.385; p < 0.005$, significant; hypothesis must be rejected

   c. $c^2 = [(500-520)^2 + (540-520)^2]/520$
            $= 1.538; p > 0.10$, nonsignificant; hypothesis cannot be rejected

   d. $c^2 = [(50-52)^2 + (54-52)^2]/52$
            $= 0.154; p > 0.50$, nonsignificant; hypothesis cannot be rejected

25. Look at the Punnett square in Figure 3-4.

    a. How many genotypes are there in the 16 squares of the grid?
    b. What is the genotypic ratio underlying the 9 : 3 : 3 : 1 phenotypic ratio?
    c. Can you devise a simple formula for the calculation of the number of progeny genotypes in dihybrid, trihybrid, and so forth crosses? Repeat for phenotypes.
    d. Mendel predicted that, within all but one of the phenotypic classes in the Punnett square, there should be several different genotypes. In particular, he performed many crosses to identify the underlying genotypes of the round, yellow phenotype. Show two different ways that could be used to identify the various genotypes underlying the round, yellow phenotype. (Remember, all the round, yellow peas look identical.)

Answer:
a. This is simply a matter of counting genotypes; there are nine genotypes in the Punnett square. Alternatively, you know there are three genotypes possible per gene, for example *R/R*, *R/r*, and *r/r*, and since both genes assort independently, there are 3 ′ 3 = 9 total genotypes.

b. Again, simply count. The genotypes are

    1 *R/R* ; *Y/Y*   1 *r/r* ; *Y/Y*   1 *R/R* ; *y/y*   1 *r/r* ; *y/y*
    2 *R/r* ; *Y/Y*   2 *r/r* ; *Y/y*   2 *R/r* ; *y/y*
    2 *R/R* ; *Y/y*
    4 *R/r* ; *Y/y*

c. To find a formula for the number of genotypes, first consider the following:

Number of genes	Number of genotypes	Number of phenotypes
1	3 = 31	2 = 21
2	9 = 32	4 = 22
3	27 = 33	8 = 23

Note that the number of genotypes is 3 raised to some power in each case. In other words, a general formula for the number of genotypes is 3*n*, where *n* equals the number of genes.

For allelic relationships that show complete dominance, the number of phenotypes is 2 raised to some power. The general formula for the number of phenotypes observed is 2*n*, where *n* equals the number of genes.

d. The round, yellow phenotype is *R/–* ; *Y/–*. Two ways to determine the exact genotype of a specific plant are through selfing or conducting a testcross.

With selfing, complete heterozygosity will yield a 9:3:3:1 phenotypic ratio. Homozygosity at one locus will yield a 3:1 phenotypic ratio, while homozygosity at both loci will yield only one phenotypic class.

With a testcross, complete heterozygosity will yield a 1:1:1:1 phenotypic ratio. Homozygosity at one locus will yield a 1:1 phenotypic ratio, while homozygosity at both loci will yield only one phenotypic class.

26. Assuming independent assortment of all genes, develop formulas that show the number of phenotypic classes and the number of genotypic classes from selfing a plant heterozygous for n gene pairs.

Answer: Assuming independent assortment and simple dominant/recessive relationships of all genes, the number of genotypic classes expected from selfing a plant heterozygous for n gene pairs is 3^n and the number of phenotypic classes expected is 2^n.

27. **Note:** The first part of this problem was introduced in Chapter 2. The line of logic is extended here.

In the plant *Arabidopsis thaliana,* a geneticist is interested in the development of trichomes (small projections) on the leaves. A large screen turns up two mutant plants (A and B) that have no trichomes, and these mutants seem to be potentially useful in studying trichome development. (If they are determined by single-gene mutations, then finding the normal and abnormal function of these genes will be instructive.) Each plant was crossed with wild type; in both cases, the next generation (F_1) had normal trichomes. When F_1 plants were selfed, the resulting F_2's were as follows:

F_2 from mutant A: 602 normal ; 198 no trichomes
F_2 from mutant B: 267 normal ; 93 no trichomes

a. What do these results show? Include proposed genotypes of all plants in your answer.
b. Assume that the genes are located on separate chromosomes. An F_1 is produced by crossing the original mutant A with the original mutant B. This F_1 is testcrossed: What proportion of testcross progeny will have no trichomes?

Answer:
a. The data for both crosses suggest that both *A* and *B* mutant plants are homozygous for a recessive allele. Both F_2 crosses give 3:1 normal to mutant ratios of progeny. For example, let A = normal and a = mutant, then

P A/A' a/a

F$_1$ A/a
F$_2$ 1 A/A phenotype: normal
      2 A/a phenotype: normal
      1 a/a phenotype: mutant (no trichomes)

b. The cross is A/A ; b/b' a/a ; B/B to give the F$_1$ of A/a ; B/b. This is then test crossed (crossed to a/a ; b/b) to give

1/4 A/a ; B/b (normal)
1/4 A/a ; b/b (no trichomes)
1/4 a/a ; B/b (no trichomes)
1/4 a/a ; b/b (no trichomes)
or 1 normal : 3 no trichomes

28. In dogs, dark coat color is dominant over albino and short hair is dominant over long hair. Assume that these effects are caused by two independently assorting genes, and write the genotypes of the parents in each of the crosses shown here, in which D and A stand for the dark and albino phenotypes, respectively, and S and L stand for the short-hair and long-hair phenotypes.

Parental phenotypes	*Number of progeny*			
	D, S	D, L	A, S	A, L
a. D, S × D, S	89	31	29	11
b. D, S × D, L	18	19	0	0
c. D, S × A, S	20	0	21	0
d. A, S × A, S	0	0	28	9
e. D, L × D, L	0	32	0	10
f. D, S × D, S	46	16	0	0
g. D, S × D, L	30	31	9	11

Use the symbols C and c for the dark and albino coat-color alleles and the symbols S and s for the short-hair and long-hair alleles, respectively. Assume homozygosity unless there is evidence otherwise. (Problem 28 is reprinted by permission of Macmillan Publishing Co., Inc., from M. Strickberger, *Genetics.* Copyright 1968 by Monroe W. Strickberger.)

Answer:
a. C/c ; S/s' C/c ; S/s There are 3 short : 1 long, and 3 dark : 1 albino. Therefore, each gene is heterozygous in the parents.
b. C/C ; S/s' $C/-$; s/s There are no albino, and there are 1 long : 1 short indicating a testcross for this trait.
c. C/c ; S/S' c/c ; $S/-$ There are no long, and there are 1 dark : 1 albino.
d. c/c ; S/s' c/c ; S/s All are albino, and there are 3 short : 1 long.
e. C/c ; s/s' C/c ; s/s All are long, and there are 3 dark : 1 albino.

f. C/C ; S/s' $C/-$; S/s There are no albino, and there are 3 short : 1 long.

g. C/c ; S/s' C/c ; s/s There are 3 dark : 1 albino and 1 short : 1 long.

29. In tomatoes, two alleles of one gene determine the character difference of purple (P) versus green (G) stems, and two alleles of a separate, independent gene determine the character difference of "cut" (C) versus "potato" (Po) leaves. The results for five matings of tomato-plant phenotypes are as follows:

Mating	Parental phenotypes	Number of progeny			
		P, C	P, Po	G, C	G, Po
1	P, C × G, C	321	101	310	107
2	P, C × P, Po	219	207	64	71
3	P, C × G, C	722	231	0	0
4	P, C × G, Po	404	0	87	0
5	P, Po × G, C	70	91	86	77

a. Determine which alleles are dominant.

b. What are the most probable genotypes for the parents in each cross? (Problem 29 is from A. M. Srb, R. D. Owen, and R. S. Edgar, *General Genetics,* 2nd ed. Copyright 1965 by W. H. Freeman and Company.)

Answer:

a. Cross 2 indicates that purple (G) is dominant to green (g), and cross 1 indicates cut (P) is dominant to potato (p).

b.

Cross 1: G/g ; P/p' g/g ; P/p There are 3 cut : 1 potato, and 1 purple : 1 green.

Cross 2: G/g ; P/p' G/g ; p/p There are 3 purple : 1 green and 1 cut : 1 potato.

Cross 3: G/G ; P/p' g/g ; P/p There are no green, and there are 3 cut : 1 potato.

Cross 4: G/g ; P/P' g/g ; p/p There are no potatoes, and there are 1 purple : 1 green.

Cross 5: G/g ; p/p' g/g ; P/p There are 1 cut : 1 potato, and there are 1 purple : 1 green.

30. A mutant allele in mice causes a bent tail. Six pairs of mice were crossed. Their phenotypes and those of their progeny are given in the following table. N is normal phenotype; B is bent phenotype. Deduce the mode of inheritance of this phenotype.

Cross	Parents		Progeny	
	♀	♂	♀	♂
1	N	B	All B	All N

2	B	N	1/2B, 1/2N	1/2B, 1/2N
3	B	N	All B	All B
4	N	N	All N	All N
5	B	B	All B	All B
6	B	B	All B	1/2B, 1/2N

a. Is it recessive or dominant?
b. Is it autosomal or sex-linked?
c. What are the genotypes of all parents and progeny?

Answer:
a. From cross 6, bent (*B*) is dominant to normal (*b*). Both parents are "bent," yet some progeny are "normal."

b. From cross 1, it appears that the trait is inherited in a sex-specific manner, in this case as X-linked (since sons always inherit one of the mother's X chromosomes).

c. In the following table, the Y chromosome is stated; the X is implied.

	Parents		*Progeny*	
Cross	♀	♂	♀	♂
1	*b/b*	*B/Y*	*B/b*	*b/Y*
2	*B/b*	*b/Y*	*B/b, b/b*	*B/Y, b/Y*
3	*B/B*	*b/Y*	*B/b*	*B/Y*
4	*b/b*	*b/Y*	*b/b*	*b/Y*
5	*B/B*	*B/Y*	*/B*	*B/Y*
6	*B/b*	*B/Y*	*B/B, B/b*	*B/Y, b/Y*

31. The normal eye color of *Drosophila* is red, but strains in which all flies have brown eyes are available. Similarly, wings are normally long, but there are strains with short wings. A female from a pure line with brown eyes and short wings is crossed with a male from a normal pure line. The F_1 consists of normal females and short-winged males. An F_2 is then produced by intercrossing the F_1. Both sexes of F_2 flies show phenotypes as follows:

3/8 red eyes, long wings
3/8 red eyes, short wings
1/8 brown eyes, long wings
1/8 brown eyes, short wings

Deduce the inheritance of these phenotypes; use clearly defined genetic symbols of your own invention. State the genotypes of all three generations and the genotypic proportions of the F_1 and F_2.

Unpacking Problem 31

Before attempting a solution to this problem, try answering the following questions:

1. What does the word "normal" mean in this problem?

   Answer: *Normal* is used to mean wild type, or red eye color and long wings.

2. The words "line" and "strain" are used in this problem. What do they mean, and are they interchangeable?

   Answer: Both *line* and *strain* are used to denote pure-breeding fly stocks, and the words are interchangeable.

3. Draw a simple sketch of the two parental flies showing their eyes, wings, and sexual differences.

   Answer: Your choice.

4. How many different characters are there in this problem?

   Answer: Three characters are being followed: eye color, wing length, and sex.

5. How many phenotypes are there in this problem, and which phenotypes go with which characters?

   Answer: For eye color, there are two phenotypes: red and brown. For wing length, there are two phenotypes: long and short. For sex, there are two phenotypes: male and female.

6. What is the full phenotype of the F_1 females called "normal"?

   Answer: The F_1 females designated normal have red eyes and long wings.

7. What is the full phenotype of the F_1 males called "short winged"?

   Answer: The F_1 males that are called short-winged have red eyes and short wings.

8. List the F_2 phenotypic ratios for each character that you came up with in answer to question 4.

   Answer: The F_2 ratio is:
   > 3/8 red eyes, long wings
   > 3/8 red eyes, short wings
   > 1/8 brown eyes, long wings
   > 1/8 brown eyes, short wings

9. What do the F_2 phenotypic ratios tell you?

   Answer: Because there is not the expected 9:3:3:1 ratio, one of the factors that distorts the expected dihybrid ratio must be present. Such factors can be sex linkage, epistasis, genes on the same chromosome, environmental effect, reduced penetrance, or a lack of complete dominance in one or both genes.

10. What major inheritance pattern distinguishes sex-linked inheritance from autosomal inheritance?

    Answer: With sex linkage, traits are inherited in a sex-specific way. With autosomal inheritance, males and females have the same probabilities of inheriting the trait.

11. Do the F_2 data show such a distinguishing criterion?

    Answer: The F_2 does not indicate sex-specific inheritance.

12. Do the F_1 data show such a distinguishing criterion?

    Answer: The F_1 data does show sex-specific inheritance—all males are short-winged, like their mothers, while all females are normal-winged, like their fathers.

13. What can you learn about dominance in the F_1? The F_2?

    Answer: The F_1 suggests that long is dominant to short and red is dominant to brown. The F_2 data show a 3 red : 1 brown ratio indicating the dominance of red but a 1 : 1 long : short ratio indicative of a testcross. Without the F_1 data, it is not possible to determine which form of the wing character is dominant.

14. What rules about wild-type symbolism can you use in deciding which allelic symbols to invent for these crosses?

Answer: If Mendelian notation is used, then the red and long alleles need to be designated with uppercase letters, for example R and L, while the brown (r) and short (l) alleles need to be designated with lowercase letters. If *Drosophila* notation is used, then the brown allele may be designated with a lowercase b and the wild-type (red) allele with a b^+; the short wing-length gene with an s and the wild-type (long) allele with an s^+. (Genes are often named after their mutant phenotype.)

15. What does "deduce the inheritance of these phenotypes" mean?

Answer: To deduce the inheritance of these phenotypes means to provide all genotypes for all animals in the three generations discussed and account for the ratios observed.

Now try to solve the problem. If you are unable to do so, make a list of questions about the things that you do not understand. Inspect the key concepts at the beginning of the chapter and ask yourself which are relevant to your questions. If this approach doesn't work, inspect the messages of this chapter and ask yourself which might be relevant to your questions.

Solution to the Problem

Start this problem by writing the crosses and results so that all the details are clear.

P   brown, short female′ red, long male
F_1 red, long females
       red, short males

These results tell you that red-eyed is dominant to brown-eyed, and since both females and males are red-eyed, this gene is autosomal. Since males differ from females in their genotype with regard to wing length, this trait is sex-linked. Knowing that *Drosophila* females are XX and males are XY, the long-winged females tell us that long is dominant to short and that the gene is X-linked. Let B = red, b = brown, S = long, and s = short. The cross can be rewritten as follows:

P b/b ; s/s' B/B ; S/Y

F_1 $1/2$ B/b ; S/s females
       $1/2$ B/b ; s/Y males

F_2 $1/16$ B/B ; S/s red, long, female
       $1/16$ B/B ; s/s red, short, female
       $1/8$ B/b ; S/s red, long, female

1/8 *B/b* ; *s/s*   red, short, female
1/16 *b/b* ; *S/s*   brown, long, female
1/16 *b/b* ; *s/s*   brown, short, female
1/16 *B/B* ; *S/Y*   red, long, male
1/16 *B/B* ; *s/Y*   red, short, male
1/8 *B/b* ; *S/Y*   red, long, male
1/8 *B/b* ; *s/Y*   red, short, male
1/16 *b/b* ; *S/Y*   brown, long, male
1/16 *b/b* ; *s/Y*   brown, short, male

The final phenotypic ratio is
3/8 red, long
3/8 red, short
1/8 brown, long
1/8 brown, short
with equal numbers of males and females in all classes.

32. In a natural population of annual plants, a single plant is found that is sickly looking and has yellowish leaves. The plant is dug up and brought back to the laboratory. Photosynthesis rates are found to be very low. Pollen from a normal dark-green-leaved plant is used to fertilize emasculated flowers of the yellowish plant. A hundred seeds result, of which only 60 germinate. All the resulting plants are sickly yellow in appearance.

a. Propose a genetic explanation for the inheritance pattern.
b. Suggest a simple test for your model.
c. Account for the reduced photosynthesis, sickliness, and yellowish appearance.

Answer:
a. Because photosynthesis is affected and the plants are yellow rather than green, it is likely that the chloroplasts are defective. If the defect maps to the DNA of the chloroplast, the trait will be maternally inherited. This fits the data that all progeny have the phenotype of the female parent and not the phenotype of the male (pollen-donor) parent.

b. If the defect maps to the DNA of the chloroplast, the trait will be maternally inherited. Use pollen from sickly, yellow plants and cross to emasculated flowers of a normal dark green-leaved plant. All progeny should have the normal dark green phenotype.

c. The chloroplasts contain the green pigment chlorophyll and are the site of photosynthesis. A defect in the production of chlorophyll would give rise to all the stated defects.

33. What is the basis for the green-and-white color variegation in the leaves of *Mirabilis*? If the following cross is made,

variegated ♀ × green ♂

what progeny types can be predicted? What about the reciprocal cross?

Answer: Maternal inheritance of chloroplasts results in the green-white color variegation observed in *Mirabilis*.

Cross 1: variegated female ♀ green male ♂ variegated, green, or white progeny
Cross 2: green female ♀ variegated male ♂ green progeny

In both crosses, the pollen (male contribution) contains no chloroplasts and thus does not contribute to the inheritance of this phenotype. Eggs from a variegated female plant can be of three types : contain only "green" chloroplasts, contain only "white" chloroplasts, or contain both (variegated). The offspring will have the phenotype associated with the egg's chloroplasts.

34. In *Neurospora*, the mutant *stp* exhibits erratic stop-and-start growth. The mutant site is known to be in the mtDNA. If an *stp* strain is used as the female parent in a cross with a normal strain acting as the male, what type of progeny can be expected? What about the progeny from the reciprocal cross?

Answer: The crosses are
Cross 1: stop-start female ♀ wild-type male ♂ all stop-start progeny
Cross 2: wild-type female ♀ stop-start male ♂ all wild-type progeny mtDNA is inherited only from the "female" in *Neurospora*

35. Two corn plants are studied. One is resistant (R) and the other is susceptible (S) to a certain pathogenic fungus. The following crosses are made, with the results shown:

S ♀ × R ♂ → all progeny S
R ♀ × S ♂ → all progeny R

What can you conclude about the location of the genetic determinants of R and S?

Answer: The genetic determinants of R and S are showing maternal inheritance and are therefore cytoplasmic. It is possible that the gene that confers resistance maps either to the mtDNA or cpDNA.

36. A presumed dihybrid in *Drosophila, B/b ; F/f* is testcrossed with *b/b ; f/f.* (*B =*

black body ; b = brown body; F = forked bristles; f = unforked bristles.) The results are

Black, forked	230
Black, unforked	210
Brown, forked	240
Brown, unforked	250

Use the χ^2 test to determine if these results fit the results expected from testcrossing the hypothesized dihybrid.

Answer: The hypothesis is that the organism being tested is a dihybrid with independently assorting genes and that all progeny are of equal viability. The expected values would be that phenotypes occur with equal frequency. There are four phenotypes so there are 3 degrees of freedom.

$$c^2 = \sum (\text{observed-expected})^2/\text{expected}$$

$$c^2 = [(230–233)^2 + (210–233)^2 + (240–233)^2 + (250–233)^2]/233$$
$$= 3.75; \ p \text{ value between 0.1 and 0.5, nonsignificant; hypothesis cannot be rejected}$$

37. Are the following progeny numbers consistent with the results expected from selfing a plant presumed to be a dihybrid of two independently assorting genes, H/h ; R/r? (H = hairy leaves; h = smooth leaves; R = round ovary; r = elongated ovary.) Explain your answer.

hairy, round	178
hairy, elongated	62
smooth, round	56
smooth, elongated	24

Answer: The hypothesis is that the organism being tested is a dihybrid with independently assorting genes and that all progeny are of equal viability. The expected values would be that phenotypes occur in a 9 : 3 : 3 : 1 ratio. There are four phenotypes so there are 3 degrees of freedom.

$$c^2 = \sum (\text{observed-expected})^2/\text{expected}$$

$$c^2 = (178–180)^2/180 + (62–60)^2/60 + (56–60)^2/60 + (24–20)2^{/}20$$
$$= 1.156; p > 0.50, \text{ nonsignificant; hypothesis cannot be rejected}$$

38. A dark female moth is crossed with a dark male. All the male progeny are dark, but half the female progeny are light and the rest are dark. Propose an explanation for this pattern of inheritance.

Answer: When results of a cross are sex-specific, sex linkage should be considered. In moths, the heterogametic sex is actually the female while the male is the homogametic sex. Assuming that dark (D) is dominant to light (d), then the data can be explained by the dark male being heterozygous (D/d) and the dark female being hemizygous (D). All male progeny will inherit the D allele from their mother and therefore be dark, while half the females will inherit D from their father, (and be dark) and half will inherit d (and be light).

39. In *Neurospora*, a mutant strain called stopper (*stp*) arose spontaneously. Stopper showed erratic "stop and start" growth, compared with the uninterrupted growth of wild-type strains. In crosses, the following results were found:

$♀$ stopper $× ♂$ wild type $→$ progeny all stopper
$♀$ wild type $× ♂$ stopper $→$ progeny all wild type

a. What do these results suggest regarding the location of the stopper mutation in the genome?
b. According to your model for part *a*, what progeny and proportions are predicted in octads from the following cross, including a mutation *nic*^{3} located on chromosome VI?

$♀$ *stp · nic3* $×$ wild type $♂$

Answer:
a. This inheritance pattern is diagnostic for cytoplasmic organelle inheritance. These crosses indicate the mutant gene resides in the mitochondria. (Neurospora is a fungi and does not have chloroplasts.)
b. All progeny from this cross will have the "maternal" trait stopper but the *nic3* allele should segregate 1:1 in the octad—four spores will be *stp* ; *nic3* and four spores will be *stp* ; *nic3*$^+$.

40. In polygenic systems, how many phenotypic classes corresponding to number of polygene "doses" are expected in selfs

a. of strains with four heterozygous polygenes?
b. of strains with six heterozygous polygenes?

Answer:
a. There should be nine classes—0, 1, 2, 3, 4, 5, 6, 7, 8 "doses."
b. There should be 13 classes—0, 1, 2, 3, 4, 5, 6, 7, 8, 9, 10, 11, 12 "doses."

41. In the self of a polygenic trihybrid R_1/r_1 ; R_2/r_2 ; R_3/r_3, use the product and sum rules to calculate the proportion of progeny with just one polygene "dose."

Answer: There are three ways for a self cross of this genotype to give rise to just one dose — one dose of R_1 and none of R_2 or R_3; one dose of R_2 and none of R_1 or R_3; or one dose of R_3 and none of R_1 or R_2. The chance of inheriting one dose of R_1 (from $R_1/r_1 \times R_1/r_1$) is 1/2. The chance of no doses of R_2 (from $R_2/r_2 \times R_2/r_2$) is 1/4 as is the chance of no doses of R_3 (from $R_3/r_3 \times R_3/r_3$). Therefore, the desired outcome of just one dose is $3 \times 1/2 \times 1/4 \times 1/4 = 3/32$.

42. Reciprocal crosses and selfs were performed between the two moss species *Funaria mediterranea* and *F. hygrometrica*. The sporophytes and the leaves of the gametophytes are shown in the accompanying diagram.

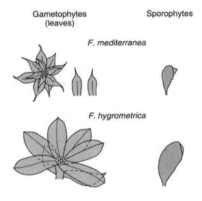

The crosses are written with the female parent first.

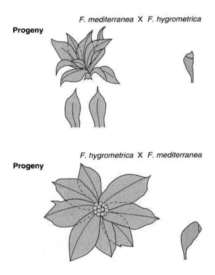

a. Describe the results presented, summarizing the main findings.
b. Propose an explanation of the results.
c. Show how you would test your explanation; be sure to show how it could be distinguished from other explanations.

Answer:

a. Both the gametophyte and the sporophyte are closer in shape to the mother than the father. Note that a size increase occurs in each type of cross.

b. Gametophyte and sporophyte morphology might be affected by cytoplasmic inheritance (from eggs) or similar extranuclear factors. Leaf size may be a function of the interplay between nuclear genome contributions.

c. If extranuclear factors are affecting morphology while nuclear factors are affecting leaf size, then repeated backcrosses could be conducted, using the hybrid as the female. This would result in the cytoplasmic information remaining constant while the nuclear information becomes increasingly like that of the backcross parent. Leaf morphology should therefore remain constant while leaf size would decrease toward the size of the backcross parent.

43. Assume that diploid plant A has a cytoplasm genetically different from that of plant B. To study nuclear–cytoplasmic relations, you wish to obtain a plant with the cytoplasm of plant A and the nuclear genome predominantly of plant B. How would you go about producing such a plant?

Answer: The goal here is to generate a plant with the cytoplasm of plant A and the nuclear genome predominantly of plant B. Remember that the cytoplasm is contributed by the egg only. So using plant A as the maternal parent, cross to B (as the paternal parent) and then backcross the progeny of this cross using plant B again as the paternal parent. Repeat for several generations until virtually the entire nuclear genome is from the B parent.

44. You are studying a plant with tissue comprising both green and white sectors. You wish to decide whether this phenomenon is due (1) to a chloroplast mutation of the type considered in this chapter or (2) to a dominant nuclear mutation that inhibits chlorophyll production and is present only in certain tissue layers of the plant as a mosaic. Outline the experimental approach that you would use to resolve this problem.

Answer: If the variegation is due to a chloroplast mutation, then the phenotype of the offspring will be controlled solely by the phenotype of the maternal parent. Look for flowers on white branches and test to see if they produce seeds that grow into all white plants regardless of the source of the pollen. To test whether a dominant nuclear mutation is responsible for the variegation, cross pollen from flowers on white branches to green plants (or flowers on green branches of same plant) and see if half the progeny are white and half are green (assuming the dominant mutation is heterozygous) or all white, if the dominant mutation is homozygous.

45. Early in the development of a plant, a mutation in cpDNA removes a specific BgIII restriction site (B) as follows:

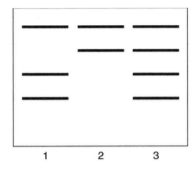

In this species, cpDNA is inherited maternally. Seeds from the plant are grown, and the resulting progeny plants are sampled for cpDNA. The cpDNAs are cut with BgIII, and Southern blots are hybridized with the probe P shown. The autoradiograms show three patterns of hybridization:

Explain the production of these three seed types.

Answer: Progeny plants inherited only normal chloroplast, cpDNA (lane 1); only mutant cpDNA (lane 2); or both (lane 3). In order to get homoplasmic cpDNA (all chloroplasts containing the same DNA), seen in lanes 1 and 2, segregation of chloroplasts had to occur.

CHALLENGING PROBLEMS

46. You have three jars containing marbles, as follows:

jar 1	600 red	and	400 white
jar 2	900 blue	and	100 white
jar 3	10 green	and	990 white

a. If you blindly select one marble from each jar, calculate the probability of obtaining

(1) a red, a blue, and a green.
(2) three whites.
(3) a red, a green, and a white.
(4) a red and two whites.
(5) a color and two whites.
(6) at least one white.

b. In a certain plant, R = red and r = white. You self a red *R/r* heterozygote with the express purpose of obtaining a white plant for an experiment. What minimum number of seeds do you have to grow to be at least 95 percent certain of obtaining at least one white individual?

c. When a woman is injected with an egg fertilized in vitro, the probability of its implanting successfully is 20 percent. If a woman is injected with five eggs simultaneously, what is the probability that she will become pregnant? (Part c is from Margaret Holm.)

Answer:

a. Before beginning the specific problems, calculate the probabilities associated with each jar.

jar 1 p(R) 600/(600 + 400) = 0.6
   p(W) = 400/(600 + 400) = 0.4

jar 2 p(B) = 900/(900 + 100) = 0.9
   p(W) 100/(900 + 100) = 0.1

jar 3 p(G) 10/(10 + 990) = 0.01
   p(W) 990/(10 + 990) = 0.99

(1) p(R, B, G) = (0.6)(0.9)(0.01) = 0.0054

(2) p(W, W, W) = (0.4)(0.1)(0.99) = 0.0396

(3) Before plugging into the formula, you should realize that, while white can come from any jar, red and green must come from specific jars (jar 1 and jar 3). Therefore, white must come from jar 2:
   p(R, W, G) = (0.6)(0.1)(0.01) = 0.0006

(4) p(R, W, W) = (0.6)(0.1)(0.99) = 0.0594

(5) There are three ways to satisfy this:
   R, W, W   or   W, B, W   or   W, W, G
   = (0.6)(0.1)(0.99) + (0.4)(0.9)(0.99) + (0.4)(0.1)(0.01)
   = 0.0594 + 0.3564 + 0.0004 = 0.4162

(6) At least one white is the same as 1 minus no whites:
   p(at least 1 W) = 1 − p(no W) = 1 − p(R, B, G)
   = 1 − (0.6)(0.9)(0.01) = 1 − 0.0054 = 0.9946

b. The cross is *R/r* × *R/r*. The probability of red (*R/–*) is 3/4, and the probability of white (*r/r*) is 1/4. Because only one white is needed, the only unacceptable result is all red.

In n trials, the probability of all red is $(3/4)n$. Because the probability of failure must be no greater than 5 percent:

$(3/4)n < 0.05$

$n > 10.41$, or 11 seeds

c. The $p(\text{failure}) = 0.8$ for each egg. Since all eggs are implanted simultaneously, the $p(5 \text{ failures}) = (0.8)5$. The $p(\text{at least one success}) = 1 - (0.8)5 = 1 - 0.328 = 0.672$

47. In tomatoes, red fruit is dominant over yellow, two-loculed fruit is dominant over many-loculed fruit, and tall vine is dominant over dwarf. A breeder has two pure lines: (1) red, two-loculed, dwarf and (2) yellow, many-loculed, tall. From these two lines, he wants to produce a new pure line for trade that is yellow, two-loculed, and tall. How exactly should he go about doing so? Show not only which crosses to make, but also how many progeny should be sampled in each case.

a. One of the genes is obviously quite distant from the other three, which appear to be tightly (closely) linked. Which is the distant gene?
b. What is the probable order of the three tightly linked genes?

(Problem 47 is from Franklin Stahl, *The Mechanics of Inheritance,* 2nd ed. Copyright 1969, Prentice Hall, Englewood Cliffs, N.J. Reprinted by permission.)

Answer: Use the following symbols:

Gene function	Dominant allele	Recessive allele
color	R = red	r = yellow
loculed	L = two	l = many
height	H = tall	h = dwarf

The starting plants are pure-breeding, so their genotypes are:

red, two-loculed, dwarf R/R ; L/L ; h/h
and
yellow, many-loculed, tall r/r ; l/l ; H/H

The farmer wants to produce a pure-breeding line that is yellow, two-loculed, and tall, which would have the genotype r/r ; L/L ; H/H.

The two pure-breeding starting lines will produce an F_1 that will be R/r ; L/l ; H/h. By doing an $F_1 \times F_1$ cross, $1/64$ of the F_2 progeny should have the correct genotype ($1/4$ r/r × $1/4$ L/L × $1/4$ H/H). The probability of NOT getting that is $(1 - 1/64)^n$, where n is number of progeny scored. For that to be less than 5 percent, n

= 191. So we need at least 191 progeny to start with, and by selecting yellow, two-loculed, and tall plants from these progeny, the known genotype will be r/r ; $L/–$; $H/–$. To identify how many of these are required for further testing (by test cross): the probability of being homozygous dominant for both (given that we are selecting only from those plants with dominant phenotypes) is $^1/_3 \times {}^1/_3 = {}^1/_9$. Therefore, the probability of a plant not being homozygous for both is $^8/_9$. We want the probability of all plants tested not being homozygous for both to be less than 5 percent, or $(8/9)^n < 0.05$. If $n = 26$, $p = .047$. So at least 26 of the yellow, two-loculed, and tall progeny should be testcrossed to a l/l; h/h parent to determine which are homozygous for the two dominant traits. (**Note**: Several l/l ; h/h testers are likely to be recovered among the 191 F$_2$ progeny generated. So you will only need one testcross for each candidate.)

For each testcross, the plant will obviously be discarded if the testcross reveals a heterozygous state for the gene in question. If no recessive allele is detected, then the minimum number of progeny that must be examined to be 95 percent confident that the plant is homozygous is based on the frequency of the dominant phenotype if heterozygous, which is 1/2. In n progeny, the probability of obtaining all dominant progeny in a testcross, given that the plant is heterozygous, is $(1/2)n$. To be 95 percent confident of homozygosity, the following formula is used, where 5 percent is the probability that it is not homozygous:

$(1/2)n = 0.05$

$n = 4.3$, or 5 phenotypically dominant progeny must be obtained from each testcross to be 95 percent confident that the plant is homozygous.

48. We have dealt mainly with only two genes, but the same principles hold for more than two genes. Consider the following cross:

$$A/a \; ; B/b \; ; C/c \; ; D/d \; ; E/e \times a/a \; ; B/b; \; c/c \; ; D/d \; ; e/e$$

a. What proportion of progeny will phenotypically resemble (1) the first parent, (2) the second parent, (3) either parent, and (4) neither parent?
b. What proportion of progeny will be genotypically the same as (1) the first parent, (2) the second parent, (3) either parent, and (4) neither parent? Assume independent assortment.

Answer:
a. Because each gene assorts independently, each probability should be considered separately and then multiplied together for the answer.

For (1) 1/2 will be A, 3/4 will be B, 1/2 will be C, 3/4 will be D, and 1/2 will be E.

$1/2 \times 3/4 \times 1/2 \times 3/4 \times 1/2 = 9/128$

For (2) 1/2 will be *a*, 3/4 will be *B*, 1/2 will be *c*, 3/4 will be *D*, and 1/2 will be *e*.

$$1/2 \times 3/4 \times 1/2 \times 3/4 \times 1/2 = 9/128$$

For (3) it is the sum of (1) and (2) = 9/128 + 9/128 = 9/64

For (4) it is 1 − (part 3) = 1 − 9/64 = 55/64

b. For (1) 1/2 will be *A/a*, 1/2 will be *B/b*, 1/2 will be *C/c*, 1/2 will be *D/d*, and 1/2 will be *E/e*.

$$1/2 \times 1/2 \times 1/2 \times 1/2 \times 1/2 = 1/32$$

For (2) 1/2 will be *a/a*, 1/2 will be *B/b*, 1/2 will be *c/c*, 1/2 will be *D/d*, and 1/2 will be *e/e*.

$$1/2 \times 1/2 \times 1/2 \times 1/2 \times 1/2 = 1/32$$

For (3) it is the sum of (1) and (2) = 1/16

For (4) it is 1 − (part 3) = 1 − 1/16 = 15/16

49. The accompanying pedigree shows the pattern of transmission of two rare human phenotypes: cataract and pituitary dwarfism. Family members with cataract are shown with a solid left half of the symbol; those with pituitary dwarfism are indicated by a solid right half.

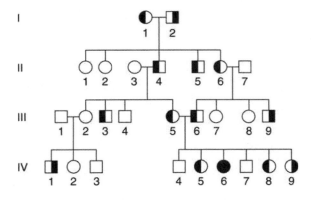

a. What is the most likely mode of inheritance of each of these phenotypes? Explain.
b. List the genotypes of all members in generation III as far as possible.
c. If a hypothetical mating took place between IV-1 and IV-5, what is the probability of the first child's being a dwarf with cataracts? A phenotypically normal child?

(Problem 49 is after J. Kuspira and R. Bhambhani, *Compendium of Problems in Genetics.* Copyright 1994 by Wm. C. Brown.)

Answer:

a. Cataracts appear to be caused by a dominant allele because affected people have affected parents. Dwarfism appears to be caused by a recessive allele because affected people have unaffected parents. Both traits appear to be autosomal.

b. Using *A* for cataracts, *a* for no cataracts, *B* for normal height, and *b* for dwarfism, the genotypes are:

III : *a/a* ; *B/b*, *a/a* ; *B/b*, *A/a* ; *B/–*, *a/a* ; *B/–*, *A/a* ; *B/b*, *A/a* ; *B/b*, *a/a* ; *B/–*, *a/a* ; *B/–*, *a/a* ; *b/b*

c. The mating is *a/a* ; *b/b* (IV-1) × *A/–* ; *B/–* (IV-5). Recall that the probability of a child's being affected by any disease is a function of the probability of each parent carrying the allele in question and the probability that one parent (for a dominant disorder) or both parents (for a recessive disorder) donate it to the child. Individual IV-1 is homozygous for these two genes, therefore, the only task is to determine the probabilities associated with individual IV-5.

The probability that individual IV-5 is heterozygous for dwarfism is 2/3. Thus the probability that she has the *b* allele and will pass it to her child is 2/3 × 1/2 = 1/3.

The probability that individual IV-5 is homozygous for cataracts is 1/3; the probability that she is heterozygous is 2/3. If she is homozygous for the allele that causes cataracts, she must pass it to her child or if she is heterozygous for cataracts, she has a probability of 1/2 of passing it to her child.

The probability that the first child is a dwarf with cataracts is the probability that the child inherits the *A* and *b* alleles from its mother which is (1/3 × 1)(2/3 × 1/2) + (2/3 × 1/2)(2/3 × 1/2) = 2/9. Alternatively, you can calculate the chance of inheriting the *b* allele (2/3 × 1/2) and not inheriting the *a* allele (1– 1/3) or 1/3 × 2/3 = 2/9.

The probability of having a phenotypically normal child is the probability that the mother donates the *a* and *B* (or not *b*) alleles, which is (2/3 × 1/2)(1– 1/3) = 2/9.

50. A corn geneticist has three pure lines of genotypes *a/a* ; *B/B* ; *C/C*, *A/A* ; *b/b* ; *C/C*, and *A/A* ; *B/B* ; *c/c*. All the phenotypes determined by *a*, *b*, and *c* will increase the market value of the corn; so, naturally, he wants to combine them all in one pure line of genotype *a/a* ; *b/b* ; *c/c*.

a. Outline an effective crossing program that can be used to obtain the *a/a* ; *b/b* ; *c/c* pure line.

b. At each stage, state exactly which phenotypes will be selected and give their expected frequencies.

c. Is there more than one way to obtain the desired genotype? Which is the best way?

Assume independent assortment of the three gene pairs. (**Note:** Corn will self- or cross-pollinate easily.)

Answer:

a. and b. Begin with any two of the three lines and cross them. If, for example, you began with *a/a* ; *B/B* ; *C/C* × *A/A* ; *b/b* ; *C/C*, the progeny would all be *A/a* ; *B/b* ; *C/C*. Crossing two of these would yield:

9	*A/–* ; *B/–* ; *C/C*
3	*a/a* ; *B/–* ; *C/C*
3	*A/–* ; *b/b* ; *C/C*
1	*a/a* ; *b/b* ; *C/C*

The *a/a* ; *b/b* ; *C/C* genotype has two of the genes in a homozygous recessive state and occurs in 1/16 of the offspring. If that were crossed with *A/A* ; *B/B* ; *c/c*, the progeny would all be *A/a* ; *B/b* ; *C/c*. Crossing two of them (or "selfing") would lead to a 27:9:9:9:3:3:3:1 ratio, and the plant occurring in 1/64 of the progeny would be the desired *a/a* ; *b/b* ; *c/c*.

There are several different routes to obtaining *a/a* ; *b/b* ; *c/c*, but the one outlined above requires only four crosses.

51. In humans, color vision depends on genes encoding three pigments. The *R* (red pigment) and *G* (green pigment) genes are close together on the X chromosome, whereas the *B* (blue pigment) gene is autosomal. A recessive mutation in any one of these genes can cause color blindness. Suppose that a color-blind man married a woman with normal color vision. The four sons from this marriage were color-blind, and the five daughters were normal. Specify the most likely genotypes of both parents and their children, explaining your reasoning. (A pedigree drawing will probably be helpful.) (Problem 51 is by Rosemary Redfield.)

Answer: First, draw the pedigree.

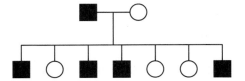

Let the genes be designated by the pigment produced by the normal allele : red pigment, *R;* green pigment, *G*; and blue pigment, *B*.

Recall that the sole X in males comes from the mother, while females obtain an X from each parent. Also recall that a difference in phenotype between sons and daughters is usually due to an X-linked gene. Because all the sons are colorblind and neither the mother nor the daughters are, the mother must carry a different allele for colorblindness on each X chromosome. In other words, she is heterozygous for both X-linked genes, and they are in repulsion: *R g/r G*. With regard to the autosomal gene, she must be *B/–*.

Because all the daughters are normal, the father, who is color-blind, must be able to complement the defects in the mother with regard to his X chromosome. Because he has only one X with which to do so, his genotype must be *R G/Y ; b/b*. Likewise, the mother must be able to complement the father's defect, so she must be *B/B*.

The original cross is therefore:

P   *R g/r G ; B/B × R G/Y ; b/b*

F$_1$ Females   Males
          R g/R G ; B/b   *R g/Y ; B/b*
          r G/R G ; B/b   *r G/Y ; B/b*

52.   Consider the accompanying pedigree for a rare human muscle disease.

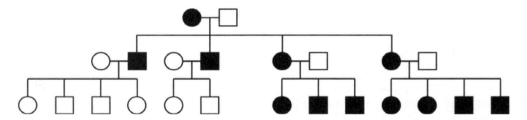

   a. What unusual feature distinguishes this pedigree from those studied earlier in this chapter?
   b. Where do you think the mutant DNA responsible for this phenotype resides in the cell?

Answer:
a. The pedigree clearly shows maternal inheritance.
b. Most likely, the mutant DNA is mitochondrial.

53.   The plant *Haplopappus gracilis* has a 2n of 4. A diploid cell culture was established and, at premitotic S phase, a radioactive nucleotide was added and was incorporated into newly synthesized DNA. The cells were then removed

from the radioactivity, washed, and allowed to proceed through mitosis.
Radioactive chromosomes or chromatids can be detected by placing
photographic emulsion on the cells; radioactive chromosomes or chromatids
appeared covered with spots of silver from the emulsion. (The chromosomes
"take their own photograph.") Draw the chromosomes at prophase and
telophase of the first and second mitotic divisions after the radioactive
treatment. If they are radioactive, show it in your diagram. If there are several
possibilities, show them, too.

Answer: In the following schematic drawings, chromosomes (or chromatids)
that are radioactive are indicated by the grains that would be observed after
radioautography. After the second mitotic division, a number of outcomes are
possible due to the random alignment and separation of the radioactive and non-
radioactive chromatids.

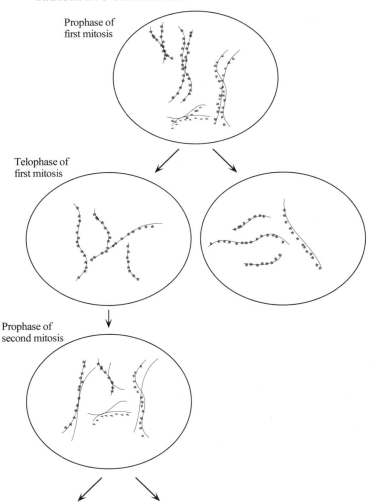

Prophase of
first mitosis

Telophase of
first mitosis

Prophase of
second mitosis

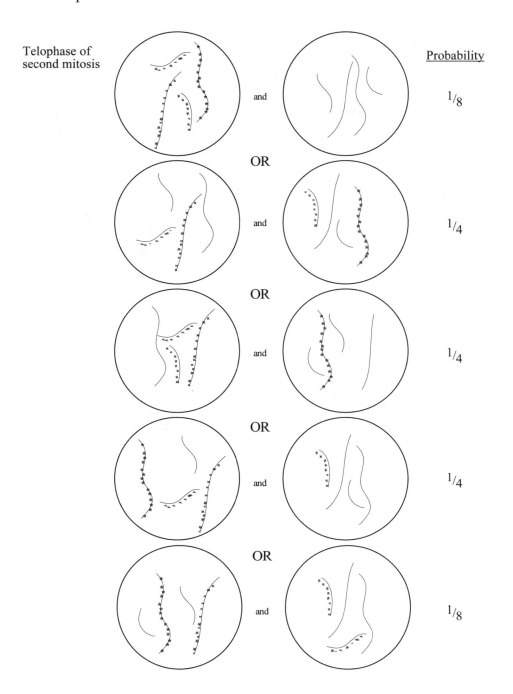

Telophase of second mitosis

Probability

and   1/8

OR

and   1/4

OR

and   1/4

OR

and   1/4

OR

and   1/8

54.   In the species of Problem 53, you can introduce radioactivity by injection into the anthers at the S phase before meiosis. Draw the four products of meiosis with their chromosomes and show which are radioactive.

Answer:

Prophase of
first meiotic
division

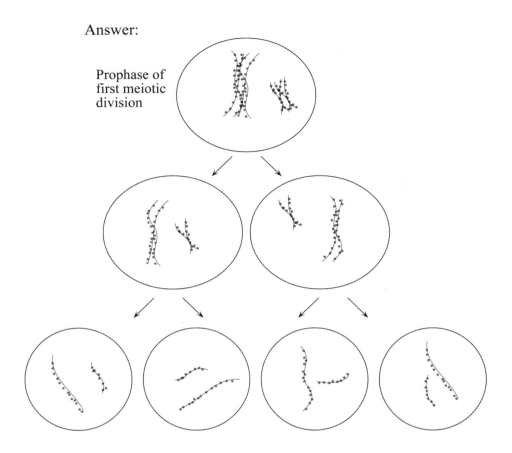

55. The DNA double helices of chromosomes can be partly unwound in situ by special treatments. What pattern of radioactivity is expected if such a preparation is bathed in a radioactive probe for

  a. a unique gene?
  b. dispersed repetitive DNA?
  c. ribosomal DNA?
  d. telomeric DNA?
  e. simple-repeat heterochromatic DNA?

Answer:
  a. In a diploid cell, expect two chromosomes (a pair of homologs) to each have a single locus of radioactivity.

  b. Expect many regions of radioactivity scattered throughout the chromosomes. The exact number and pattern would be dependent on the specific sequence in question, and where and how often it is present within the genome.

  c. The multiple copies of the genes for ribosomal RNA are organized into large tandem arrays called nucleolar organizers (NO). Therefore, expect broader areas of radioactivity compared to (a). The number of these regions would equal the number of NO present in the organism.

d. Each chromosome end would be labeled by telomeric DNA.

e. The multiple repeats of this heterochromatic DNA are organized into large tandem arrays. Therefore, expect broader areas of radioactivity compared to (a). Also, there may be more than one area in the genome of the same simple repeat.

56. If genomic DNA is cut with a restriction enzyme and fractionated by size by electrophoresis, what pattern of Southern hybridization is expected for the probes cited in Problem 55?

Answer: The following is meant to be examples of what is possible. It is also possible, for instance, that more than one band would be present in (a), depending on the position of the restriction sites within the sequence complementary to the probe used.

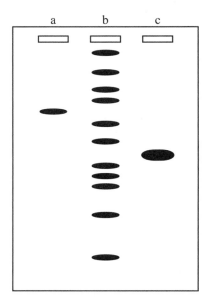

For (d) and (e), the specifics or where the DNA is cut relative to the telomeric DNA or heterochromatic DNA will effect what is observed. Assuming the restriction sites are not within the telomeric or heterochromatic DNA, then (d) will be similar to (b), and (c) will have one or several very large bands.

57. The plant *Haplopappus gracilis* is diploid and $2n = 4$. There are one long pair and one short pair of chromosomes. The diagrams below (numbered 1 through 12) represent anaphases ("pulling apart" stages) of individual cells in meiosis or mitosis in a plant that is genetically a dihybrid (A/a ; B/b) for genes on different chromosomes. The lines represent chromosomes or chromatids, and the points of the V's represent centromeres. In each case, indicate if the diagram represents a cell in meiosis I, meiosis II, or mitosis. If a diagram shows an impossible situation, say so.

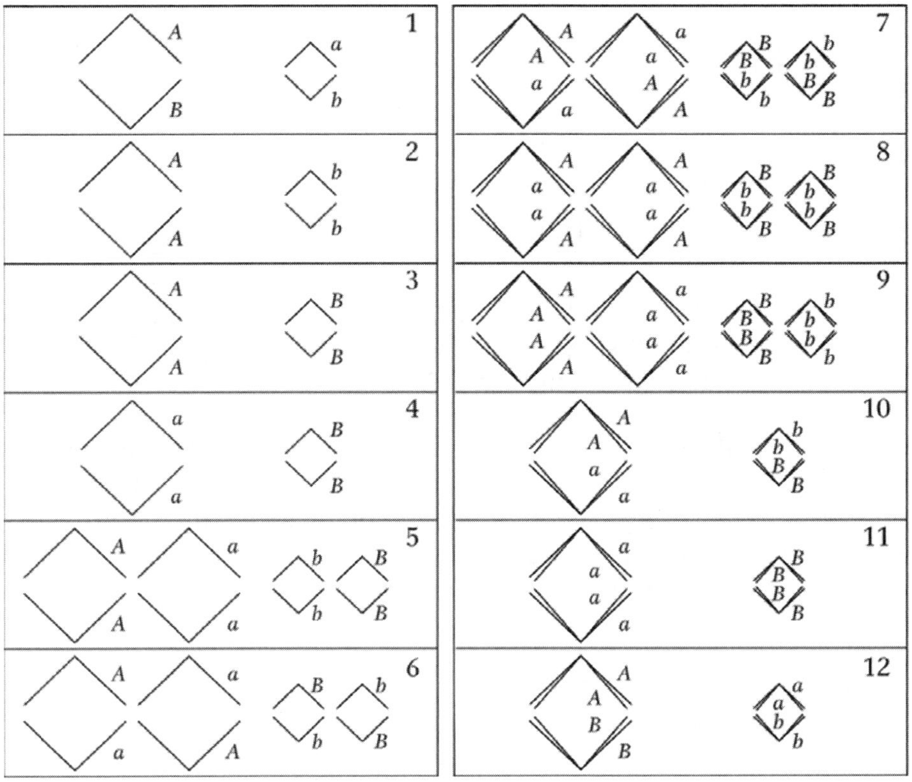

Answer:

(1) Impossible: the alleles of the same genes are on nonhomologous chromosomes

(2) Meiosis II

(3) Meiosis II

(4) Meiosis II

(5) Mitosis

(6) Impossible: appears to be mitotic anaphase but alleles of sister chromatids are not identical

(7) Impossible: too many chromosomes

(8) Impossible: too many chromosomes

(9) Impossible: too many chromosomes

(10) Meiosis I

(11) Impossible: appears to be meiosis of homozygous *a/a* ; *B/B*

(12) Impossible: the alleles of the same genes are on nonhomologous chromosomes

58. The pedigree below shows the recurrence of a rare neurological disease (large black symbols) and spontaneous fetal abortion (small black symbols) in one family. (A slash means that the individual is deceased.) Provide an explanation for this pedigree in regard to the cytoplasmic segregation of defective mitochondria.

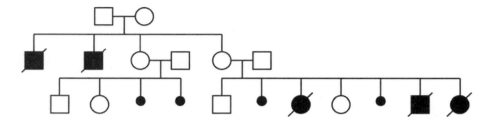

Answer: Recall that each cell has many mitochondria, each with numerous genomes. Also recall that cytoplasmic segregation is routinely found in mitochondrial mixtures within the same cell.

The best explanation for this pedigree is that the mother in generation I experienced a mutation in a single cell that was a progenitor of her egg cells (primordial germ cell). By chance alone, the two males with the disorder in the second generation were from egg cells that had experienced a great deal of cytoplasmic segregation prior to fertilization, while the two females in that generation received a mixture.

The spontaneous abortions that occurred for the first woman in generation II were the result of extensive cytoplasmic segregation in her primordial germ cells: aberrant mitochondria were retained. The spontaneous abortions of the second woman in generation II also came from such cells. The normal children of this woman were the result of extensive segregation in the opposite direction: normal mitochondria were retained. The affected children of this woman were from egg cells that had undergone less cytoplasmic segregation by the time of fertilization, so that they developed to term but still suffered from the disease.

59. A man is brachydactylous (very short fingers; rare autosomal dominant) and his wife is not. Both can taste the chemical phenylthiocarbamide (autosomal dominant; common allele), but their mothers could not.

    a. Give the genotypes of the couple.
       If the genes assort independently and the couple has four children, what is the probability of
    b. all of them being brachydactylous?
    c. none being brachydactylous?
    d. all of them being tasters?
    c. ribosomal DNA?
    d. telomeric DNA?
    e. simple-repeat heterochromatic DNA?

f. all of them being brachydactylous tasters?

g. none being brachydactylous tasters?

h. at least one being a brachydactylous taster?

Answer:

a. Let B = brachydactylous, b = normal, T = taster and t = nontaster. The genotypes of the couple are B/b ; T/t for the male and b/b ; T/t for the female.

b. For all four children to be brachydactylous, the chance is $(1/2)^4 = 1/16$.

c. For none of the four children to be brachydactylous, the chance is $(1/2)^4 = 1/16$.

d. For all to be tasters, the chance is $(3/4)^4 = 81/256$.

e. For all to be nontasters, the chance is $(1/4)^4 = 1/256$.

f. For all to be brachydactylous tasters, the chance is $(1/2 \times 3/4)^4 = 81/4096$.

g. Not being a brachydactylous taster is the same $1-$ (the chance of being a brachydactylous taster) or $1 - (1/2 \times 3/4) = 5/8$. The chance that all four children are not brachydactylous tasters is $(5/8)^4 = 625/4096$.

h. The chance that at least one is a brachydactylous taster is $1 -$ (the chance of none being a brachydactylous taster) or $1 - (5/8)^4$.

60. One form of male sterility in corn is maternally transmitted. Plants of a male-sterile line crossed with normal pollen give male-sterile plants. In addition, some lines of corn are known to carry a dominant nuclear restorer allele (Rf) that restores pollen fertility in male-sterile lines.

   a. Research shows that the introduction of restorer alleles into male-sterile lines does not alter or affect the maintenance of the cytoplasmic factors for male sterility. What kind of research results would lead to such a conclusion?

   b. A male-sterile plant is crossed with pollen from a plant homozygous for Rf. What is the genotype of the F_1? The phenotype?

   c. The F_1 plants from part b are used as females in a testcross with pollen from a normal plant (rf/rf). What are the results of this testcross? Give genotypes and phenotypes, and designate the kind of cytoplasm.

   d. The restorer allele already described can be called Rf-1. Another dominant restorer, Rf-2, has been found. Rf-1 and Rf-2 are located on different chromosomes. Either or both of the restorer alleles will give pollen fertility. With the use of a male-sterile plant as a tester, what will be the result of a cross in which the male parent is

(i)   heterozygous at both restorer loci?

(ii)   homozygous dominant at one restorer locus and homozygous recessive at the other?

(iii)   heterozygous at one restorer locus and homozygous recessive at the other?

(iv)   heterozygous at one restorer locus and homozygous dominant at the other?

Answer: For the following, S will signify cytoplasm of a male-sterile line and N will signify cytoplasm of a non-male-sterile line. *Rf* will signify the dominant nuclear restorer allele, and *rf* is the recessive non-restorer allele.

a. If S *rf/rf* (male-sterile plants) are crossed with pollen from N *Rf/Rf* plants, the offspring will all be S *Rf/rf* and male fertile. If these offspring are then crossed with pollen from N *rf/rf* plants, half the offspring will be S *Rf/rf* (male-fertile) and half will be S *rf/rf* (male-sterile). The S cytoplasm will not be altered or affected even though the maternal offspring parent plant was *Rf/rf*.

b. The cross is S *rf/rf* × N *Rf/Rf* so all the progeny will be S *Rf/rf* and male-fertile.

c. The cross is S *Rf/rf* × N *rf/rf* so half the progeny will be S *Rf/rf* (male-fertile) and half will be S *rf/rf* (male-sterile).

d.

i.   The cross is S *rf-1/rf-1* ; *rf-2/rf-2* × N *Rf-1/rf-1* ; *Rf-2/rf-2*
        The progeny will be: 1/4 S *Rf-1/rf-1* ; *Rf-2/rf-2* (male-fertile)
                                1/4 S *Rf-1/rf-1* ; *rf-2/rf-2* (male-fertile)
                                1/4 S *rf-1/rf-1* ; *Rf-2/rf-2* (male-fertile)
                                1/4 S *rf-1/rf-1* ; *rf-2/rf-2* (male-sterile)

ii.   The cross is S *rf-1/rf-1* ; *rf-2/rf-2* × N *Rf-1/Rf-1* ; *rf-2/rf-2*
        The progeny will all be: S *Rf-1/rf-1* ; *rf-2/rf-2* (male-fertile)

iii.   The cross is S *rf-1/rf-1* ; *rf-2/rf-2* × N *Rf-1/rf-1* ; *rf-2/rf-2*
        The progeny will be: 1/2 S *Rf-1/rf-1* ; *rf-2/rf-2* (male-fertile)
                                1/2 S *rf-1/rf-1* ; *rf-2/rf-2* (male-sterile)

vi.   The cross is S *rf-1/rf-1* ; *rf-2/rf-2* × N *Rf-1/rf-1* ; *Rf-2/Rf-2*
        The progeny will be: 1/2 S *Rf-1/rf-1* ; *Rf-2/rf-2* (male-fertile)
                                1/2 S *rf-1/rf-1* ; *Rf-2/rf-2* (male-fertile)

4

Mapping Eukaryotic Chromosomes by Recombination

WORKING WITH THE FIGURES

(The first 11 problems require inspection of text figures.)

1.   In Figure 4-3, would there be any noncrossover meiotic products in the meiosis illustrated? If so, what colors would they be in the color convention used?

   Answer: Figure 4-3 is drawn to show how crossing over can produce new combinations of alleles, so only the crossover meiotic products are shown for simplicity. In addition to those, there would be two noncrossover products: *pr vg* designated with brown, and *pr⁺ vg⁺* designated yellow.

2.   In Figure 4-6, why does the diagram not show meioses in which two crossovers occur between the same two chromatids (such as the two inner ones)?

   Answer: Figure 4-6 is drawn to show the result of double crossovers involving more than two chromatids (three or four). A double crossover between the inner two chromatids only would not meet this condition.

3.   In Figure 4-8, some meiotic products are labeled parental. Which parent is being referred to in this terminology?

   Answer: The parental meiotic products are chromosomes that have descended intact from one of the two original parents (P₁) in the cross. In this case, the parent for the *AB/ab* genotype would be the *AB/AB* parent (homozygous brown) and the parent for the *ab/ab* genotype would be the *ab/ab* (homozygous yellow).

4.   In Figure 4-9, why is only locus A shown in a constant position?

Answer: Locus A is held in a constant position for a reference to the other loci. In the first two panels, this allows a direct comparison of the relative distances of B and C from A. In the third panel, it allows a visualization of the two ways that B and C can be positioned around A.

5.   In Figure 4-10, what is the mean frequency of crossovers per meiosis in the region A–B? The region B–C?

Answer: Figure 4-10 shows the four products of one meiosis, depicting crossovers at various locations along the chromosome. There are five total crossovers in the A-C region of this chromosome. One crossover occurs in the A-B region, for a frequency of 0.20. Four crossovers (two singles and a three-stranded double) occur in the B-C region for a frequency of 0.80. Because A and C are near the ends of the chromosome, but not at the ends, crossovers could occur outside the A-C interval. To calculate the overall frequency of crossovers that occurred in the two designated regions, the number of meiocytes for which crossovers occurred outside the A-C region would have to be known.

6.   In Figure 4-11, is it true to say that from such a cross the product $v\ cv^+$ can have two different origins?

Answer: Yes. Ignoring the middle gene ct, the product $v\ cv^+$ can arise from either a nonrecombinant parental chromosome or a double crossover. If only the two outside genes, v and cv were involved in the cross, the nonparental double crossover product would not be detectable phenotypically and would appear as a parental chromosome.

7.   In Figure 4-14, in the bottom row four colors are labeled SCO. Why are they not all the same size (frequency)?

Answer: Figure 4-14 represents a trihybrid testcross with linked genes, so there are two genetic intervals to consider. In a typical three-point testcross, those intervals will be different sizes with correspondingly different frequencies of SCOs. The colored boxes depicting SCOs are different sizes to reflect a difference in the number of single crossovers. Note that the two adjacent colors in each pair (green/brown, purple/gray) are about the same size, reflecting the roughly equal number of reciprocal products in each crossover.

8.   Using the conventions of Figure 4-15, draw parents and progeny classes from a cross

$P\ M''''/p\ M' \times p\ M'/p\ M''''$

Answer: In this cross, the dominant disease gene *P* is linked to the microsatellite allele M‴. The progeny classes of the cross shown would be ¼ *P* M‴/*p* M′ (affected by the disease), ¼ *P* M‴/*p* M″″ (affected), ¼ *p* M′/*p* M′, (unaffected) and ¼ *p* M′/*p* M″″ (unaffected).

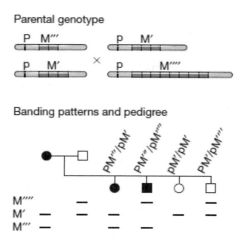

9.   In Figure 4-17, draw the arrangements of alleles in an octad from a similar meiosis in which the upper product of the first division segregated in an upside-down manner at the second division.

Answer: A reversal in the orientation of the upper product would produce a second division pattern of aAAa, and an octad of aaAAAAaa.

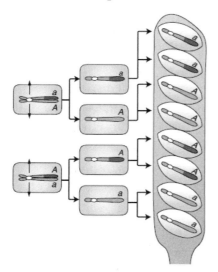

10.   In Figure 4-19, what would be the RF between *A/a* and *B/b* in a cross in which purely by chance all meioses had four-strand double crossovers in that region?

Answer: Figure 4-19 shows that a single four-stranded double crossover produces 100% RF for that meiosis. If by chance all meioses had four-stranded double crossovers, the overall RF would be 100%.

11. **a.** In Figure 4-21, let GC = *A* and AT = *a*, then draw the fungal octad that would result from the final structure (5).

b. (Challenging) Insert some closely linked flanking markers into the diagram, say *P/p* to the left and *Q/q* to the right (assume either cis or trans arrangements). Assume neither of these loci show non-Mendelian segregation. Then draw the final octad based on the structure in part 5.

Answer:

a. The heteroduplex DNA in the upper chromatid would produce a non-identical spore pair with GC (= *A*) on top and AT (= *a*) on bottom. The bottom chromatid would replicate to produce a normal identical spore pair. Thus, the final octad would be A-A A-a a-a a-a.

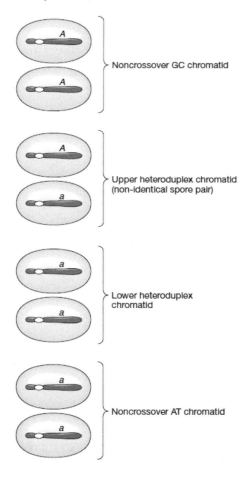

b. That the two loci show no non-Mendelian segregation means that both lie outside the heteroduplex region. If *P* and *Q* were in *cis* and positioned to the left and right, respectively, of the gene *A*/crossover region, they would recombine due to the crossover. The final octad would be *PQ PQ Pq Pq pQ pQ pq pq*.

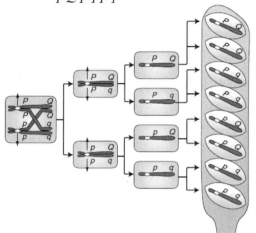

BASIC PROBLEMS

12. A plant of genotype

$$\frac{A \qquad B}{a \qquad b}$$

is testcrossed with

$$\frac{a \qquad b}{a \qquad b}$$

If the two loci are 10 m.u. apart, what proportion of progeny will be *AB/ab*?

Answer: You perform the following cross and are told that the two genes are 10 m.u. apart.

A B/a b ′ a b/a b

Among their progeny, 10 percent should be recombinant (*A b/a b* and *a B/a b*) and 90 percent should be parental (*A B/a b* and *a b/a b*). Therefore, *A B/a b* should represent 1/2 of the parentals or 45 percent.

13. The *A* locus and the *D* locus are so tightly linked that no recombination is ever observed between them. If *Ad/Ad* is crossed with *aD/aD* and the F$_1$ is intercrossed, what phenotypes will be seen in the F$_2$ and in what proportions?

Answer:

P $A\,d\,/\,A\,d$ $'$ $a\,D\,/\,a\,D$

F$_1$ $A\,d\,/\,a\,D$

F$_2$ 1 $A\,d\,/\,A\,d$ phenotype: A d

2    $A\,d\,/\,a\,D$ phenotype: A D

1    $a\,D\,/\,a\,D$ phenotype: a D

14. The *R* and *S* loci are 35 m.u. apart. If a plant of genotype

$$\frac{R \qquad\qquad S}{r \qquad\qquad s}$$

is selfed, what progeny phenotypes will be seen and in what proportions?

Answer:

P $R\,S/r\,s$ $'$ $R\,S/r\,s$

gametes 1/2 $(1 - 0.35)$ $R\,S$
             1/2 $(1 - 0.35)$ $r\,s$
             1/2 (0.35) $R\,s$
             1/2 (0.35) $r\,S$

F$_1$ genotypes   0.1056 $R\,S/R\,S$ 0.1138 $r\,s/r\,S$
             0.1056 $r\,s/r\,s$ 0.1138 $r\,s/R\,s$
             0.2113 $R\,S/r\,s$ 0.0306 $R\,s/R\,s$
             0.1138 $R\,S/r\,S$ 0.0306 $r\,S/r\,S$
             0.1138 $R\,S/R\,s$ 0.0613 $R\,s/r\,S$

F$_1$ phenotypes   0.6058 R S
             0.1056 r s
             0.1444 R s
             0.1444 r S

15. The cross $E/E \cdot F/F \times e/e \cdot f/f$ is made, and the F$_1$ is then backcrossed with the recessive parent. The progeny genotypes are inferred from the phenotypes. The progeny genotypes, written as the gametic contributions of the heterozygous parent, are in the following proportions:

$E \cdot F$ 2/6
$E \cdot f$ 1/6
$e \cdot F$ 1/6
$e \cdot f$ 2/6

Explain these results.

Answer: The cross is $E/e \cdot F/f \times e/e \cdot f/f$. If independent assortment exists, the progeny should be in a 1:1:1:1 ratio, which is not observed. Therefore, there is linkage. $E\,f$ and $e\,F$ are recombinants equaling one-third of the progeny. The two genes are 33.3 map units (m.u.) apart.

RF = 100% × 1/3 = 33.3%

16. A strain of *Neurospora* with the genotype $H \cdot I$ is crossed with a strain with the genotype $h \cdot i$. Half the progeny are $H \cdot I$, and the other half are $h \cdot i$. Explain how this outcome is possible.

Answer: Because only parental types are recovered, the two genes must be tightly linked and recombination must be very rare. Knowing how many progeny were looked at would give an indication of how close the genes are.

17. A female animal with genotype $A/a \cdot B/b$ is crossed with a double-recessive male ($a/a \cdot b/b$). Their progeny include 442 $A/a \cdot B/b$, 458 $a/a \cdot b/b$, 46 $A/a \cdot b/b$, and 54 $a/a \cdot B/b$. Explain these results.

Answer: The problem states that a female that is $A/a \cdot B/b$ is testcrossed. If the genes are unlinked, they should assort independently, and the four progeny classes should be present in roughly equal proportions. This is clearly not the case. The $A/a \cdot B/b$ and $a/a \cdot b/b$ classes (the parentals) are much more common than the $A/a \cdot b/b$ and $a/a \cdot B/b$ classes (the recombinants). The two genes are on the same chromosome and are 10 map units apart.

RF = 100% × (46 + 54)/1000 = 10%

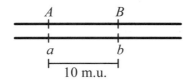

18. If $A/A \cdot B/B$ is crossed with $a/a \cdot b/b$ and the F_1 is testcrossed, what percentage of the testcross progeny will be $a/a \cdot B/b$ if the two genes are **(a)** unlinked; **(b)** completely linked (no crossing over at all); **(c)** 10 m.u. apart; **(d)** 24 m.u. apart?

Answer: The cross is $A/A \cdot B/B \times a/a \cdot a/a$. The F_1 would be $A/a \cdot B/b$.

a. If the genes are unlinked, all four progeny classes from the testcross (including a/a ; b/b) would equal 25 percent.

b. With completely linked genes, the F_1 would produce only $A\,B$ and $a\,b$ gametes. Thus, there would be a 50 percent chance of having $a\,b/a\,b$ progeny from a testcross of this F_1.

c. If the two genes are linked and 10 map units apart, 10 percent of the test-cross progeny should be recombinants. Since the F_1 is *A B/a b*, *a b* is one of the parental classes (*A B* being the other) and it should equal 1/2 of the total parentals or 45 percent.

d. 38 percent (see part c).

19. In a haploid organism, the *C* and *D* loci are 8 m.u. apart. From a cross *C d* × *c D*, give the proportion of each of the following progeny classes: **(a)** *C D*; **(b)** *c d*; **(c)** *C d*; **(d)** all recombinants.

Answer: Meiosis is occurring in an organism that is *C d/c D*, ultimately producing haploid spores. The parental genotypes are *C d* and *c D*, in equal frequency. The recombinant types are *C D* and *c d*, in equal frequency. Eight map units means 8 percent recombinants. Thus, *C D* and *c d* will each be present at a frequency of 4 percent, and *C d* and *c D* will each be present at a frequency of (100% − 8%)/2 = 46%.

a. 4 percent
b. 4 percent
c. 46 percent
d. 8 percent

20. A fruit fly of genotype *B R/b r* is testcrossed with *b r/b r*. In 84 percent of the meioses, there are no chiasmata between the linked genes; in 16 percent of the meioses, there is one chiasma between the genes. What proportion of the progeny will be *B r/b r*?

Answer: To answer this question, you must realize that

(1) One chiasma involves two of the four chromatids of the homologous pair so if 16 percent of the meioses have one chiasma, it will lead to 8 percent recombinants observed in the progeny (one half of the chromosomes of such a meiosis are still parental), and
(2) Half of the recombinants will be *B r,* so the correct answer is 4 percent.

21. A three-point testcross was made in corn. The results and a recombination analysis are shown in the display below, which is typical of three-point testcrosses (*p* = purple leaves, + = green; *v* = virus-resistant seedlings, + = sensitive; *b* = brown midriff to seed, + = plain). Study the display and answer parts *a* through *c*.

P   +/+ · +/+ · +/+ × *p/p* · *v/v* · *B/b*

Gametes $+ \cdot + \cdot +$ $p \cdot v \cdot b$

a. Determine which genes are linked.
b. Draw a map that shows distances in map units.
c. Calculate interference, if appropriate.

Class	Progeny phenotypes	F$_1$ gametes	Numbers	Recombinant for		
				p–b	p–v	v–b
1	gre sen pla	$+ \cdot + \cdot +$	3,210			
2	pur res bro	$p \cdot v \cdot b$	3,222			
3	gre res pla	$+ \cdot v \cdot +$	1,024		R	R
4	pur sen bro	$p \cdot + \cdot b$	1,044		R	R
5	pur res pla	$p \cdot v \cdot +$	690	R		R
6	gre sen bro	$+ \cdot + \cdot b$	678	R		R
7	gre res bro	$+ \cdot v \cdot b$	72	R	R	
8	pur sen pla	$p \cdot + \cdot +$	60	R	R	
			Total 10,000	1,500	2,200	3,436

Unpacking Problem 21

1. Sketch cartoon drawings of the P, F$_1$, and tester corn plants, and use arrows to show exactly how you would perform this experiment. Show where seeds are obtained.

   Answer: There is no correct drawing; any will do. Pollen from the tassels is placed on the silks of the females. The seeds are the F$_1$ corn kernels.

2. Why do all the +'s look the same, even for different genes? Why does this not cause confusion?

   Answer: The +'s all look the same because they signify wild type for each gene. The information is given in a specific order, which prevents confusion, at least initially. However, as you work the problem, which may require you to reorder the genes, errors can creep into your work if you do not make sure that you reorder the genes for each genotype in exactly the same way. You may find it easier to write the complete genotype, p^+ instead of +, to avoid confusion.

3. How can a phenotype be purple and brown, for example, at the same time?

   Answer: The phenotype is purple leaves and brown midriff to seeds. In other words, the two colors refer to different parts of the organism.

4.   Is it significant that the genes are written in the order *p-v-b* in the problem?

Answer: There is no significance in the original sequence of the data.

5.   What is a tester and why is it used in this analysis?
Answer: A tester is a homozygous recessive for all genes being studied. It is used so that the meiotic products in the organism being tested can be seen directly in the phenotype of the progeny.

6.   What does the column marked "Progeny phenotypes" represent? In class 1, for example, state exactly what "gre sen pla" means.

Answer: The progeny phenotypes allow you to infer the genotypes of the plants. For example, *gre* stands for "green," the phenotype of $p^+/–$; *sen* stands for "virus-sensitive," the phenotype of $v^+/–$; and *pla* stands for "plain seed," the phenotype of $b^+/–$. In this testcross, all progeny have at least one recessive allele so the "gre sen pla" progeny are actually $p^+/p \cdot v^+/v \cdot b^+/b$.

7.   What does the line marked "Gametes" represent, and how is it different from the column marked "F$_1$ gametes"?

In what way is comparison of these two types of gametes relevant to recombination?

Answer: *Gametes* refers to the gametes of the two pure-breeding parents. F$_1$ *gametes* refers to the gametes produced by the completely heterozygous F$_1$ progeny. They indicate whether crossing-over or independent assortment have occurred. In this case, because there is either independent assortment or crossing-over, or both, the data indicate that the three genes are not so tightly linked that zero recombination occurred.

8.   Which meiosis is the main focus of study? Label it on your drawing.

Answer: The main focus is meiosis occurring in the F$_1$ parent.

9.   Why are the gametes from the tester not shown?

Answer: The gametes from the tester are not shown because they contribute nothing to the phenotypic differences seen in the progeny.

10.   Why are there only eight phenotypic classes? Are there any classes missing?

Answer: Eight phenotypic classes are expected for three autosomal genes, whether or not they are linked, when all three genes have simple dominant-recessive relationships among their alleles. The general formula for the number of expected phenotypes is 2^n, where n is the number of genes being studied.

11. What classes (and in what proportions) would be expected if all the genes are on separate chromosomes?

Answer: If the three genes were on separate chromosomes, the expectation is a 1:1:1:1:1:1:1:1 ratio.

12. To what do the four pairs of class sizes (very big, two intermediates, very small) correspond?

Answer: The four classes of data correspond to the parentals (largest), two groups of single crossovers (intermediate), and double crossovers (smallest).

13. What can you tell about gene order simply by inspecting the phenotypic classes and their frequencies?

Answer: By comparing the parentals with the double crossovers, gene order can be determined. The gene in the middle flips with respect to the two flanking genes in the double-crossover progeny. In this case, one parental is +++ and one double crossover is p++. This indicates that the gene for leaf color (p) is in the middle.

14. What will be the expected phenotypic class distribution if only two genes are linked?

Answer: If only two of the three genes are linked, the data can still be grouped, but the grouping will differ from that mentioned in (12) above. In this situation, the unlinked gene will show independent assortment with the two linked genes. There will be one class composed of four phenotypes in approximately equal frequency, which combined will total more than half the progeny. A second class will be composed of four phenotypes in approximately equal frequency, and the combined total will be less than half the progeny. For example, if the cross were $a\ b/+ +\ ;\ c/+ \times a\ b/a\ b\ ;\ c/c$, then the parental class (more frequent class) would have four components: $a\ b\ c$, $a\ b\ +$, $+ + c$, and $+ + +$. The recombinant class would be $a + c$, $a + +$, $+ b\ c$, and $+ b +$.

15.   What does the word "point" refer to in a three-point testcross? Does this word usage imply linkage? What would a four-point testcross be like?

Answer: *Point* refers to locus. The usage does not imply linkage but rather a testing for possible linkage. A four-point testcross would look like the following: $a/+ \cdot b/+ \cdot c/+ \cdot d/+ \times a/a \cdot b/b \cdot c/c \cdot d/d$.

16.   What is the definition of *recombinant*, and how is it applied here?

Answer: A *recombinant* refers to an individual who has alleles inherited from two different grandparents, both of whom were the parents of the individual's heterozygous parent. Another way to think about this term is that in the recombinant individual's heterozygous parent, recombination took place among the genes that were inherited from his or her parents. In this case, the recombination took place in the F_1, and the recombinants are among the F_2 progeny.

17.   What do the "Recombinant for" columns mean?

Answer: The "recombinant for" columns refer to specific gene pairs and progeny that exhibit recombination between those gene pairs.

18.   Why are there only three "Recombinant for" columns?

Answer: There are three "recombinant for" columns because three genes can be grouped in three different gene pairs.

19.   What do the *R*'s mean, and how are they determined?

Answer: *R* refers to recombinant progeny, and they contain different configurations of alleles than were present in their heterozygous parent.

20.   What do the column totals signify? How are they used?

Answer: Column totals indicate the number of progeny that experience crossing- over between the specific gene pairs. They are used to calculate map units between the two genes.

21.   What is the diagnostic test for linkage?

Answer: The diagnostic test for linkage is a recombination frequency of significantly less than 50 percent.

22.   What is a map unit? Is it the same as a centimorgan?

Answer: A map unit represents 1 percent crossing over and is the same as a centimorgan.

23.   In a three-point testcross such as this one, why aren't the F_1 and the tester considered to be parental in calculating recombination? (They are parents in one sense.)

Answer: In the tester, recombination cannot be detected in the gamete contribution to the progeny because the tester is homozygous. The F_1 individuals have genotypes fixed by their parents' homozygous state and, again, recombination cannot be detected in their phenotypes. Recombination between the P configurations occurs when the F_1 forms gametes and is detected in the phenotypes of the F_2 progeny.

24.   What is the formula for interference? How are the "expected" frequencies calculated in the coefficient-of-coincidence formula?

Answer: Interference I = 1 − coefficient of coincidence = 1 − (observed double crossovers/expected double crossovers). The expected double crossovers are equal to the (frequency of crossing over in the first region, in this case between v and p) × (frequency of crossing over in the second region, between p and b) × number of progeny.

25.   Why does part c of the problem say "if appropriate"?

Answer: If the three genes are not all linked, then interference cannot be calculated.

26.   How much work is it to obtain such a large progeny size in corn? Which of the three genes would take the most work to score? Approximately how many progeny are represented by one corncob?

Answer: A great deal of work is required to obtain 10,000 progeny in corn because each seed on a cob represents one progeny. Each cob may contain as many as 200 seeds. While seed characteristics can be assessed at the cob stage, for other characteristics such as leaf color, viral sensitivity, and midriff color, each seed must be planted separately and assessed after germination and growth. The bookkeeping task is also enormous.

Solution to the Problem

a. The three genes are linked.

b. Comparing the parentals (most frequent) with the double crossovers (least frequent), the gene order is v p b. There were 2200 recombinants between v and p, and 1500 between p and b. The general formula for map units is:

m.u. = 100% (number of recombinants)/total number of progeny

Therefore, the map units between v and p = 100% (2200)/10,000 = 22 m.u., and the map units between p and b = 100%(1500)/10,000 = 15 m.u. The map is

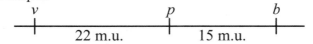

c. I = 1 – observed double crossovers/expected double crossovers
= 1 – 132/(0.22)(0.15)(10,000)
= 1 – 0.4 = 0.6

22. You have a *Drosophila* line that is homozygous for autosomal recessive alleles *a, b,* and *c,* linked in that order. You cross females of this line with males homozygous for the corresponding wild-type alleles. You then cross the F₁ heterozygous males with their heterozygous sisters. You obtain the following F₂ phenotypes (where letters de note recessive phenotypes and pluses denote wild-type phenotypes): 1364 + + +, 365 *a b c*, 87 *a b* +, 84 + + *c*, 47 *a* + +, 44 + *b c*, 5 *a* + *c*, and 4 + *b* +.

a. What is the recombinant frequency between *a* and *b*? Between *b* and *c*? (Remember, there is no crossing over in *Drosophila* males.)

b. What is the coefficient of coincidence?

Answer:
P a b c/a b c × a^+ b^+ c^+/a^+ b^+ c^+

F₁ a^+ b^+ c^+/a b c × a^+ b^+ c^+/a b c

F₂ 1364 a^+ b^+ c^+
      365   a b c
       87   a b c^+
       84   a^+ b^+ c
       47   a b^+ c^+
       44   a^+ b c

$$5 \qquad a\ b^+\ c$$
$$4 \qquad a^+\ b\ c^+$$

This problem is somewhat simplified by the fact that recombination does not occur in male *Drosophila*. Also, only progeny that received the *a b c* chromosome from the male will be distinguishable among the F_2 progeny.

a. Because you cannot distinguish between $a^+\ b^+\ c^+/a^+\ b^+\ c^+$ and $a^+\ b^+\ c^+/a\ b\ c$, use the frequency of $a\ b\ c/a\ b\ c$ to estimate the frequency of $a^+\ b^+\ c^+$ (parental) gametes from the female.

parentals   730   (2 × 365)
CO *a–b*:   91   ($a\ b^+\ c^+$, $a^+\ b\ c = 47 + 44$)
CO *b–c*:   171   ($a\ b\ c^+$, $a^+\ b^+\ c = 87 + 84$)
DCO: 9   ($a^+\ b\ c^+$, $a\ b+\ c = 4 + 5$)
    1001

a–b: 100%(91 + 9)/1001 = 10 m.u.
b–c: 100%(171 + 9)/1001 = 18 m.u.

b. Coefficient of coincidence = (observed DCO)/(expected DCO)
= 9/[(0.1)(0.18)(1001)] = 0.5

23. R. A. Emerson crossed two different pure-breeding lines of corn and obtained a phenotypically wild-type F_1 that was heterozygous for three alleles that determine recessive phenotypes: *an* determines anther; *br*, brachytic; and *f*, fine. He testcrossed the F_1 with a tester that was homozygous recessive for the three genes and obtained these progeny phenotypes: 355 anther; 339 brachytic, fine; 88 completely wild type; 55 anther, brachytic, fine; 21 fine; 17 anther, brachytic; 2 brachytic; 2 anther, fine.

a. What were the genotypes of the parental lines?
b. Draw a linkage map for the three genes (include map distances).
c. Calculate the interference value.

Answer:
a. By comparing the two most frequent classes (parentals: $an\ br^+\ f^+$, $an^+\ br\ f$) to the least frequent classes (DCO: $an^+\ br\ f^+$, $an\ br+\ f$), the gene order can be determined. The gene in the middle switches with respect to the other two (the order is *an f br*). Now the crosses can be written fully.

P $an\ f^+\ br^+/an\ f^+\ br^+ \times an^+\ f\ br/an^+\ f\ br$

F_1 $an^+\ f\ br/an\ f^+\ br^+ \times an\ f\ br/an\ f\ br$

F_2 355 $an\ f^+\ br^+/an\ f\ br$ parental

339   *an⁺ f br/an f br*   parental
88   *an⁺ f⁺ br⁺/an f br*   CO *an–f*
55   *an f br/an f br*   CO *an–f*
21   *an⁺ f br⁺/an f br*   CO *f–br*
17   *an f⁺ br/an f br*   CO *f–br*
2   *an⁺ f⁺ br/an f br*   DCO
2   *an f br⁺/an f br*   DCO
879

b. *an–f*:   100% $(88 + 55 + 2 + 2)/879 = 16.72$ m.u.
    f–br: 100% $(21 + 17 + 2 + 2)/879 = 4.78$ m.u.

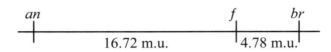

c.   Interference $= 1 - $ (observed DCO/expected DCO)
                $= 1 - 4/(0.1672)(0.0478)(879) = 0.431$

24. Chromosome 3 of corn carries three loci (*b* for plant-color booster, *v* for virescent, and *lg* for liguleless). A testcross of triple recessives with F_1 plants heterozygous for the three genes yields progeny having the following genotypes: 305 + *v lg*, 275 *b* + +, 128 *b* + *lg*, 112 + *v* +, 74 + + *lg*, 66 *b v* +, 22 + + +, and 18 *b v lg*. Give the gene sequence on the chromosome, the map distances between genes, and the coefficient of coincidence.

Answer: By comparing the most frequent classes (parental: + *v lg*, *b* + +) with the least frequent classes (DCO: + + +, *b v lg*) the gene order can be determined. The gene in the middle switches with respect to the other two, yielding the following sequence: *v b lg*. Now the cross can be written:

P   *v b⁺ lg/v+ b lg⁺ × v b lg/v b lg*

F_1 305   *v b⁺ lg/v b lg*   parental
     275   *v+ b lg⁺/v b lg⁺*   parental
     128   *v⁺ b lg/v b lg*   CO *b–lg*
     112   *v b⁺ lg⁺/v b lg*   CO *b–lg*
     74   *v⁺ b⁺ lg/v b lg*   CO *v–b*
     66   *v b lg⁺/v b lg*   CO *v–b*
     22   *v⁺ b⁺ lg⁺/v b lg*   DCO
     18   *v b lg/v b lg*   DCO

v–b: 100%$(74 + 66 + 22 + 18)/1000 = 18.0$ m.u.
b–lg: 100%$(128 + 112 + 22 + 18)/1000 = 28.0$ m.u.

c.c. = observed DCO/expected DCO $= (22 + 18)/(0.28)(0.18)(1000) = 0.79$

25. Groodies are useful (but fictional) haploid organisms that are pure genetic tools. A wild-type groody has a fat body, a long tail, and flagella. Mutant lines are known that have thin bodies, are tailless, or do not have flagella. Groodies can mate with one another (although they are so shy that we do not know how) and produce recombinants. A wild-type groody mates with a thin-bodied groody lacking both tail and flagella. The 1000 baby groodies produced are classified as shown in the illustration here. Assign genotypes, and map the three genes. (Problem 25 is from Burton S. Guttman.)

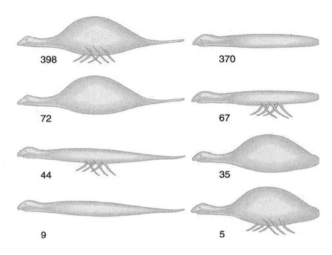

Answer: Let F = fat, L = long tail, and Fl = flagella. The gene sequence is $F\ L\ Fl$ (compare most frequent with least frequent). The cross is:

P $F\ L\ Fl/f\ l\ fl \times f\ l\ fl/f\ l\ fl$

F_1	398	$F\ L\ Fl/f\ l\ fl$	parental
	370	$f\ l\ fl/f\ l\ fl$	parental
	72	$F\ L\ fl/f\ l\ fl$	CO $L–Fl$
	67	$f\ l\ Fl/f\ l\ fl$	CO $L–Fl$
	44	$f\ L\ Fl/f\ l\ fl$	CO $F–L$
	35	$F\ l\ fl/f\ l\ fl$	CO $F–L$
	9	$f\ L\ fl/f\ l\ fl$	DCO
	5	$F\ l\ Fl/f\ l\ fl$	DCO

$L–Fl$: $100\%(72 + 67 + 9 + 5)/1000 = 15.3$ m.u.
$F–L$: $100\%(44 + 35 + 9 + 5)/1000 = 9.3$ m.u.

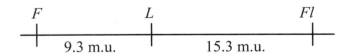

26. In *Drosophila*, the allele dp^+ determines long wings and dp determines short ("dumpy") wings. At a separate locus, e^+ determines gray body and e

determines ebony body. Both loci are autosomal. The following crosses were made, starting with pure-breeding parents:

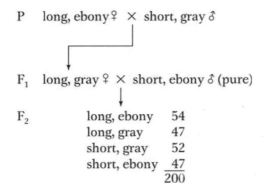

P   long, ebony ♀ × short, gray ♂

F$_1$ long, gray ♀ × short, ebony ♂ (pure)

F$_2$ long, ebony   54
                long, gray   47
                short, gray   52
                short, ebony   47
                               ———
                               200

Use the χ^2 test to determine if these loci are linked. In doing so, indicate **(a)** the hypothesis, **(b)** calculation of χ^2, **(c)** *p* value, **(d)** what the *p* value means, **(e)** your conclusion, **(f)** the inferred chromosomal constitutions of parents, F$_1$, tester, and progeny.

Answer:
a. The hypothesis is that the genes are not linked. Therefore, a 1:1:1:1 ratio is expected.

b. χ^2 $= (54-50\ )^2/50 + (47-50\ )^2/50 + (52-50\ )^2/50 + (47-50)^2/50$
  $= 0.32 + 0.18 + 0.08 + 0.18 = 0.76$

c. With three degrees of freedom, the *p* value is between 0.50 and 0.90.

d. Between 50 percent and 90 percent of the time values this extreme from the prediction would be obtained by chance alone.

e. Accept the initial hypothesis.

f. Because the χ^2 value was insignificant, we conclude the two genes are assorting independently. The genotypes of all individuals are

P dp^+/dp^+ ; $e/e \times dp/dp$; e^+/e^+

F$_1$ dp^+/dp ; e^+/e

tester dp/dp ; e/e

progeny   long, ebony dp^+/dp ; e/e
          long, gray dp^+/dp ; e^+/e
          short, gray dp/dp ; e^+/e
          short, ebony dp/dp ; e/e

27. The mother of a family with 10 children has blood type Rh⁺. She also has a very rare condition (elliptocytosis, phenotype E) that causes red blood cells to be oval rather than round in shape but that produces no adverse clinical effects. The father is Rh⁻ (lacks the Rh⁺ antigen) and has normal red blood cells (phenotype e). The children are 1 Rh⁺ e, 4 Rh⁺ E, and 5 Rh⁻ e. Information is available on the mother's parents, who are Rh⁺ E and Rh⁻ e. One of the 10 children (who is Rh⁺ E) marries someone who is Rh⁺ e, and they have an Rh⁺ E child.

   a. Draw the pedigree of this whole family.
   b. Is the pedigree in agreement with the hypothesis that the Rh⁺ allele is dominant and Rh⁻ is recessive?
   c. What is the mechanism of transmission of elliptocytosis?
   d. Could the genes governing the E and Rh phenotypes be on the same chromosome? If so, estimate the map distance between them, and comment on your result.

Answer:

a.

   b. Yes

   c. Dominant

   d. As drawn, the pedigree hints at linkage. If unlinked, expect that the phenotypes of the 10 children should be in a 1:1:1:1 ratio of Rh⁺ E, Rh⁺ e, Rh⁻ E, and Rh⁻ e. There are actually five Rh⁻ e, four Rh⁺ E, and one Rh⁺ e. If linked, this last phenotype would represent a recombinant, and the distance between the two genes would be 100% (1/10) = 10 m.u. However, there is just not enough data to strongly support that conclusion.

28. From several crosses of the general type $A/A \cdot B/B \times a/a \cdot B/b$ the F_1 individuals of type $A/a \cdot B/b$ were testcrossed with $a/a \cdot B/b$. The results are as follows:

Testcross of F₁ from cross	Testcross progeny			
	$A/a \cdot B/b$	$a/a \cdot B/b$	$A/a \cdot B/b$	$a/a \cdot B/b$
1	310	315	287	288
2	36	38	23	23
3	360	380	230	230
4	74	72	50	44

For each set of progeny, use the χ^2 test to decide if there is evidence of linkage.

Answer: Assume there is no linkage. (This is your hypothesis. If it can be rejected, the genes are linked.) The expected values would be that genotypes occur with equal frequency. There are four genotypes in each case ($n = 4$) so there are 3 degrees of freedom.

$$\chi^2 = \sum (\text{observed} - \text{expected})^2/\text{expected}$$

Cross 1: $\chi^2 = [(310–300)^2 + (315–300)^2 + (287–300)^2 + (288–300)^2]/300$
= 2.1266; p > 0.50, nonsignificant; hypothesis cannot be rejected

Cross 2: $\chi^2 = [(36–30)^2 + (38–30)^2 + (23–30)^2 + (23–30)^2]/30$
= 6.6; p > 0.10, nonsignificant; hypothesis cannot be rejected

Cross 3: $\chi^2 = [(360–300)^2 + (380–300)^2 + (230–300)^2 + (230–300)^2]/300$
= 66.0; p < 0.005, significant; hypothesis must be rejected

Cross 4: $\chi^2 = [(74–60)^2 + (72–60)^2 + (50–60)^2 + (44–60)^2]/60$
= 11.60; p < 0.01, significant; hypothesis must be rejected

29. In the two pedigrees diagrammed here, a vertical bar in a symbol stands for steroid sulfatase deficiency, and a horizontal bar stands for ornithine transcarbamylase deficiency.

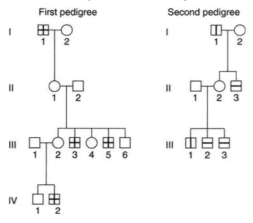

a. Is there any evidence in these pedigrees that the genes determining the deficiencies are linked?
b. If the genes are linked, is there any evidence in the pedigree of crossing over between them?
c. Draw genotypes of these individuals as far as possible.

Answer:

a. Both disorders must be recessive to yield the patterns of inheritance that are observed. Notice that only males are affected, strongly suggesting X linkage for both disorders. In the first pedigree, there is a 100 percent correlation between the presence or absence of both disorders, indicating close linkage. In the second pedigree, the presence and absence of both disorders are inversely correlated, again indicating linkage. In the first pedigree, the two defective alleles must be *cis* within the heterozygous females to show 100 percent linkage in the affected males, while in the second pedigree the two defective alleles must be trans within the heterozygous females.

b. and c. Let *a* stand for the allele giving rise to steroid sulfatase deficiency (vertical bar) and *b* stand for the allele giving rise to ornithine transcarbamylase deficiency (horizontal bar). Crossing over cannot be detected without attaching genotypes to the pedigrees. When this is done, it can be seen that crossing over need not occur in either of the pedigrees to give rise to the observations.

First pedigree:

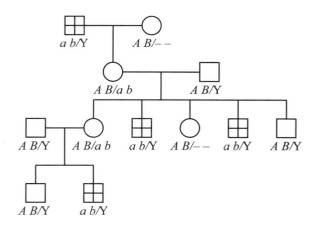

Second pedigree:

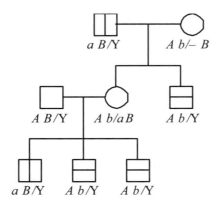

30. In the accompanying pedigree, the vertical lines stand for protan color blindness, and the horizontal lines stand for deutan color blindness. These are separate conditions causing different misperceptions of colors; each is determined by a separate gene.

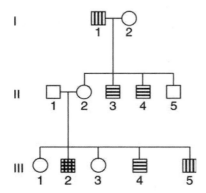

    a. Does the pedigree show any evidence that the genes are linked?

    b. If there is linkage, does the pedigree show any evidence of crossing over? Explain your answers to parts a and b with the aid of the diagram.

    c. Can you calculate a value for the recombination between these genes? Is this recombination by independent assortment or by crossing over?

Answer:

    a. Note that only males are affected by both disorders. This suggests that both are X-linked recessive disorders. Using *p* for protan and *P* for non-protan, and *d* for deutan and *D* for non-deutan, the inferred genotypes are listed on the pedigree below. The Y chromosome is shown, but the X is represented by the alleles carried.

    b. Individual II-2 must have inherited both disorders in the trans configuration (on separate chromosomes). Therefore, individual III-2 inherited both traits as the result of recombination (crossing over) between his mother's X chromosomes.

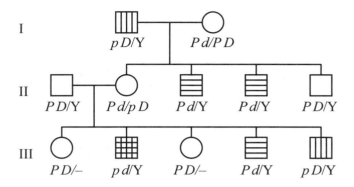

c. Because both genes are X-linked, this represents crossing over. The progeny size is too small to give a reliable estimate of recombination.

31. In corn, a triple heterozygote was obtained carrying the mutant alleles *s* (shrunken), *w* (white aleurone), and *y* (waxy endosperm), all paired with their normal wild-type alleles. This triple heterozygote was testcrossed, and the progeny contained 116 shrunken, white; 4 fully wild type; 2538 shrunken; 601 shrunken, waxy; 626 white; 2708 white, waxy; 2 shrunken, white, waxy; and 113 waxy.

   a. Determine if any of these three loci are linked and, if so, show map distances.
   b. Show the allele arrangement on the chromosomes of the triple heterozygote used in the testcross.
   c. Calculate interference, if appropriate.

   Answer:
   a. and b. Again, the best way to determine whether there is linkage is through chi-square analysis, which indicates that it is highly unlikely that the three genes assort independently. To determine linkage by simple inspection, look at gene pairs. Because this is a testcross, independent assortment predicts a 1:1:1:1 ratio.

   Comparing shrunken and white, the frequencies are:
   + +   (113 + 4)/total
   s wh   (116 + 2)/total
   + *wh*   (2708 + 626)/total
   s +   (2538 + 601)/total

   There is not independent assortment between shrunken and white, which means that there is linkage.

   Comparing shrunken and waxy, the frequencies are:

+	+	(626 + 4)/total
s	wa	(601 + 2)/total
+	wa	(2708 + 113)/total
s	+	(2538 + 116)total

There is not independent assortment between shrunken and waxy, which means that there is linkage.

Comparing white and waxy, the frequencies are:

+	+	(2538 + 4)/total
wh	wa	(2708 + 2)/total
wh	+	(626 + 116)/total
+	wa	(601 + 113)/total

There is not independent assortment between waxy and white, which means that there is linkage.

Because all three genes are linked, the strains must be + s +/wh + wa and wh s wa/wh s wa (compare most frequent, parentals, to least frequent, double crossovers, to obtain the gene order). The cross can be written as:

P   + s +/wh + wa × wh s wa/wh s wa

F₁   as in problem

Crossovers between white and shrunken and shrunken and waxy are:

113	601
116	626
4	4
2	2
235	1233

Dividing by the total number of progeny and multiplying by 100 percent yields the following map:

white   *shrunken*   *waxy*

        3.5 m.u.   18.4 m.u.

c.   Interference = 1 – (observed double crossovers/expected double crossovers)
           = 1 – 6/(0.035)(0.184)(6,708) = 0.86

32.   a.   A mouse cross A/a · B/b x a/a   · b/b is made, and in the progeny there are

25% *A/a · B/b*, 25% *a/a · b/b*,
25% *A/a · b/b*, 25% *a/a · B/b*

Explain these proportions with the aid of simplified meiosis diagrams.

b. A mouse cross *C/c · D/d ʹ/c   · d/d* is made, and in the progeny there are

45% *C/c · d/d*, 45% *c/c · D/d*,
5% *c/c · d/d*, 5% *C/c · D/d*

Explain these proportions with the aid of simplified meiosis diagrams.

Answer:

a. The results of this cross indicate independent assortment of the two genes. This might be diagrammed as

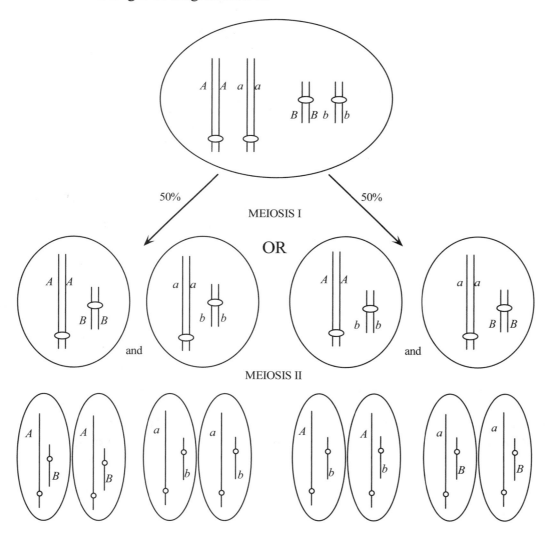

b. The results of this cross indicate that the two genes are linked and 10 m.u. apart. Further, the recessive alleles are in repulsion in the dihybrid (*C d/c D* × *c d/c d*). This might be diagrammed as

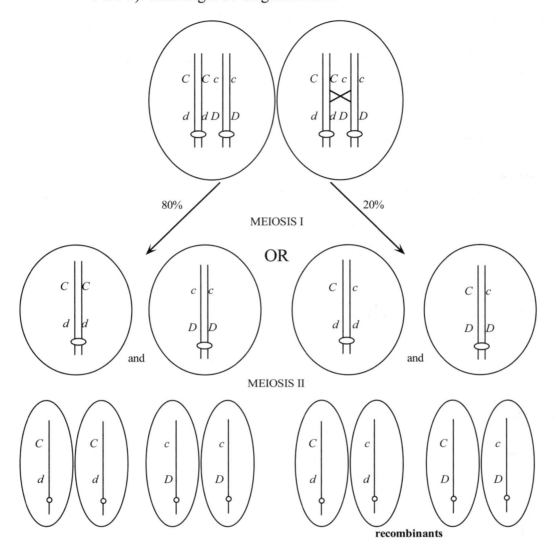

33. In the tiny model plant *Arabidopsis*, the recessive allele *hyg* confers seed resistance to the drug hygromycin, and *her*, a recessive allele of a different gene, confers seed resistance to herbicide. A plant that was homozygous *hyg/hyg · her/her* was crossed with wild type, and the F₁ was selfed. Seeds resulting from the F₁ self were placed on petri dishes containing hygromycin and herbicide.

  a. If the two genes are unlinked, what percentage of seeds are expected to grow?

b. In fact, 13 percent of the seeds grew. Does this percentage support the hypothesis of no linkage? Explain. If not, calculate the number of map units between the loci.

c. Under your hypothesis, if the F_1 is testcrossed, what proportion of seeds will grow on the medium containing hygromycin and herbicide?

Answer:

a. If the genes are unlinked, the cross becomes:

P $\quad$ *hyg/hyg ; her/her* × *hyg$^+$/hyg$^+$; her$^+$/her$^+$*

F_1 $\quad$ *hyg$^+$/hyg ; her$^+$/her* × *hyg$^+$/hyg ; her$^+$/her*

F_2 $\quad$ 9/16 $\quad$ *hyg$^+$/– ; her$^+$/–*
$\quad\quad\quad$ 3/16 $\quad$ *hyg$^+$/– ; her/her*
$\quad\quad\quad$ 3/16 $\quad$ *hyg/hyg ; her$^+$/–*
$\quad\quad\quad$ 1/16 $\quad$ *hyg/hyg ; her/her*

So only 1/16 (or 6.25 percent) of the seeds would be expected to germinate.

b. and c. No. More than twice the expected seeds germinated so assume the genes are linked. The cross then becomes:

P $\quad$ *hyg her/hyg her* × *hyg$^+$ her$^+$/hyg$^+$ her$^+$*

F_1 $\quad$ *hyg$^+$ her$^+$/hyg her* × *hyg$^+$ her$^+$/hyg her*

F_2 $\quad$ 13 percent $\quad$ *hyg her/hyg her*

Because this class represents the combination of two parental chromosomes, it is equal to:

$$p(hyg\ her) \times p(hyg\ her) = (1/2\ \text{parentals})^2 = 0.13$$
and
$\quad$ parentals = 0.72 $\quad\quad$ so recombinants = $1 - 0.72 = 0.28$

Therefore, a testcross of *hyg$^+$ her$^+$/hyg her* should give:

$\quad$ 36% $\quad$ *hyg$^+$ her$^+$/hyg her*
$\quad$ 36% $\quad$ *hyg her/hyg her*
$\quad$ 14% $\quad$ *hyg$^+$ her/hyg her*
$\quad$ 14% $\quad$ *hyg her$^+$/hyg her*

and 36 percent of the progeny should grow (the *hyg her/hyg her* class).

34. In a diploid organism of genotype *A/a* ; *B/b* ; D/d, the allele pairs are all on different chromosome pairs. The two diagrams in the next column purport to show anaphases ("pulling apart" stages) in individual cells. State whether each drawing represents mitosis, meiosis I, or meiosis II or is impossible for this particular genotype.

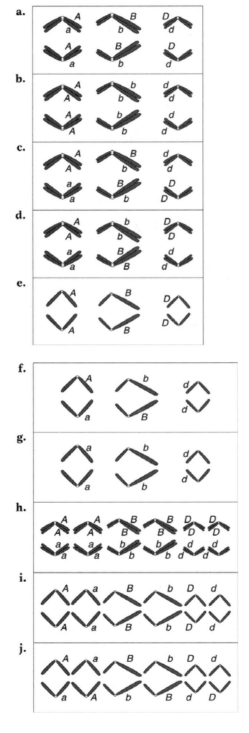

Answer:

a. Meiosis I (crossing-over has occurred between all genes and their centromeres.)

b. Impossible

c. Meiosis I (crossing-over has occurred between gene B and its centromere.)

d. Meiosis I

e. Meiosis II

f. Meiosis II (crossing-over has occurred between genes A and B and their centromeres.)

g. Meiosis II

h. Impossible

i. Mitosis

j. Impossible

35. The *Neurospora* cross *al-2$^+$* × *al-2* is made. A linear tetrad analysis reveals that the second-division segregation frequency is 8 percent.

   a. Draw two examples of second-division segregation patterns in this cross.
   b. What can be calculated by using the 8 percent value?

Answer:
a. *al-2$^+$*   *al-2*

b. The 8 percent value can be used to calculate the distance between the gene and the centromere. That distance is ½ the percentage of second-division segregation, or 4 percent.

36. From the fungal cross *arg-6* ⊕ *al-2* × *arg-6$^+$* ⊕ *al-2$^+$*, what will the spore genotypes be in unordered tetrads that are **(a)** parental ditypes? **(b)** tetratypes? **(c)** nonparental ditypes?

Answer:
a. *arg-6* · *al-2*, *arg-6* · *al-2*, *arg-6$^+$* · *al-2$^+$*, *arg-6$^+$* · *al-2$^+$*

b. *arg-6$^+$* · *al-2$^+$*, *arg-6$^+$* · *al-2*, *arg-6* · *al-2$^+$*, *arg-6* · *al-2*

c. *arg-6$^+$ · al-2, arg-6$^+$ · al-2, arg-6 · al-2$^+$, arg-6 · al-2$^+$*

37. For a certain chromosomal region, the mean number of crossovers at meiosis is calculated to be two per meiosis. In that region, what proportion of meioses are predicted to have **(a)** no crossovers? **(b)** one crossover? **(c)** two crossovers?

Answer: The formula for this problem is $f(i) = e^{-m}m^i/i!$ where $m = 2$ and $i = 0$, 1, or 2.

a. $f(0) = e^{-2}2^0/0! = e^{-2} = 0.135$ or 13.5%

b. $f(1) = e^{-2}2^1/1! = e^{-2}(2) = 0.27$ or 27%

c. $f(2) = e^{-2}2^2/2! = e^{-2}(2) = 0.27$ or 27%

38. A *Neurospora* cross was made between a strain that carried the mating-type allele *A* and the mutant allele *arg-1* and another strain that carried the mating-type allele *a* and the wild-type allele for *arg-1* (+). Four hundred linear octads were isolated, and they fell into the seven classes given in the table below. (For simplicity, they are shown as tetrads.)

a. Deduce the linkage arrangement of the mating-type locus and the *arg-1* locus. Include the centromere or centromeres on any map that you draw. Label *all* intervals in map units.
b. Diagram the meiotic divisions that led to class 6. Label clearly.

1	2	3	4	5	6	7
A · arg	*A · +*	*A · arg*	*A · arg*	*A · arg*	*A · +*	*A · +*
A · arg	*A · +*	*A · +*	*a · arg*	*a · +*	*a · arg*	*a · arg*
a · +	*a · arg*	*a · arg*	*A · +*	*A · arg*	*A · +*	*A · arg*
a · +	*a · arg*	*a · +*	*a · +*	*a · +*	*a · arg*	*a · +*
127	125	100	36	2	4	6

Unpacking Problem 38

1. Are fungi generally haploid or diploid?

Answer: Fungi are generally haploid.

2. How many ascospores are in the ascus of Neuro spora? Does your answer match the number presented in this problem? Explain any discrepancy.

Answer: There are four pairs, or eight ascospores, in each ascus. One member of each pair is presented in the data.

3.   What is mating type in fungi? How do you think it is determined experimentally?

      Answer: A mating type in fungi is analogous to gender in humans, in that the mating types of two organisms must differ in order to have a mating that produces progeny. Mating type is determined experimentally simply by seeing if progeny result from specific crosses.

4.   Do the symbols *A* and *a* have anything to do with dominance and recessiveness?

      Answer: The mating types *A* and *a* do not indicate dominance and recessiveness. They simply symbolize the mating-type difference.

5.   What does the symbol *arg-1* mean? How would you test for this genotype?

      Answer: *arg-1* indicates that the organism requires arginine for growth. Testing for the genotype involves isolating nutritional mutants and then seeing if arginine supplementation will allow for growth.

6.   How does the *arg-1* symbol relate to the symbol +?

      Answer: *arg-1*$^+$ indicates that the organism is wild type and does not require supplemental arginine for growth.

7.   What does the expression *wild type* mean?

      Answer: *Wild type* refers to the common form of an organism in its natural population.

8.   What does the word *mutant* mean?

      Answer: *Mutant* means that, for the trait being studied, an organism differs from the wild type.

9.   Does the biological function of the alleles shown have anything to do with the solution of this problem?

      Answer: The actual function of the alleles in this problem does not matter in solving the problem.

10.   What does the expression *linear octad analysis* mean?

Answer: *Linear octad analysis* refers to the fact that the ascospores in each ascus are in a linear arrangement that reflects the order in which the two meiotic divisions and a subsequent mitotic division occurred to produce them. By tracking traits and correlating them with position, it is possible to detect crossing-over events that occurred at the tetrad (four-strand, homologous pairing) stage prior to the two meiotic divisions. Since the mitotic division occurs last, and mitotic sister spores are typically identical, the mitotic sisters may be treated as a pair of identical twins and are listed only once in the diagram, so the octad is treated as a tetrad.

11.   In general, what more can be learned from linear tetrad analysis that cannot be learned from unordered tetrad analysis?

Answer: Linear tetrad analysis allows for the mapping of centromeres in relation to genes, which cannot be done with unordered tetrad analysis.

12.   How is a cross made in a fungus such as *Neurospora*? Explain how to isolate asci and individual ascospores. How does the term *tetrad* relate to the terms *ascus* and *octad*?

Answer: A cross is made in *Neurospora* by placing the two organisms in the same test tube or Petri dish and allowing them to grow. Gametes develop and fertilization, followed by meiosis, mitosis, and ascus formation, occurs. The asci are isolated, and the ascospores are dissected out of them with the aid of a microscope. The ascus has an octad, or eight spores, within it, and the spores are arranged in four (tetrad) pairs.

13.   Where does meiosis take place in the *Neurospora* life cycle? (Show it on a diagram of the life cycle.)

Answer: Meiosis occurs immediately following fertilization in *Neurospora*.

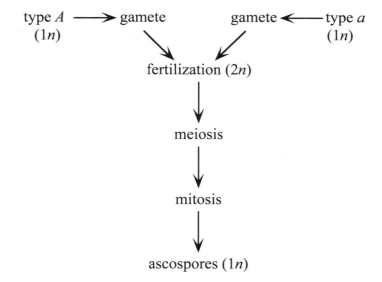

14. What does Problem 38 have to do with meiosis?

    Answer: Meiosis produced the ascospores that were analyzed.

15. Can you write out the genotypes of the two parental strains?

    Answer: The cross is $A \cdot arg\text{-}1 \times a \cdot arg\text{-}1^+$.

16. Why are only four genotypes shown in each class?

    Answer: Although there are eight ascospores, they occur in pairs. Each pair represents one chromatid of the originally paired chromosomes. By convention, both members of a pair are represented by a single genotype.

17. Why are there only seven classes? How many ways have you learned for classifying tetrads generally? Which of these classifications can be applied to both linear and unordered tetrads? Can you apply these classifications to the tetrads in this problem? (Classify each class in as many ways as possible.) Can you think of more possibilities in this cross? If so, why are they not shown?

    Answer: The seven classes represent the seven types of outcomes. The specific outcomes can be classified as follows:

Class	1	2	3	4	5	6	7
outcome	PD	NPD	T	T	PD	NPD	T
A/a	I	I	I	II	II	II	II
$arg\text{-}1^+/arg\text{-}1$	I	I	II	I	II	II	II

where PD = parental ditype, NPD = nonparental ditype, T = tetratype, I = first-division segregation, and II = second-division segregation.

Other classes can be detected, but they indicate the same underlying process. For example, the following three asci are equivalent:

1	2	3
$A\ arg$	$A\ arg^+$	$A\ arg^+$
$A\ arg^+$	$A\ arg$	$A\ arg$
$a\ arg$	$a\ arg$	$a\ arg^+$
$a\ arg^+$	$a\ arg^+$	$a\ arg$

In the first ascus, a crossover occurred between chromatids 2 and 3, while in the second ascus it occurred between chromatids 1 and 3, and in the third ascus the crossover was between chromatids 1 and 4. A fourth equivalent ascus would contain a crossover between chromatids 2 and 3. All four indicate a crossover between the second gene and its centromere and all are tetratypes.

18.   Do you think there are several different spore orders within each class? Why would these different spore orders not change the class?

Answer: This is exemplified in the answer to (17) above.

19.   Why is the following class not listed?

$a \cdot +$
$a \cdot +$
$A \cdot arg$
$A \cdot arg$

Answer: The class is identical with class 1 in the problem, but inverted. These are included in class 1 in the table.

20.   What does the expression *linkage arrangement* mean?

Answer: *Linkage arrangement* refers to the relative positions of the two genes and the centromere along the length of the chromosome.

21.   What is a genetic interval?

Answer: A genetic interval refers to the region between two loci, whose size is measured in map units.

22. Why does the problem state "centromere or centromeres" and not just "centromere"? What is the general method for mapping centromeres in tetrad analysis?

Answer: The problem does not specify whether the two loci are on separate chromosomes or are on the same chromosome. The general formula for calculating the distance of a locus to its centromere is to measure the percentage of tetrads that show second-division segregation patterns for that locus and divide by two. You are supposed to determine whether the genes share a common centromere or are associated with different centromeres.

23. What is the total frequency of $A \oplus +$ ascospores? (Did you calculate this frequency by using a formula or by inspection? Is this a recombinant genotype? If so, is it the only recombinant genotype?)

Answer: Recall that there are eight ascospores per ascus. By inspection, the frequency of recombinant $A\ arg\text{-}1^+$ ascospores is $4(125) + 2(100) + 2(36) + 4(4) + 2(6) = 800$. There is also the reciprocal recombinant genotype $a\ arg\text{-}1$.

24. The first two classes are the most common and are approximately equal in frequency. What does this information tell you? What is their content of parental and recombinant genotypes?

Answer: Class 1 is parental; class 2 is nonparental ditype. Because they occur at equal frequencies, the two genes are not linked.

Solution to the Problem

a. The cross is $A \cdot arg\text{-}1 \times a \cdot arg\text{-}1^+$. Use the classification of asci in part (17) above. First, decide if the two genes are linked by using the formula PD>>NPD, when the genes are linked, while PD = NPD when they are not linked. PD = 127 + 2 = 129 and NPD = 125 + 4 = 129, which means that the two genes are not linked. Alternatively,

RF $= 100\%\ (1/2T + NPD)/\text{total asci}$

$= 100\%\ [(1/2)(100 + 36 + 6) + (125 + 4)]/400 = 50\%.$

Next, calculate the distance between each gene and its centromere using the formula RF = 100%(1/2 number of tetrads exhibiting MII segregation)/(total number of asci):

$A\text{–centromere} = 100\%(1/2)(36 + 2 + 4 + 6)/400$

$$= 100\%(24/400) = 6 \text{ m.u.}$$

$$arg^+\text{–centromere} \quad = 100\%(1/2)(100 + 2 + 4 + 6)/400$$

$$= 100\%(56/400) = 14 \text{ m.u.}$$

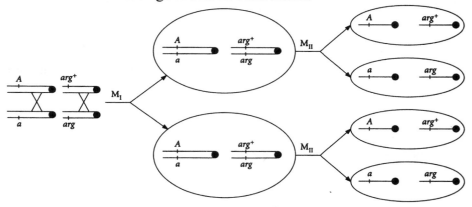

b. Class 6 can be obtained if a single crossover occurred between chromatids 2 and 3 between each gene and its centromere.

39. A geneticist studies 11 different pairs of *Neurospora* loci by making crosses of the type $a \cdot b \times a^+ \cdot b^+$ and then analyzing 100 linear asci from each cross. For the convenience of making a table, the geneticist organizes the data as if all 11 pairs of genes had the same designation—*a* and *b*—as shown below. For each cross, map the loci in relation to each other and to centromeres.

Number of asci of type

Cross	$a \cdot b$ $a \cdot b$ $a^+ \cdot b^+$ $a^+ \cdot b^+$	$a \cdot b^+$ $a \cdot b^+$ $a^+ \cdot b$ $a^+ \cdot b$	$a \cdot b$ $a \cdot b^+$ $a^+ \cdot b^+$ $a^+ \cdot b$	$a \cdot b$ $a^+ \cdot b$ $a^+ \cdot b^+$ $a \cdot b^+$	$a \cdot b$ $a^+ \cdot b^+$ $a^+ \cdot b^+$ $a \cdot b$	$a \cdot b^+$ $a^+ \cdot b$ $a^+ \cdot b$ $a \cdot b^+$	$a \cdot b^+$ $a^+ \cdot b$ $a^+ \cdot b^+$ $a \cdot b$
1	34	34	32	0	0	0	0
2	84	1	15	0	0	0	0
3	55	3	40	0	2	0	0
4	71	1	18	1	8	0	1
5	9	6	24	22	8	10	20
6	31	0	1	3	61	0	4
7	95	0	3	2	0	0	0
8	6	7	20	22	12	11	22
9	69	0	10	18	0	1	2
10	16	14	2	60	1	2	5
11	51	49	0	0	0	0	0

Answer: Before beginning this problem, classify all asci as PD, NPD, or T and determine whether there is M_I or M_{II} segregation for each gene:

			Asci type				
	1	2	3	4	5	6	7
type	PD	NPD	T	T	PD	NPD	T
gene a	I	I	I	II	II	II	II
gene b	I	I	II	I	II	II	II

If PD >> NPD, linkage is indicated. The distance between a gene and its centromere = 100% (1/2)(MII)/total. The distance between two genes = 100% (1/2T + NPD)/total.

Cross 1: PD = NPD and RF = 50%; the genes are not linked.
a–centromere: 100% (1/2)(0)/100 = 0 m.u. Gene *a* is close to the centromere.
b–centromere: 100% (1/2)(32)/100 = 16 m.u.

Cross 2: PD >> NPD; the genes are linked.
a–*b*: 100% [1/2(15) + 1]/100 = 8.5 m.u.
a–centromere: 100% (1/2)(0)/100 = 0 m.u. Gene *a* is close to the centromere.
b–centromere: 100% (1/2)(15)/100 = 7.5 m.u.

Cross 3: PD >> NPD; the genes are linked.
a–*b*: 100% [1/2(40) + 3]/100 = 23 m.u.
a–centromere: 100% (1/2)(2)/100 = 1 m.u.
b–centromere: 100% (1/2)(40 + 2)/100 = 21 m.u.

Cross 4: PD > > NPD; the genes are linked.
a–*b*: 100% [1/2(20) + 1]/100 = 11 m.u.
a–centromere: 100% (1/2)(10)/100 = 5 m.u.
b–centromere: 100% (1/2)(18 + 8 + 1)/100 = 13.5 m.u.

Cross 5: PD = NPD (and RF = 49%); the genes are not linked.
a–centromere: 100% (1/2)(22 + 8 + 10 + 20)/99 = 30.3 m.u.
b–centromere: 100% (1/2)(24 + 8 + 10 + 20)/99 = 31.3 m.u.

These values are approaching the 67 percent theoretical limit of loci exhibiting M_{II} patterns of segregation and should be considered cautiously.

Cross 6: PD >> NPD; the genes are linked.
a–*b*: 100% [1/2(1 + 3 + 4) + 0]/100 = 4 m.u.
a–centromere: 100% (1/2)(3 + 61 + 4)/100 = 34 m.u.
b–centromere: 100% (1/2)(1 + 61 + 4)/100 = 33 m.u.

These values are at the 67 percent theoretical limit of loci exhibiting M_{II} patterns of segregation, and therefore, both loci can be considered unlinked to the centromere.

Cross 7: PD >> NPD; the genes are linked.
a–b: 100% [1/2(3 + 2) + 0]/100 = 2.5 m.u.
a–centromere: 100% (1/2)(2)/100 = 1 m.u.
b–centromere: 100% (1/2)(3)/100 = 1.5 m.u.

Cross 8: PD = NPD; the genes are not linked.
a–centromere: 100% (1/2)(22 + 12 + 11+ 22)/100 = 33.5 m.u.
b–centromere: 100% (1/2)(20 + 12 + 11 + 22)/100 = 32.5 m.u.
Same as cross 5.

Cross 9: PD >> NPD; the genes are linked.
a–b: 100% [1/2(10 + 18 + 2) + 1]/100 = 16 m.u.
a–centromere: 100% (1/2)(18 + 1 + 2)/100 = 10.5 m.u.
b–centromere: 100% (1/2)(10 + 1 + 2)/100 = 6.5 m.u.

Cross 10: PD = NPD; the genes are not linked.
a–centromere: 100% (1/2)(60 + 1+ 2 + 5)/100 = 34 m.u.
b–centromere: 100% (1/2)(2 + 1 + 2 + 5)/100 = 5 m.u.

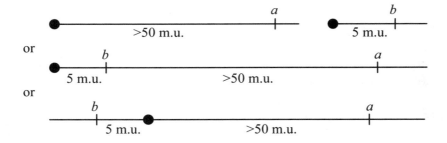

Cross 11: PD = NPD; the genes are not linked.
a–centromere: 100% (1/2)(0)/100 = 0 m.u.
b–centromere: 100% (1/2)(0)/100 = 0 m.u.

40.   Three different crosses in *Neurospora* are analyzed on the basis of unordered tetrads. Each cross combines a different pair of linked genes. The results are shown in the following table:

Cross	Parents (%)	Parental ditypes (%)	Tetra- types (%)	Non-parental ditypes (%)
1	$a \cdot b^+ \times a^+ \cdot b$	51	45	4
2	$c \cdot d^+ \times c^+ \cdot d$	64	34	2
3	$e \cdot f^+ \times e^+ \cdot f$	45	50	5

For each cross, calculate:

a. the frequency of recombinants (RF).
b. the uncorrected map distance, based on RF.
c. the corrected map distance, based on tetrad frequencies.
d. the corrected map distance, based on the mapping function.

Answer: The number of recombinants is equal to NPD + 1/2T. The uncorrected map distance is based on RF = (NPD + ½T)/total. The corrected map distance, based on the Perkins formula, is RF = 50(T + 6NPD)/total. The formula for the Haldane mapping function is RF = ½(1 – e^{-m}). To convert to map distance, m, the measure of crossover frequency is multiplied by 50 percent to give the corrected map distance.

Cross 1:
recombinant frequency = 4% + ½ (45%) = 26.5%
uncorrected map distance = [4% + ½ (45%)]/100% = 26.5 m.u.
corrected map distance, Perkins' formula = 50[45% + 6(4%)]/100% = 34.5 m.u.

The formula for the Haldane mapping function is RF = ½(1 – e^{-m}). Solving for this situation, e^{-m} = 1 – (2 × 0.265) = 0.47 so m = 0.755. To convert to map distance, m, the measure of crossover frequency is multiplied by 50 percent to give a corrected map distance of 37.8%.

Cross 2:
recombinant frequency = 2% + 1/2(34%) = 19%
uncorrected map distance = [2% + 1/2(34%)]/100% = 19 m.u.
corrected map distance, Perkins formula = 50[34% + 6(2%)]/100% = 23 m.u.
corrected map distance, using mapping function = 23.9 m.u.

Cross 3:
recombinant frequency = 5% + ½ (50%) = 30%
uncorrected map distance = [5% + ½ (50%)]/100% = 30 m.u.
corrected map distance, Perkins formula = 50[50% + 6(5%)]/100% = 40 m.u.
corrected map distance, using mapping function = 45.8 m.u.

41. On *Neurospora* chromosome 4, the *leu3* gene is just to the left of the centromere and always segregates at the first division, whereas the *cys2* gene is to the right of the centromere and shows a second-division segregation frequency of 16 percent. In a cross between a *leu3* strain and a *cys2* strain, calculate the predicted frequencies of the following seven classes of linear tetrads where *l* = *leu3* and *c* = *cys2*. (Ignore double and other multiple crossovers.)

(i) *l c*	(ii) *l+*	(iii) *l c*	(iv) *l c*	(v) *l c*	(vi) *l+*	(vii) *l+*
l c	*l +*	*l +*	*+ c*	*+ +*	*+ c*	*+ c*
+ +	*+ c*	*+ +*	*+ +*	*+ +*	*+ c*	*+ +*
+ +	*+ c*	*+ c*	*l +*	*l c*	*l +*	*l c*

Answer: The *leu3* gene is centromere-linked and always segregates at the first division (M$_I$ patterns). The *cys2* gene is at a distance from its centromere and recombination between these respective locations will result in a second-division segregation (M$_{II}$) pattern. For this, there are four patterns of spores, all equally likely:

cys		*cys*		*+*		*+*
+	and	*+*	and	*cys*	and	*cys*
cys		*+*		*+*		*cys*
+		*cys*		*cys*		*+*

If recombination between the centromere and *cys2* does not occur, then the alleles will show first-division segregation (M$_I$).

i. Given the cross *leu3 +* × *+ cys2*, this tetrad is classified as a nonparental ditype. For linked genes, NPDs are the result of double crossover events, and since you are told to ignore multiple crossovers, the expected frequency for this would be 0 percent.

ii. This tetrad is a parental ditype (PD). Due to random alignment and segregation during meiosis I, two linear tetrads, both classified as PD, are equally likely:

l +	and	*+ c*
l +		*+ c*
+ c		*l +*
+ c		*l +*

Therefore, you would expect 100 percent (M$_I$ segregration of *leu3*) × 84 percent (M$_I$ segregation of *cys2* ' ½) = 42 percent of this class of tetrad.

iii. This tetrad is a tetratype (T). It shows one of the two M$_I$ patterns for *leu3* and one of the four M$_{II}$ patterns for *cys2*. You expect 50 percent × 16/4 percent = 2 percent of this class of tetrad.

iv. – vii. 0 percent. In all these tetrads, *leu3* is shown to have a second-division segregation pattern, and you are told that it always segregates at the first division.

42. A rice breeder obtained a triple heterozygote carrying the three recessive alleles for albino flowers (al), brown awns **(b)**, and fuzzy leaves (fu), all paired with

their normal wild-type alleles. This triple heterozygote was testcrossed. The progeny phenotypes were

    170 wild type
    150 albino, brown, fuzzy
      5 brown
      3 albino, fuzzy
    710 albino
    698 brown, fuzzy
     42 fuzzy
     38 albino, brown

a. Are any of the genes linked? If so, draw a map labeled with map distances. (Don't bother with a correction for multiple crossovers.)

b. The triple heterozygote was originally made by crossing two pure lines. What were their genotypes?

Answer:

a. Yes, the data indicate that the three genes are linked. The most common classes of progeny, albino ($al\ b^+\ fu^+$) and brown, fuzzy ($al^+\ b\ fu$) represent the "parental" chromosomes. (Gene order is not specified.) The least common classes of progeny, brown ($al^+\ b\ fu^+$) and albino, fuzzy ($al\ b^+\ fu$) represent the outcomes of double crossover events and can be used to deduce gene order. In comparing the most common to the least common, the gene "in the middle" can be determined to be *fu*. (Compare $al\ b^+\ fu^+$ to $al\ b^+$ *fu* and $al^+\ b\ fu$ to $al^+\ b\ fu^+$.) To calculate map distances, you must now determine the various recombinant classes.

		al – fu	*fu – b*
710	$al\ fu^+\ b^+$	P	P
698	$al^+\ fu\ b$		P P
170	$al^+\ fu^+\ b^+$	R	P
150	$al\ fu\ b$		R P
42	$al^+\ fu\ b^+$	P	R
38	$al\ fu^+\ b$		P R
5	$al^+\ fu^+\ b$	R	R
3	$al\ fu\ b^+$		R R

So there are $170 + 150 + 5 + 3 = 328$ recombinants between *al* and *fu* for a map distance of (328/progeny) × 100% = 18.1 m.u. and $42 + 38 + 5 + 3 = 88$ recombinants between *fu* and *b* for a map distance of 4.8 m.u.

The most common classes of progeny (the parentals) tell you the original genotypes:

$$al\ fu^+\ b^+\ /\ al\ fu^+\ b^+\ \times\ al^+\ fu\ b\ /\ al^+\ fu\ b$$

43. In a fungus, a proline mutant (*pro*) was crossed with a histidine mutant (*his*). A nonlinear tetrad analysis gave the following results:

+	+	+	+	+	*his*
+	+	+	*his*	+	*his*
pro	*his*	*pro*	+	*pro*	+
pro	*his*	*pro*	*his*	*pro*	+
6		82		112	

a. Are the genes linked or not?

b. Draw a map (if linked) or two maps (if not linked), showing map distances based on straightforward recombinant frequency where appropriate.

c. If there is linkage, correct the map distances for multiple crossovers (choose one approach only).

Answer:

a. The cross was *pro* + × + *his*. This makes the first tetrad class NPD (6 of these), the second tetrad class T (82 of these) and the third tetrad class PD (112 of these). When PD >> NPD, you know the two genes are linked.

b. Map distance can be calculated using the formula RF = (NPD + 1/2 T)100%. In this case, the frequency of NPD is 6/200 or 3 percent, and the frequency of T is 82/200 or 41 percent. Map distance is therefore 23.5 m.u. between these two loci.

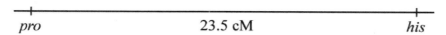

pro	23.5 cM	*his*

c. To correct for multiple crossovers, the Perkins formula may be used. Thus, map distance = (T + 6NPD)50% or (0.41 + 0.18)50% = 29.5 m.u.

44. In the fungus *Neurospora*, a strain that is auxotrophic for thiamine (mutant allele *t*) was crossed with a strain that is auxotrophic for methionine (mutant allele *m*). Linear asci were isolated and classified into the following groups.

Spore pair	Ascus types					
1 and 2	*t* +	*t* +	*t* +	*t* +	*t m*	*t m*
3 and 4	*t* +	*t m*	+ *m*	+ +	*t m*	+ +
5 and 6	+ *m*	+ +	*t* +	*t m*	+ +	*t* +
7 and 8	+ *m*	+ *m*	+ *m*	+ *m*	+ +	+ *m*
Number	260	76	4	54	1	5

a. Determine the linkage relations of these two genes to their centromere(s) and to each other. Specify distances in map units.

b. Draw a diagram to show the origin of the ascus type with only one single representative (second from right).

Answer:

a. The cross is $t + \times + m$. Knowing this, you can then determine that column one (260) is PD, column two (76) is T, column three (4) is PD, column four (54) is T, column five (1) is NPD, and column six (5) is T. Given that PD >> NPD, you know the two genes are linked. For determining the distance of a gene from its centromere, you need to determine what percentage of asci show M_{II} segregation for that gene and then divide that in half to find map distance. For gene t, columns three, four, and six show M_{II} segregation (63/400) and for gene m, columns two, three, and six show M_{II} segregation (85/400). Therefore, gene t is 7.88 m.u. and gene m is 10.63 m.u. from their centromere. What remains to be determined is whether the two genes reside on the same side of their centromere or on opposite sides. The map distance between the two genes will determine which option is correct. For simple map distance, the formula to use is RF = 1/2 T + NPD, so (1/2(135) + 1)/400 or 17.13 m.u.. This best fits the following map:

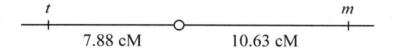

$$t \qquad\qquad\qquad\qquad\qquad\qquad\qquad m$$

7.88 cM   10.63 cM

b. When two genes are linked, it requires a four-stranded double crossover to generate an NPD type ascus. Also, since neither gene shows M_{II} segregation, both crossovers must have occurred on the same side of the centromere.

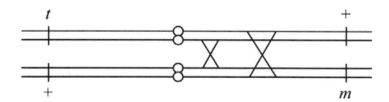

45. A corn geneticist wants to obtain a corn plant that has the three dominant phenotypes: anthocyanin (A), long tassels (L), and dwarf plant (D). In her collection of pure lines, the only lines that bear these alleles are *AA LL dd* and *aa ll DD*. She also has the fully recessive line *aa ll dd*. She decides to intercross the first two and testcross the resulting hybrid to obtain in the progeny 0 plant of the desired phenotype (which would have to be *Aa Ll Dd* in this case). She knows that the three genes are linked in the order written and that the distance between the *A/a* and the *L/l* loci is 16 map units and that the distance between the *L/l* and the *D/d* loci is 24 map units.

    a. Draw a diagram of the chromosomes of the parents, the hybrid, and the tester.

    b. Draw a diagram of the crossover(s) necessary to produce the desired genotype.

    c. What percentage of the testcross progeny will be of the phenotype that she needs?

    d. What assumptions did you make (if any)?

Answer:

a. A diagram of the *AA LL dd* parent's chromosomes would be

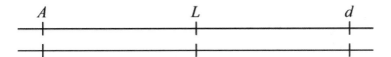

A diagram of the *aa ll dd* parent's chromosomes would be

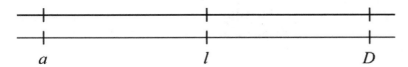

A diagram of the hybrid's chromosomes would be

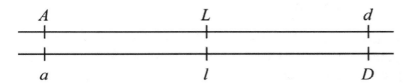

A diagram of the tester's chromosomes would be

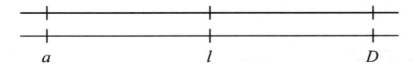

b. A diagram of the crossover(s) necessary to produce the desired genotype would be

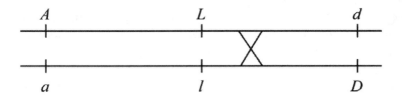

    Please note, this diagram shows a location of crossover that would give the desired genotype but does not take into consideration that each chromosome would actually consist of two chromatids.

c. The genotype of the desired crossover product is *A L D*. This is a result of a single crossover event between the *L* and *D* genes. The map distance between these genes includes both these single crossover events and also any double crossover events (simultaneous events between genes *A* and *L* and between genes *L* and *D*). The double crossovers will not give the desired outcome and therefore must be subtracted to calculate the percentage of test- cross progeny with the desired phenotype. The expected double crossover is simply the chance of recombination in one interval multiplied by the chance of recombination in another interval, or in this case $0.16 \times 0.24 = 0.038$. Therefore, the 24 percent recombination between genes *L* and *D* is the sum of 3.8 percent double crossover events and 20.2 percent single crossover events. Thus, the percentage of *A L D/a l d* progeny from this testcross should be $1/2 \times 20.2 = 10.1$ percent of the total.

d. To calculate the frequency of double crossovers, we assumed that the two recombinations were independent and that there was no interference.

46. In the model plant *Arabidopsis thaliana* the following alleles were used in a cross:

T = presence of trichomes	*t* = absence of trichomes
D = tall plants	*d* = dwarf plants
W = waxy cuticle	*w* = nonwaxy
A = presence of purple anthocyanin pigment	*a* = absence (white)

The *T/t* and *D/d* loci are linked 26 m.u. apart on chromosome 1, whereas the *W/w* and *A/a* loci are linked 8 m.u. apart on chromosome 2.

A pure-breeding double-homozygous recessive trichome less nonwaxy plant is crossed with another pure-breeding double-homozygous recessive dwarf white plant.

a. What will be the appearance of the F_1?

b. Sketch the chromosomes 1 and 2 of the parents and the F_1, showing the arrangement of the alleles.

c. If the F_1 is testcrossed, what proportion of the progeny will have all four recessive phenotypes?

Answer:

a. The F_1 would have all the dominant phenotypes: presence of trichomes, tall, waxy cuticle, and presence of purple pigment.

b. A sketch of the parents' chromosomes would be

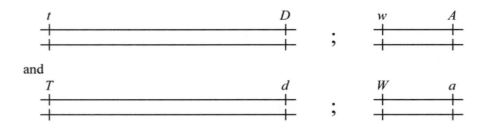

And a sketch of the F$_1$'s chromosomes would be

c. For the progeny of a test cross to have all four recessive phenotypes, it must inherit a recombinant *t d* chromosome 1 and a *w a* recombinant chromosome 2 from its F$_1$ parent. Since the *T/t* and *D/d* loci are 26 m.u. apart, 26 percent of the progeny are expected to be recombinant for these genes and of these, half, or 13 percent, will be *t d*. Similarly, 8 percent of the progeny are expected to be recombinant for the *W/w* and *A/a* loci and half of these, or 4 percent, will be *w a*. Since these are independent events, 13 percent × 4 percent = 0.52 percent of the total progeny will have all four recessive phenotypes.

47. In corn, the cross *WW ee FF* × *ww EE ff* is made. The three loci are linked as follows:

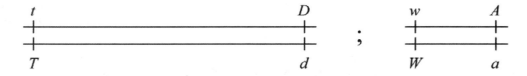

Assume no interference.

a. If the F$_1$ is testcrossed, what proportion of progeny will be *ww ee ff*?
b. If the F$_1$ is selfed, what proportion of progeny will be *ww ee ff*?

Answer:

a. The cross is *W e F / W e F* × *w E f / w E f* and the F$_1$ are *W e F / w E f*. Progeny that are *ww ee ff* from a testcross of this F$_1$ must have inherited one of the double crossover recombinant chromosomes (*w e f*). Assuming no interference, the expected percentage of double crossovers is 8 percent × 24 percent = 1.92 percent, and half of this is 0.96 percent.

b. To obtain a *ww ee ff* progeny from a self cross of this F$_1$ requires the independent inheritance of two doubly recombinant *w e f* chromosomes. The chance of this, based on the answer to part (a) of this question would be 0.96 × 0.96 = 0.009 percent.

48. The fungal cross $+ \cdot + \times c \cdot m$ was made, and nonlinear (unordered) tetrads were collected. The results were

+ +	+ +	+ m
+ +	+ m	+ m
c m	c +	c +
c m	c m	c +
Total 112	82	6

a. From these results, calculate a simple recombinant frequency.
b. Compare the Haldane mapping function and the Perkins formula in their conversions of the RF value into a "corrected" map distance.
c. In the derivation of the Perkins formula, only the possibility of meioses with zero, one, and two crossovers was considered. Could this limit explain any discrepancy in your calculated values? Explain briefly (no calculation needed).

Answer: Since the cross is $+ + \times c\ m$, the first class of tetrad is PD, the second class is T, and the third class is NPD. That PD >> NPD tells you the genes are linked.

a. RF = (1/2T + NPD)100% = (0.205 + 0.03)100% = 23.5%

b. The Perkins formula is RF = (T + 6 NPD)50% = 29.5%

The formula for the Haldane mapping function is RF = $1/2(1 - e^{-m})$. Solving for this situation, $e^{-m} = 1 - (2 \times 0.235) = 0.53$ so $m = 0.635$.

To convert to map distance, m, the measure of crossover frequency is multiplied by 50 percent, to give a corrected map distance of 31.74%.

c. Yes. The Perkins formula only takes into account 0, 1, and 2 crossovers while the Haldane mapping formula has no limitations on the number of crossover events. It predicts the chance of any nonzero number of crossovers.

49. In mice, the following alleles were used in a cross:

W = waltzing gait w = nonwaltzing gait
G = normal gray color g = albino
B = bent tail b = straight tail

A waltzing gray bent-tailed mouse is crossed with a nonwaltzing albino straight-tailed mouse and, over several years, the following progeny totals are obtained:

waltzing	gray	bent	18
waltzing	albino	bent	21
nonwaltzing	gray	straight	19
nonwaltzing	albino	straight	22
waltzing	gray	straight	4
waltzing	albino	straight	5
nonwaltzing	gray	bent	5
nonwaltzing	albino	bent	6
Total			100

a. What were the genotypes of the two parental mice in the cross?
b. Draw the chromosomes of the parents.
c. If you deduced linkage, state the map unit value or values and show how they were obtained.

Answer: There are three patterns of possible outcomes from a testcross of a triply heterozygous parent. If all genes are linked, you expect pairs of roughly equal numbers of progeny in four different frequencies (for example, see problem 20 of this chapter). If all genes are unlinked, the expectation is eight classes of progeny in roughly equal numbers due to the independent assortment of all genes. The last possibility is that two of the genes are linked, but the third is unlinked. In that case, the expectation is two groups of four of two different frequencies. This final pattern is observed in the data of this problem.

In reviewing the data, the four most common classes are either waltzing bent or nonwaltzing straight with gray or albino segregating equally among those two groups. This observation tells you that the W/w and B/b genes are linked and both are unlinked to the G/g gene.

a. $W/w \cdot G/G \cdot A/A \times w/w \cdot g/g \cdot a/a$

b.

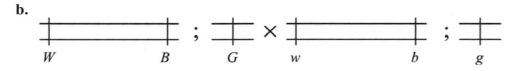

c. For this cross, $W\,B$ and $w\,b$ begin linked to each other so any progeny of the testcross that are $W\,b$ or $w\,B$ are recombinant.

$$\text{map distance} = (4 + 5 + 5 + 6)/100 \times 100\% = 20 \text{ map units}$$

50. Consider the *Neurospora* cross + ; + × *f* ; *p*

It is known that the $+/f$ locus is very close to the centromere on chromosome 7—in fact, so close that there are never any second-division segregations. It is also known that the $+/p$ locus is on chromosome 5, at such a distance that there is usually an average of 12 percent second- division segregations. With this information, what will be the proportion of octads that are

a. parental ditypes showing M_I patterns for both loci?
b. nonparental ditypes showing M_I patterns for both loci?
c. tetratypes showing an M_I pattern for $+/f$ and an M_{II} pattern for $+/p$?
d. tetratypes showing an M_{II} pattern for $+/f$ and an M_I pattern for $+/p$?

Answer: The cross is $+$; $+ \times f$; p and you know that f is linked to its centromere and only segregates at the first division (M_I). The p gene, on average, segregates at the second divisions (M_{II}) 12 percent of the time (and therefore segregates at the first division 88 percent of the time).

a. All octads will show M_I segregation for the $+/f$ gene and 88 percent will show M_I segregation for the $+/p$ gene. Since the genes are unlinked, PD = NPD so half of all octads that meet these conditions will be PD, or in this case, 44 percent.

b. 44 percent (see part (a)).

c. Since all octads will show M_I segregation for $+/f$, all M_{II} patterns for $+/p$ will be tetratypes or, in this case, 12 percent.

d. 0 percent. $+/f$ does not segregate at the second division.

51. In a haploid fungus, the genes *al-2* and *arg-6* are 30 map units apart on chromosome 1, and the genes *lys-5* and *met-1* are 20 map units apart on chromosome 6. In a cross

al-2 + ; *+ met-1* × *+ arg-6* ; *lys-5 +*

what proportion of progeny would be prototrophic $+ +$; $+ +$?

Answer: The cross is *al-2 +* ; *+ met-l* × *+ arg-6* ; *lys-5 +*. The transient diploid will be *al-2 +/+ arg-6* ; *+ met-1/lys-5 +*. From this, one-half of the recombinant chromosomes 1 will be $+ +$ and one-half of the recombinant chromosomes 6 will be $+ +$. Since these are independent events, ½(30 percent) × 1/2(20 percent) = 1.5 percent of the progeny will be $+ +$; $+ +$.

52. The recessive alleles k (kidney-shaped eyes instead of wild-type round), c (cardinal-colored eyes instead of wild-type red), and e (ebony body instead of wild-type gray) identify three genes on chromosome 3 of *Drosophila*. Females with kidney-shaped, cardinal-colored eyes were mated with ebony males. The F$_1$ was wild type. When F$_1$ females were testcrossed with *kk cc ee* males, the following progeny phenotypes were obtained:

k	*c*	*e*	3
k	*c*	+	876
k	+	*e*	67
k	+	+	49
+	*c*	*e*	44
+	*c*	+	58
+	+	*e*	899
+	+	+	4
Total			2000

a. Determine the order of the genes and the map distances between them.
b. Draw the chromosomes of the parents and the F$_1$.
c. Calculate interference and say what you think of its significance.

Answer:

a. The data include that the three genes are linked. The most common classes of progeny, *k c +* and *+ + e* represent the "parental" chromosomes. (Gene order is not specified.) The least common classes of progeny, *k c e* and *+ + +* represent the outcomes of double crossover events and can be used to deduce gene order. In comparing the most common to the least common, the gene "in the middle" can be determined to be *e*. (Compare *k c +* to *k c e* and *+ + e* to *+ + +*.) To calculate map distances, you must now determine the various recombinant classes.

		k − e	*e − c*
899	*+ + e*	P	P
876	*k c +*	P	P
67	*k + e*	R	P
58	*+ c +*	R	P
49	*k + +*	P	R
44	*+ c e*	P	R
4	*+ + +*	R	R
3	*k c e*	R	R

So there are $67 + 58 + 4 + 3 = 132$ recombinants between *k* and *e* for a map distance of (132/total progeny) $\times$ 100% = 6.6 m.u. and $49 + 44 + 4 + 3 = 100$ recombinants between *e* and *c* for a map distance of 5 m.u.

b.

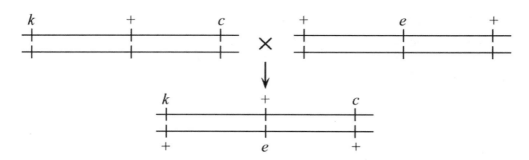

c. The expected frequency of double crossovers is 6.6 percent × 5 percent = 0.33 percent. The actual frequency of double crossovers is (4 + 3)/2000 × 100% = 0.35 percent. To calculate interference, the formula is I = 1 − (observed DCO/expected DCO) = -0.06. In this case, the interference is "negative." The occurrence of one recombination event appears to slightly increase the chances of another. Of course, the difference between observed (7) and expected (6.6) is not statistically meaningful. While there are observed instances of negative interference, generally, interference values are between 0 (no interference) and 1 (complete interference).

53. From parents of genotypes *A/A · B/B* and *a/a · B/b*, a dihybrid was produced. In a testcross of the dihybrid, the following seven progeny were obtained:

A/a · B/b, a/a · b/b, A/a · B/b, A/a · b/b,
a/a · b/b, A/a · B/b, and a/a · B/b

Do these results provide convincing evidence of linkage?

Answer: The results tell us little about linkage. Although the number of recombinants (3) is less than the number of parentals (5), one can have no confidence in the fact that the RF is <50%. The main problem is that the sample size is small, so just one individual more or less in a genotypic class can dramatically affect the ratios. Even the Chi-square test is unreliable at such small sample sizes. It *is* probably safe to say that there is not tight linkage because several recombinants were found in a relatively small sample. However, one cannot distinguish between more distant linkage and independent assortment. A larger sample size is required.

CHALLENGING PROBLEMS

54. Use the Haldane map function to calculate the corrected map distance in cases where the measured RF = 5%, 10%, 20%, 30%, and 40%. Sketch a graph of RF against corrected map distance and use it to answer the question, When should one use a map function?

Answer: The formula for the Haldane mapping function is RF = $1/2(1 - e^{-m})$. Solving for a measured RF = 5%, $e^{-m} = 1 - (2 \times 0.05) = 0.90$, so $m = 0.105$. To convert to map distance, m, the measure of crossover frequency is multiplied by 50 percent, to give a corrected map distance of 5.27%. Similarly, an RF = 10% would be corrected to 11.2%, RF = 20% to 25.5%, RF = 30% to 46% and RF = 40% to 80%! A graph of this data is presented below.

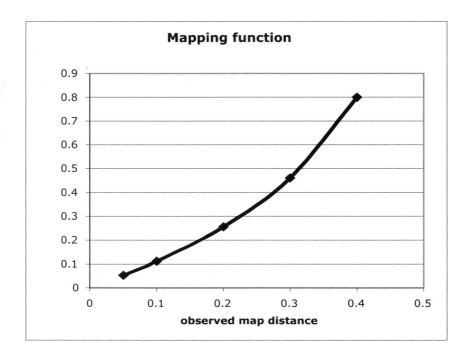

As can be observed, when map distances are small, the observed and corrected recombination frequencies are much the same. So the larger the RF, the more important it is to correct for underestimation of map distance by use of the mapping function.

Unpacking the Problem

55. An individual heterozygous for four genes, $A/a \cdot B/b \cdot C/c \cdot D/d$, is testcrossed with $a/a \cdot B/b \cdot c/c \cdot d/d$, and 1000 progeny are classified by the gametic contribution of the heterozygous parent as follows:

$a \cdot B \cdot C \cdot D$ 42
$A \cdot b \cdot c \cdot d$ 43
$A \cdot B \cdot C \cdot d$ 140
$a \cdot b \cdot c \cdot D$ 145
$a \cdot B \cdot c \cdot D$ 6
$A \cdot b \cdot C \cdot d$ 9
$A \cdot B \cdot c \cdot d$ 305

$a \cdot b \cdot C \cdot D$ *310*

a. Which genes are linked?
b. If two pure-breeding lines had been crossed to produce the heterozygous individual, what would their genotypes have been?
c. Draw a linkage map of the linked genes, showing the order and the distances in map units.
d. Calculate an interference value, if appropriate.

Answer:

a. All of these genes are linked. To determine this, each gene pair is examined separately. For example, are A and B linked?

$A\,B = 140 + 305 = 445$
$a\,b = 145 + 310 = 455$
$a\,B = 42 + 6 = 48$
$A\,b = 43 + 9 = 52$

Conclusion: the two genes are linked and 10 m.u. apart.

Are A and D linked?

$A\,D = 0$
$a\,d = 0$
$A\,d = 43 + 140 + 9 + 305 = 497$
$a\,D = 42 + 145 + 6 + 310 = 503$

Conclusion: the two genes show no recombination, and at this resolution are 0 m.u. apart.

Are B and C linked?

$B\,C = 42 + 140 = 182$
$b\,c = 43 + 145 = 188$
$B\,c = 6 + 305 = 311$
$b\,C = 9 + 310 = 319$

Conclusion: the two genes are linked and 37 m.u. apart.

Are C and D linked?

$C\,D = 42 + 310 = 350$
$c\,d = 43 + 305 = 348$
$C\,d = 140 + 9 = 149$
$c\,D = 145 + 6 = 151$

Conclusion: the two genes are linked and 30 m.u. apart. Therefore, all four genes are linked.

b. and c. Because A and D show no recombination, first rewrite the progeny omitting D and d (or omitting A and a).

$a B C$	42
$A b c$	43
$A B C$	140
$a b c$	145
$a B c$	6
$A b C$	9
$A B c$	305
$a b C$	310
	1000

Note that the progeny now look like those of a typical three-point testcross, with $A B c$ and $a b C$ the parental types (most frequent) and $a B c$ and $A b C$ the double recombinants (least frequent). The gene order is $B A C$. This is determined either by the map distances or by comparing double recombinants with the parentals; the gene that switches in reference with the other two is the gene in the center ($B A c \rightarrow B a c$, $b a C \rightarrow b A C$).

Next, rewrite the progeny again, this time putting the genes in the proper order, and classify the progeny.

$B a C$	42	CO A–B
$b A c$	43	CO A–B
$B A C$	140	CO A–C
$b a c$	145	CO A–C
$B a c$	6	DCO
$b A C$	9	DCO
$B A c$	305	parental
$b a C$	310	parental

To construct the map of these genes, use the following formula:

$$\text{distance between two genes} = \frac{(100\%)\ (\text{number of single CO} + \text{number of DCO})}{\text{total number of progeny}}$$

For the A to B distance

$$= \frac{(100\%)(42 + 43 + 6 + 9)}{1000} = 10 \text{ m.u.}$$

For the A to C distance

$$= \frac{(100\%)(140 + 145 + 6 + 9)}{} = 30 \text{ m.u.}$$

1000

The map is

The parental chromosomes actually were *B (A,d) c/b (a,D) C*, where the parentheses indicate that the order of the genes within is unknown.

d. Interference = 1 – (observed DCO/expected DCO)

$$= 1 - (6 + 9)/[(0.10)(0.30)(1000)]$$
$$= 1 - 15/30 = 0.5$$

56. An autosomal allele N in humans causes abnormalities in nails and patellae (kneecaps) called the nail–patella syndrome. Consider marriages in which one partner has the nail–patella syndrome and blood type A and the other partner has normal nails and patellae and blood type O. These marriages produce some children who have both the nail–patella syndrome and blood type A. Assume that unrelated children from this phenotypic group mature, intermarry, and have children. Four phenotypes are observed in the following percentages in this second generation:

nail–patella syndrome, blood type A	66%
normal nails and patellae, blood type O	16%
normal nails and patellae, blood type A	9%
nail–patella syndrome, blood type O	9%

Fully analyze these data, explaining the relative frequencies of the four phenotypes. (See page 214-215 for the genetic basis of these blood types.)

Answer: The verbal description indicates the following cross and result:

P   *N/– · A/– × n/n · O/O*

F₁   *N/n · A/O × N/n · A/O*

The results indicate linkage, so the cross and results can be rewritten:
P   *N A/– – × n O/n O*

F₁   *N A/ n O × N A/n O*

F₂   66%   *N A/– – or N –/– A*
       16%   *n O/n O*

9%   *n A/n –*
9%   *N O/– O*

Only one genotype is fully known: 16 percent *n O/n O*, a combination of two parental gametes. The frequency of two parental gametes coming together is the frequency of the first times the frequency of the second. Therefore, the frequency of each *n O* gamete is the square root of 0.16, or 0.4. Within an organism the two parental gametes occur in equal frequency. Therefore, the frequency of *N A* is also 0.4. The parental total is 0.8, leaving 0.2 for all recombinants. Therefore, *N O* and *n A* occur at a frequency of 0.1 each. The two genes are 20 m.u. apart. Complete frequencies for all genotypes contributing to the four phenotypes can be obtained from a Punnett square using the gamete frequencies provided above.

57. Assume that three pairs of alleles are found in *Drosophila*: x^+ and x, y^+ and y, and z^+ and z. As shown by the symbols, each non-wild-type allele is recessive to its wild-type allele. A cross between females heterozygous at these three loci and wild-type males yields progeny having the following genotypes: 1010 *x+* · *y+* · *z+* females, 430 *x* · *y+* · *z* males, 441 *x+* · *y* · *z+* males, 39 *x* · *y* · *z* males, 32 *x+* · *y+* · *z* males, 30 *x+* · *y+* · *z+* males, 27 *x* · *y* · *z+* males, 1 *x+* · *y* · *z* male, and 0 *x* · *y+* · *z+* males.

  a. On what chromosome of *Drosophila* are the genes carried?

  b. Draw the relevant chromosomes in the heterozygous female parent, showing the arrangement of the alleles.

  c. Calculate the map distances between the genes and the coefficient of coincidence.

Answer:

a. and b. The data indicate that the progeny males have a different phenotype than the females. Therefore, all the genes are on the X chromosome. The two most frequent phenotypes in the males indicate the genotypes of the X chromosomes in the female, and the two least frequent phenotypes in the males indicate the gene order. Gene z is in the middle. Data from only the males are used to determine map distances. The cross is:

P $x\,z\,y^+/x^+\,z^+\,y$ ′ $x^+\,z^+\,y^+/Y$

F₁ males			
	430	$x\,z\,y^+/Y$	parental
	441	$x^+\,z^+\,y/Y$	parental
	39	$x\,z\,y/Y$	CO z–y
	30	$x^+\,z^+\,y^+/Y$	CO z–y
	32	$x^+\,z\,y^+/Y$	CO x–z
	27	$x\,z^+\,y/Y$	CO x–z
	1	$x^+\,z\,y/Y$	DCO
	0	$x\,z^+\,y^+/Y$	DCO

c. $z-y$: 100%(39 + 30 + 1)/1000 = 7.0 m.u.
     $x-z$: 100%(32 + 27 + 1)/1000 = 6.0 m.u.

c.c.   = observed DCO/expected DCO
        = 1/[(0.06)(0.07)(1000)] = 0.238

58. From the five sets of data given in the following table, determine the order of genes by inspection—that is, without calculating recombination values. Recessive phenotypes are symbolized by lowercase letters and dominant phenotypes by pluses.

Phenotypes observed in 3-point testcross	Data sets 1	2	3	4	5
+ + +	317	1	30	40	305
+ + c	58	4	6	232	0
+ b +	10	31	339	84	28
+ b c	2	77	137	201	107
a + +	0	77	142	194	124
a + c	21	31	291	77	30
a b +	72	4	3	235	1
a b c	203	1	34	46	265

Answer: The data given for each of the three-point testcrosses can be used to determine the gene order by realizing that the rarest recombinant classes are the result of double crossover events. By comparing these chromosomes to the "parental" types, the alleles that have switched represent the gene in the middle.

For example, in (1), the most common phenotypes (+ + + and a b c) represent the parental allele combinations. Comparing these to the rarest phenotypes of this data set (+ b c and a + +) indicates that the a gene is recombinant and must be in the middle. The gene order is b a c.

For (2), + b c and a + + (the parentals) should be compared to + + + and a b c (the rarest recombinants) to indicate that the a gene is in the middle. The gene order is b a c.

For (3), compare + b + and a + c with a b + and + + c, which gives the gene order b a c.

For (4), compare + + c and a b + with + + + and a b c, which gives the gene order a c b.

For (5), compare + + + and a b c with + + c and a b +, which gives the gene order a c b.

59. From the phenotype data given in the following table for two three-point testcrosses for (1) *a, b,* and *c* and (2) *b, c,* and *d,* determine the sequence of the four genes *a, b, c,* and *d* and the three map distances between them. Recessive phenotypes are symbolized by lowercase letters and dominant phenotypes by pluses.

1		2	
+ + +	669	b c d	8
a b +	139	b + +	441
a + +	3	b + d	90
+ + c	121	+ c d	376
+ b c	2	+ + +	14
a + c	2280	+ + d	153
a b c	653	+ c +	65
+ b +	2215	b c +	141

Answer: The gene order is *a c b d.*

Recombination between *a* and *c* occurred at a frequency of:

$100\%(139 + 3 + 121 + 2)/(669 + 139 + 3 + 121 + 2 + 2{,}280 + 653 + 2{,}215)$
$= 100\%(265/6{,}082) = 4.36\%$

Recombination between *b* and *c* in cross 1 occurred at a frequency of:

$100\%(669 + 3 + 2 + 653)/(669 + 139 + 3 + 121 + 2 + 2{,}280 + 653 + 2{,}215)$
$= 100\%(1{,}327/6{,}082) = 21.82\%$

Recombination between *b* and *c* in cross 2 occurred at a frequency of:
$100\%(8 + 14 + 153 + 141)/(8 + 441 + 90 + 376 + 14 + 153 + 65 + 141)$
$= 100\%(316/1{,}288) = 24.55\%$

The difference between the two calculated distances between *b* and *c* is not surprising because each set of data would not be expected to yield exactly identical results. Also, many more offspring were analyzed in cross 1. Combined, the distance would be:

$100\%[(316 + 1{,}327)/(1{,}288 + 6{,}082)] = 22.3\%$

Recombination between *b* and *d* occurred at a frequency of:
$100\%(8 + 90 + 14 + 65)/(8 + 441 + 90 + 376 + 14 + 153 + 65 + 141)$
$= 100\%(177/1{,}288) = 13.68\%$

60. The father of Mr. Spock, first officer of the starship *Enterprise*, came from planet Vulcan; Spock's mother came from Earth. A Vulcan has pointed ears (determined by allele *P*), adrenals absent (determined by *A*), and a right-sided heart (determined by *R*). All these alleles are dominant to normal Earth alleles. The three loci are autosomal, and they are linked as shown in this linkage map:

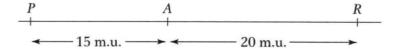

If Mr. Spock marries an Earth woman and there is no (genetic) interference, what proportion of their children will have

a. Vulcan phenotypes for all three characters?
b. Earth phenotypes for all three characters?
c. Vulcan ears and heart but Earth adrenals?
d. Vulcan ears but Earth heart and adrenals?

(Problem 60 is from D. Harrison, *Problems in Genetics*. Addison-Wesley, 1970.)

Answer: Part (a) of this problem is solved two ways, once in the standard way, once in a way that emphasizes a more mathematical approach.

The cross is:

P $\quad$ *P A R/P A R* $\times$ *p a r/p a r*

F_1 $\quad$ *P A R/p a r* $\times$ *p a r/p a r*, a three-point test cross

a. In order to find what proportion will have the Vulcan phenotype for all three characteristics, we must determine the frequency of parentals. Crossing over occurs 15 percent of the time between P and A, which means it does not occur 85 percent of the time. Crossing over occurs 20 percent of the time between A and R, which means that it does not occur 80 percent of the time.

p (no crossover between either gene)
= *p*(no crossover between *P* and *A*) × *p*(no crossover between A and *R*)
= (0.85)(0.80) = 0.68

Half the parentals are Vulcan, so the proportion that are completely Vulcan is 1/2(0.68) = 0.34

Mathematical method:

Number of parentals $\quad$ = 1 – (single CO individuals – DCO individuals)
$\qquad\qquad\qquad\qquad$ = 1 – {[0.15 + 0.20 – 2(0.15)(0.20)] – (0.15) (0.20)} \
$\qquad\qquad\qquad\qquad$ = 0.68

Because half the parentals are Earth alleles and half are Vulcan, the frequency of children with all three Vulcan characteristics is 1/2(0.68) = 0.34

b. Same as above, 0.34

c. To yield Vulcan ears and hearts and Earth adrenals, a crossover must occur in both regions, producing double crossovers. The frequency of Vulcan ears and hearts and Earth adrenals will be half the DCOs, or 1/2(0.15) (0.20) = 0.015

d. To yield Vulcan ears and an Earth heart and adrenals, a single crossover must occur between *P* and *A*, and no crossover can occur between *A* and *R*. The frequency will be:

$$p(CO\ P\text{–}A) \times p(\text{no CO } A\text{–}R) = (0.15)(0.80) = 0.12$$

Of these, 1/2 are *P a r* and 1/2 are *p A R*. Therefore, the proportion with Vulcan ears and an Earth heart and adrenals is 0.06

61. In a certain diploid plant, the three loci *A, B,* and *C* are linked as follows:

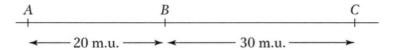

One plant is available to you (call it the parental plant). It has the constitution *A b c/a B C.*

a. With the assumption of no interference, if the plant is selfed, what proportion of the progeny will be of the genotype *a b c/a b c?*
b. Again, with the assumption of no interference, if the parental plant is crossed with the *a b c/a b c* plant, what genotypic classes will be found in the progeny? What will be their frequencies if there are 1000 progeny?
c. Repeat part *b,* this time assuming 20 percent interference between the regions.

Answer:
a. To obtain a plant that is *a b c/a b c* from selfing of *A b c/a B C,* both gametes must be derived from a crossover between *A* and *B.* The frequency of the *a b c* gamete is:

$$1/2\ p(CO\ A\text{–}B) \times p(\text{no CO } B\text{–}C) = 1/2(0.20)(0.70) = 0.07$$

Therefore, the frequency of the homozygous plant will be $(0.07)2 = 0.0049$

b. The cross is $A\ b\ c/a\ B\ C \times a\ b\ c/a\ b\ c$.

To calculate the progeny frequencies, note that the parentals are equal to all those that did not experience a crossover. Mathematically this can be stated as:

parentals $= p(\text{no CO } A\text{–}B) \times p(\text{no CO } B\text{–}C)$
$= (0.80)(0.70) = 0.56$

Because each parental should be represented equally:

$A\ b\ c = 1/2(0.56) = 0.28$
$a\ B\ C = 1/2(0.56) = 0.28$

As calculated above, the frequency of the $a\ b\ c$ gamete is:

$1/2\ p(\text{CO } A\text{–}B) \times p(\text{no CO } B\text{–}C) = 1/2(0.20)(0.70) = 0.07$

as is the frequency of $A\ B\ C$.

The frequency of the $A\ b\ C$ gamete is:

$1/2\ p(\text{CO } B\text{–}C) \times p(\text{no CO } A\text{–}B) = 1/2(0.30)(0.80) = 0.12$

as is the frequency of $a\ B\ c$.

Finally, the frequency of the $A\ B\ c$ gamete is:

$1/2\ p(\text{CO } A\text{–}B) \times p(\text{CO } B\text{–}C) = 1/2(0.20)(0.30) = 0.03$

as is the frequency of $a\ b\ C$.

So for 1,000 progeny, the expected results are

$A\ b\ c$	280
$a\ B\ C$	280
$A\ B\ C$	70
$a\ b\ c$	70
$A\ b\ C$	120
$a\ B\ c$	120
$A\ B\ c$	30
$a\ b\ C$	30

c. Interference $= 1 -$ observed DCO/expected DCO
$0.2\ = 1 -$ observed DCO/$(0.20)(0.30)$

observed DCO $= (0.20)(0.30) - (0.20)(0.20)(0.30) = 0.048$

The *A–B* distance = 20% = 100% [*p*(CO *A–B*) + *p*(DCO)]
Therefore, *p*(CO *A–B*) = 0.20 – 0.048 = 0.152

Similarly, the *B–C* distance = 30% = 100% [*p*(CO *B–C*) + *p*(DCO)]
Therefore, *p*(CO *B–C*) = 0.30 – 0.048 = 0.252

The *p*(parental) = 1 – *p*(CO *A–B*) – *p*(CO *B–C*) – *p*(observed DCO)
              = 1 – 0.152 – 0.252 – 0.048 = 0.548

So for 1,000 progeny, the expected results are:

A b c	274
a B C	274
A B C	76
a b c	76
A b C	126
a B c	126
A B c	24
a b C	24

62. The following pedigree shows a family with two rare abnormal phenotypes: blue sclerotic (a brittle-bone defect), represented by a black-bordered symbol, and hemophilia, represented by a black center in a symbol. Members represented by completely black symbols have both disorders. The numbers in some symbols are the numbers of those types.

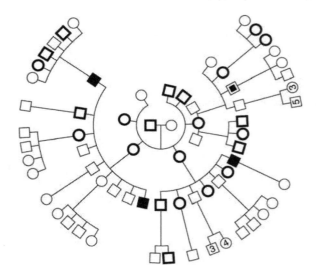

  a. What pattern of inheritance is shown by each condition in this pedigree?
  b. Provide the genotypes of as many family members as possible.
  c. Is there evidence of linkage?

   d. Is there evidence of independent assortment?

   e. Can any of the members be judged as recombinants (that is, formed from at least one recombinant gamete)?

Answer:

a. Blue sclerotic (*B*) appears to be an autosomal dominant disorder. Hemophilia (*h*) appears to be an X-linked recessive disorder.

b. If the individuals in the pedigree are numbered as generations I through IV and the individuals in each generation are numbered clockwise, starting from the top right-hand portion of the pedigree, their genotypes are:

I: *b/b* ; *H/h, B/b* ; *H/Y*

II: *B/b* ; *H/Y, B/b* ; *H/Y, b/b* ; *H/Y, B/b* ; *H/h, b/b* ; *H/Y, B/b* ; *H/h, B/b* ; *H/h, B/b* ; *H/–, b/b* ; *H/–*

III: *b/b* ; *H/–, B/b* ; *H/–, b/b* ; *h/Y, b/b* ; *H/Y, B/b* ; *H/Y, B/b* ; *H/–, B/b* ; *H/Y, B/b* ; *h/Y, B/b* ; *H/–, b/b* ; *H/Y, B/b* ; *H/–, b/b* ; *H/Y, B/b* ; *H/–, B/b* ; *H/Y, B/b* ; *h/Y, b/b* ; *H/Y, b/b* ; *H/Y, b/b* ; *H/–, b/b* ; *H/Y, b/b* ; *H/Y, B/b* ; *H/–, B/b* ; *H/Y, B/b* ; *h/Y*

IV: *b/b* ; *H/–, B/b* ; *H/–, B/b* ; *H/–, b/b* ; *H/h, b/b* ; *H/h, b/b* ; *H/Y, b/b* ; *H/H, b/b* ; *H/Y, b/b* ; *H/h, b/b* ; *H/H, b/b* ; *H/H, b/b* ; *H/Y, b/b* ; *H/Y, b/b* ; *H/H, b/b* ; *H/Y, b/b* ; *H/Y, B/b* ; *H/Y, b/b* ; *H/Y, b/b* ; *H/–, b/b* ; *H/Y, b/b* ; *H/Y, b/b* ; *H/–, b/b* ; *H/H, b/b* ; *H/–, b/b* ; *H/–, b/b* ; *H/Y, b/b* ; *H/Y, b/b* ; *H/Y, b/b* ; *H/h, B/b* ; *H/h, B/b* ; *H/Y, b/b* ; *H/Y, B/b* ; *H/Y, b/b* ; *H/h*

c. There is no evidence of linkage between these two disorders. Because of the modes of inheritance for these two genes, no linkage would be expected.

d. The two genes exhibit independent assortment.

e. No individual could be considered intrachromosomally recombinant. However, a number show interchromosomal recombination, for example, all individuals in generation III that have both disorders.

63. The human genes for color blindness and for hemophilia are both on the X chromosome, and they show a recombinant frequency of about 10 percent. The linkage of a pathological gene to a relatively harmless one can be used for genetic prognosis. Shown here is part of a bigger pedigree. Blackened symbols indicate that the subjects had hemophilia, and crosses indicate color blindness. What information could be given to women III-4 and III-5 about the likelihood of their having sons with hemophilia?

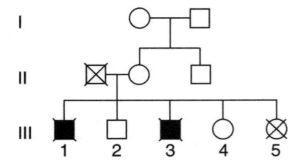

(Problem 63 is adapted from J. F. Crow, *Genetics Notes: An Introduction to Genetics*. Burgess, 1983.)

Answer: If h = hemophilia and b = colorblindness, the genotypes for individuals in the pedigree can be written as

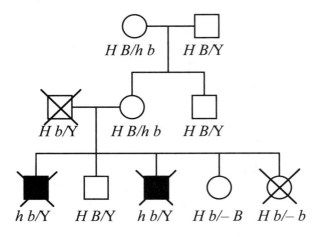

The mother of the two women in question would produce the following gametes:

0.45	$H B$
0.45	$h b$
0.05	$h B$
0.05	$H b$

Woman III-4 can be either $H b/H B$ (0.45 chance) or $H b/h B$ (0.05 chance) because she received B from her mother. If she is $H b/h B$ [0.05/(0.45 + 0.05) = 0.10 chance], she will produce the parental and recombinant gametes with the same probabilities as her mother. Thus, her child has a 45 percent chance of receiving $h B$, a 5 percent chance of receiving $h b$, and a 50 percent chance of receiving a Y from his father. The probability that her child will be a hemophiliac son is (0.1)(0.5)(0.5) = 0.025 = 2.5 percent.

Woman III-5 can be either $H b/H b$ (0.05 chance) or $H b/h b$ (0.45 chance), because she received b from her mother. If she is $H b/h b$ [0.45/(0.45 + 0.05) = 0.90 chance], she has a 50 percent chance of passing h to her child, and there is a 50 percent chance that the child will be male. The probability that she will have a son with hemophilia is $(0.9)(0.5)(0.5) = 0.225 = 22.5$ percent.

64. A geneticist mapping the genes *A, B, C, D,* and *E* makes two 3-point testcrosses. The first cross of pure lines is

$A/A \cdot B/B \cdot C/C \cdot D/D \cdot E/E$ ×/$a/a \cdot b/b \cdot C/C \cdot d/d \cdot E/E$

The geneticist crosses the F1 with a recessive tester and classifies the progeny by the gametic contribution of the F1:

$$
\begin{array}{ll}
A \cdot B \cdot C \cdot D \cdot E & 316 \\
a \cdot b \cdot C \cdot d \cdot E & 314 \\
A \cdot B \cdot C \cdot d \cdot E & 31 \\
a \cdot b \cdot C \cdot D \cdot E & 39 \\
A \cdot b \cdot C \cdot d \cdot E & 130 \\
a \cdot B \cdot C \cdot D \cdot E & 140 \\
A \cdot b \cdot C \cdot D \cdot E & 17 \\
a \cdot B \cdot C \cdot d \cdot E & \underline{\quad 13} \\
 & 1000
\end{array}
$$

The second cross of pure lines is $A/A \cdot B/B \cdot C/C \cdot D/D \cdot E/E$ ×/$a/a \cdot B/B \cdot c/c \cdot D/D \cdot e/e.$

The geneticist crosses the F1 from this cross with a recessive tester and obtains

$$
\begin{array}{ll}
A \cdot B \cdot C \cdot D \cdot E & 243 \\
a \cdot B \cdot c \cdot D \cdot e & 237 \\
A \cdot B \cdot c \cdot D \cdot e & 62 \\
a \cdot B \cdot C \cdot D \cdot E & 58 \\
A \cdot B \cdot C \cdot D \cdot e & 155 \\
a \cdot B \cdot c \cdot D \cdot E & 165 \\
a \cdot B \cdot C \cdot D \cdot e & 46 \\
A \cdot B \cdot c \cdot D \cdot E & \underline{\quad 34} \\
 & 1000
\end{array}
$$

The geneticist also knows that genes *D* and *E* assort independently.

a. Draw a map of these genes, showing distances in map units wherever possible.

b. Is there any evidence of interference?

Answer:

a. Cross 1 reduces to:

P $\qquad A/A \cdot B/B \cdot D/D \times a/a \cdot b/b \cdot d/d$

F_1 $A/a \cdot B/b \cdot D/d \times a/a \cdot b/b \cdot d/d$

The testcross progeny indicate these three genes are linked.

Testcross	A B D	316	parental
progeny	a b d	314	parental
	A B d	31	CO B–D
	a b D	39	CO B–D
	A b d	130	CO A–B
	a B D	140	CO A–B
	A b D	17	DCO
	a B d	13	DCO

A–B: 100%(130 + 140 + 17 + 13)/1000 = 30 m.u.
B–D: 100%(31 + 39 + 17 + 13)/1000 = 10 m.u.

Cross 2 reduces to:

P $A/A \cdot C/C \cdot E/E \times a/a \cdot c/c \cdot e/e$

F_1 $A/a \cdot C/c \cdot E/e \times a/a \cdot c/c \cdot e/e$

The testcross progeny indicate these three genes are linked.

Testcross	A C E	243	parental
progeny	a c e	237	parental
	A c e	62	CO A–C
	a C E	58	CO A–C
	A C e	155	CO C–E
	a c E	165	CO C–E
	a C e	46	DCO
	A c E	34	DCO

A–C: 100% (62 + 58 + 46 + 34)/1000 = 20 m.u.
C–E: 100% (155 + 165 + 46 + 34)/1000 = 40 m.u.

The map that accommodates all the data is:

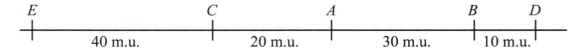

b. Interference (I) = 1 – [(observed DCO)/(expected DCO)]

For cross 1: I = 1 – {30/[(0.30)(0.10)(1000)]} = 1 – 1 = 0, no interference

For cross 2: $I = 1 - \{80/[(0.20)(0.40)(1000)]\} = 1 - 1 = 0$, no interference

65. In the plant *Arabidopsis*, the loci for pod length (*L*, long; *l*, short) and fruit hairs (*H*, hairy; *h*, smooth) are linked 16 map units apart on the same chromosome. The following crosses were made:

(i) $L\,H/L\,H \times l\,h/l\,h \;\rightarrow\; F_1$
(ii) $L\,h/L\,h \times l\,H/l\,H \;\rightarrow\; F_1$

If the F_1's from cross i and cross ii are crossed,

a. what proportion of the progeny are expected to be *l h/l h*?
b. what proportion of the progeny are expected to be *L h/l h*?

Answer:

a. The first F_1 is *L H/l h,* and the second is *l H/L h.* For progeny that are *l h/l h,* they have received a "parental" chromosome from the first F_1 and a "recombinant" chromosome from the second F_1. The genes are 16 percent apart so the chance of a parental chromosome is $1/2(100 - 16\%) = 42\%$, and the chance of a recombinant chromosome is $1/2(16\%) = 8\%$.

The chance of both events $= 42\% \times 8\% = 3.36\%$

b. To obtain *Lh/l h* progeny, either a parental chromosome from each parent was inherited *or* a recombinant chromosome from each parent was inherited. The total probability will therefore be:

$(42\% \times 42\%) + (8\% \times 8\%) = (17.6\% + 0.6\%) = 18.2\%$

66. In corn (*Zea mays*), the genetic map of part of chromosome 4 is as follows, where *w, s,* and *e* represent recessive mutant alleles affecting the color and shape of the pollen:

If the following cross is made

$+ + +/+ + + \times w\,s\,e/w\,s\,e$

and the F_1 is testcrossed with *w s e/w s e,* and if it is assumed that there is no interference on this region of the chromosome, what proportion of progeny will be of genotypes?

a. $+ + +$ **e.** $+ + e$

b. $w\,s\,e$ **f.** $w\,s+$

c. $+\,s\,e$ **g.** $w+e$

d. $w++$ **h.** $+\,s+$

Answer: Crossing over occurs 8 percent of the time between w and s, which means it does not occur 92 percent of the time. Crossing over occurs 14 percent of the time between s and e, which means that it does not occur 86 percent of the time.

a. and b. The frequency of parentals = p (no crossover between either gene)

$$= p(\text{no CO } w\text{–}s) \times p(\text{no CO } s\text{–}e) = (0.92)(0.86)$$
$$= 0.791$$
or
$$1/2(0.791) = 0.396 \text{ each}$$

c. and d. The frequency that will show recombination between w and s only

$$= p(\text{CO } w\text{–}s) \times p(\text{no CO } s\text{–}e) = (0.08)(0.86) = 0.069$$
or
$$1/2(0.069) = 0.035 \text{ each}$$

e. and f. The frequency that will show recombination between s and e only

$$= p(\text{CO } s\text{–}e) \times p(\text{no CO } w\text{–}s) = (0.14)(0.92) = 0.128$$
or
$$1/2(0.128) = 0.064 \text{ each}$$

g. and h. The frequency that will show recombination between w and s and s and e

$$= p(\text{CO } w\text{–}s) \times p(\text{CO } s\text{–}e) = (0.08)(0.14) = 0.011$$
or
$$1/2(0.011) = 0.006 \text{ each}$$

67. Every Friday night, genetics student Jean Allele, exhausted by her studies, goes to the student union's bowling lane to relax. But, even there, she is haunted by her genetic studies. The rather modest bowling lane has only four bowling balls: two red and two blue. They are bowled at the pins and are then collected and returned down the chute in random order, coming to rest at the end stop. As the evening passes, Jean notices familiar patterns of the four balls as they come to rest at the stop. Compulsively, she counts the different patterns. What patterns did she see, what were their frequencies, and what is the relevance of this matter to genetics?

Answer: This problem is analogous to meiosis in organisms that form linear tetrads. Let red = R and blue = r. This can now be compared to meiosis in an organism that is R/r. The patterns, their frequencies, and the division of segregation are given below. Notice that the probabilities change as each ball/allele is selected. This occurs when there is sampling without replacement.

1/2 R
$\times 1/3\ R \times 1/1\ r \times 1/1\ r = 1/6\ RRrr$ first division
$\times 2/3\ r$ or
$\times 1/2\ R \times 1/1\ r = 1/6\ RrRr$ second division
$\times 1/2\ r \times 1/1\ R = 1/6\ RrrR$ second division

1/2 r
$\times 1/3\ r \times 1/1\ R \times 1/1\ R = 1/6\ rrRR$ first division
$\times 2/3\ R$ or
$\times 1/2\ R \times 1/1\ r = 1/6\ rRRr$ second division
$\times 1/2\ r \times 1/1\ R = 1/6\ RrRr$ second division

These results indicate one-third first-division segregation and two-thirds second-division segregation.

68. In a tetrad analysis, the linkage arrangement of the p and q loci is as follows:

Assume that

- in region i, there is no crossover in 88 percent of meioses and there is a single crossover in 12 percent of meioses;
- in region ii, there is no crossover in 80 percent of meioses and there is a single crossover in 20 percent of meioses; and
- there is no interference (in other words, the situation in one region does not affect what is going on in the other region).

What proportions of tetrads will be of the following types? **(a)** M_IM_I, PD; **(b)** M_IM_I, NPD; **(c)** M_IM_{II}, T; **(d)** $M_{II}M_I$, T; **(e)** $M_{II}M_{II}$, PD; **(f)** $M_{II}M_{II}$, NPD; **(g)** $M_{II}M_{II}$,T. (**Note:** Here the M pattern written first is the one that pertains to the p locus.) **Hint:** The easiest way to do this problem is to start by calculating the frequencies of asci with crossovers in both regions, region i, region ii, and neither region. Then determine what M_I and M_{II} patterns result.

Answer: As the problem suggests, calculate the frequencies of the various possibilities. The percentage of tetrads without crossing over is 88% × 80% =

70.4%. The percentage of tetrads with a single crossover in region (i) and none in region (ii) is 12% × 80% = 9.6%. The percentage of tetrads with a single crossover in region (ii) and none in region (i) is 20% × 88% = 17.6%, and the percentage of tetrads with crossovers in both regions is 12% × 20% = 2.4%.

Now work out the patterns of segregation that result in each case

For no crossovers

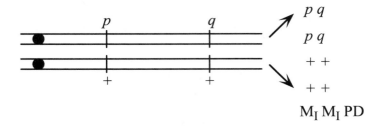

For a single crossover in region (i)

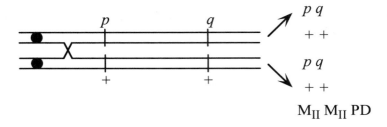

For a single crossover in region (ii)

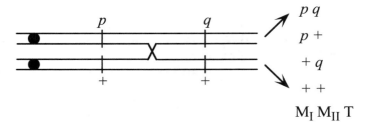

For double crossovers, there are four types, all equally likely

two-strand

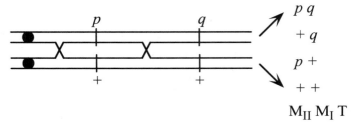

four-strand

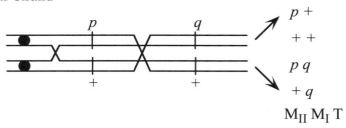

M_{II} M_I T

and two different three-strand

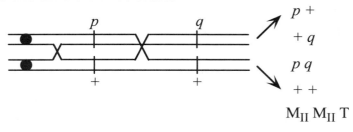

M_{II} M_{II} T

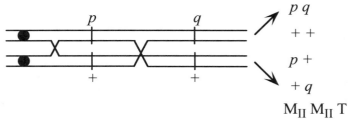

M_{II} M_{II} T

(a) M_I M_I PD is the result of no crossovers = 70.4%

(b) M_I M_I NPD is not found as a result.

(c) M_I M_{II} T is the result of a single crossover in region (ii) = 17.6%

(d) M_{II} M_I T is the result of the two- and four-strand double crossovers = 1.2%

(e) M_{II} M_{II} PD is the result of a single crossover in region (i) = 9.6%

(f) M_{II} M_{II} NPD is not found as a result.

(g) MII MII T is the result of both three-strand double crossovers = 1.2%

69. For an experiment with haploid yeast, you have two different cultures. Each will grow on minimal medium to which arginine has been added, but neither will grow on minimal medium alone. (Minimal medium is inorganic salts plus sugar.) Using appropriate methods, you induce the two cultures to mate. The diploid cells then divide meiotically and form unordered tetrads. Some of the ascospores will grow on minimal medium. You classify a large number of these tetrads for the phenotypes ARG⁻ (arginine requiring) and ARG⁺ (arginine independent) and record the following data:

Segregation of ARG⁻ : ARG⁺	Frequency (%)
4 : 0	40
3 : 1	20
2 : 2	40

   a. Using symbols of your own choosing, assign genotypes to the two parental cultures. For each of the three kinds of segregation, assign genotypes to the segregants.
   b. If there is more than one locus governing arginine requirement, are these loci linked?

Answer:

a. and b. The data support the independent assortment of two genes (call them *arg1* and *arg2*). The cross becomes $arg1 ; arg2^+ \times arg1^+ ; arg2$ and the resulting tetrads are:

$$
\begin{array}{ccc}
4:0\ (\text{PD}) & 3:1\ (\text{T}) & 2:2\ (\text{NPD}) \\
arg1\ ;\ arg2^+ & arg1\ ;\ arg2^+ & arg1\ ;\ arg2 \\
arg1\ ;\ arg2^+ & arg1^+\ ;\ arg2 & arg1\ ;\ arg2 \\
arg1^+\ ;\ arg2 & arg1\ ;\ arg2 & arg1^+\ ;\ arg2^+ \\
arg1^+\ ;\ arg2 & arg1^+\ ;\ arg2^+ & arg1^+\ ;\ arg2^+ \\
\end{array}
$$

Because PD = NPD, the genes are unlinked.

70. An RFLP analysis of two pure lines $A/A \cdot B/B$ and $a/a \cdot B/b$ showed that the former was homozygous for a long RFLP allele (l) and the latter for a short allele (s). The two were crossed to form an F1, which was then backcrossed to the second pure line. A thousand progeny were scored as follows:

Aa Bb ss	9
Aa Bb ls	362
aa bb ls	11
aa bb ss	358
Aa bb ss	43
Aa bb ls	93
aa Bb ls	37
aa Bb ss	87

   a. What do these results tell us about linkage?
   b. Draw a map if appropriate.
   c. Incorporate the RFLP fragments into your map.

Answer:

a. The cross is $A/A \cdot B/B \cdot l/l \times a/a \cdot b/b \cdot s/s$. The resulting F_1 is backcrossed to the second parent and the backcross progeny indicate that all three markers are linked.

b. and c. By comparing the most common progeny to the least (*A B l/a b s* to *A B s/a b s* and *a b s/abs* to *a b l/a b s*) we can deduce that the RFLP allele is between genes *A/a* and *B/b*. The distance between these markers is $(43 + 37 + 9 + 11)/1000 = 10$ map units between gene *A/a* and the RFLP and $(93 + 87 + 9 + 11)/1000 = 20$ map units between the RFLP and gene *B/b*. The restriction sites are not drawn to scale relative to the genetic distances.

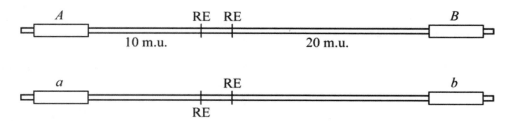

5

The Genetics of Bacteria and Their Viruses

WORKING WITH THE FIGURES

1. In Figure 5-2, in which of the four processes shown is a complete bacterial genome transferred from one cell to another?

   Answer: In none of the four processes shown is a complete bacterial genome transferred.

2. In Figure 5-3, if the concentration of bacterial cells in the original suspension is 200/ml and 0.2 ml is plated onto each of 100 petri dishes, what is the expected average number of colonies per plate?

   Answer: You would expect an average of 40 colonies per plate.

3. In Figure 5-5,
   a. Why do A$^-$ and B$^-$ cells not form colonies on the plating medium?
   b. What genetic event do the purple colonies in the middle plate represent?

   Answer:
   a. The A$^-$ and B$^-$ strains do not grow on the minimal medium because they each contain mutations in genes required to synthesize compounds lacking in the medium.
   b. Prototrophic colonies result from recombination between the A$^-$ and B$^-$ genomes following conjugation.

4. In Figure 5-10c, what do the yellow nuclei represet?

   Answer: The yellow spots represent the DNA of F$^-$ recipients that have taken up DNA from the Hfr donor strain.

5. In Figure 5-11, which donor alleles become part of the recombinant genome produced?

Answer: The donor alleles a^+, b^+, and c^+ become part of the recombinant genome.

6. In Figure 5-12,
   a. Which Hfr gene enters the recipient last? (Which diagram shows it actually entering?)
   b. What is the maximum percentage of cases of transfer of this gene?
   c. Which genes have entered at 25 min? Could they all become part of a stable exconjugant genome?

Answer:
   a. The *gal* is shown entering the donor at 25 minutes.
   b. The maximum transfer frequency of the *gal* gene is approximately 25%.
   c. The *azi, ton,* and *lac* genes have all entered and can become part of a stable transconjugant.

7. In Figure 5-14, which is the last gene to be transferred into the F⁻ from each of the five Hfr strains?

Answer: For the H strain, the last gene transferred is *thi* 2 *lac* 1 *pro* 3 *gal* 312 *gly*.

8. In Figure 5-15, how are each of the following genotypes produced?
   a. $F^+ a^-$
   b. $F^- a^-$
   c. $F^- a^+$
   d. $F^+ a^+$

Answer:
   a. $F^+ a^-$ cells result from transfer of the F plasmid into the $F^- a^-$ strain.
   b. $F^- a^-$ strains result when the recipient in an Hfr cross does not acquire the donor a^+ allele by recombination.
   c. $F^- a^+$ strains result when the recipient in an Hfr cross acquires the donor a^+ allele by recombination.
   d. $F^+ a^+$ strains result from resynthesis of the donated F plasmid in the donor strain.

9. In Figure 5-17, how many crossovers are required to produce a completely prototrophic exconjugant?

Answer: Two crossovers are required to produce a prototroph.

10. In Figure 5-18c, why is the crossover shown occurring in the orange segments of DNA?

Answer: The orange segments represent IS elements, which provide homology for crossing over.

11. In Figure 5-19, how many different bacterial species are shown as having contributed DNA to the plasmid pk214?

Answer: Ten different bacterial species have donated sequences to the R plasmid.

12. In Figure 5-25, can you point to any phage progeny that could transduce?

Answer: No, in this figure all of the phage particles contain phage DNA.

13. In Figure 5-28, what are the physical features of the plaques of recombinant phages?

Answer: The recombinant plaque types are small, clear plaques and large, cloudy plaques.

14. In Figure 5-29, do you think that b^+ could be transduced instead of a^+? As well as a^+?

Answer: The b^+ allele could be transduced instead of a^+, but they will not be co-transduced.

15. In Figure 5-30, which genes show the highest frequencies of cotransduction?

Answer: The *narC, supF, C* and *galU* genes show the highest frequencies of co-transduction.

16. In Figure 5-32, what do the half-red, half-blue segments represent?

Answer: They represent hybrid attachment sites comprised of sequences from both the phage and bacterial genomes.

17. In Figure 5-33, which is the rarest genotype produced in the initial lysate?

Answer: The rarest phage genotype produced in this lysate will be λd*gal*.

18. In Figure 5-34, precisely which gene is eventually identified from the genome sequence?

Answer: The gene is the one colored in orange.

BASIC PROBLEMS

19. Describe the state of the F factor in an Hfr, F$^+$, and F$^-$ strain.

Answer: An Hfr strain has the fertility factor F integrated into the chromosome. An F$^+$ strain has the fertility factor free in the cytoplasm. An F$^-$ strain lacks the fertility factor.

20. How does a culture of F$^+$ cells transfer markers from the host chromosome to a recipient?

Answer: All cultures of F$^+$ strains have a small proportion of cells in which the F factor is integrated into the bacterial chromosome and are, by definition, Hfr cells. These Hfr cells transfer markers from the host chromosome to a recipient during conjugation.

21. With respect to gene transfer and the integration of the transferred gene into the recipient genome, compare
  a. Hfr crosses by conjugation and generalized transduction.
  b. F′ derivatives such as F′ *lac* and specialized transduction.

Answer:
  a. Hfr cells involved in conjugation transfer host genes in a linear fashion. The genes transferred depend on both the Hfr strain and the length of time during which the transfer occurred. Therefore, a population containing several different Hfr strains will appear to have an almost random transfer of host genes. This is similar to generalized transduction, in which the viral protein coat forms around a specific amount of DNA rather than specific genes. In generalized transduction, any gene can be transferred.
  b. F′ factors arise from improper excision of an Hfr from the bacterial chromosome. They can have only specific bacterial genes on them because the integration site is fixed for each strain. Specialized transduction resembles this in that the viral particle integrates into a specific region of the bacterial chromosome and then, upon improper excision, can take with it

only specific bacterial genes. In both cases, the transferred gene exists as a second copy.

22. Why is generalized transduction able to transfer any gene, but specialized transduction is restricted to only a small set?

Answer: Generalized transduction occurs with lytic phages that enter a bacterial cell, fragment the bacterial chromosome, and then, while new viral particles are being assembled, improperly incorporate some bacterial DNA within the viral protein coat. Because the amount of DNA, not the information content of the DNA, is what governs viral particle formation, any bacterial gene can be included within the newly formed virus. In contrast, specialized transduction occurs with improper excision of viral DNA from the host chromosome in lysogenic phages. Because the integration site is fixed, only those bacterial genes very close to the integration site will be included in a newly formed virus.

23. A microbial geneticist isolates a new mutation in *E. coli* and wishes to map its chromosomal location. She uses interrupted-mating experiments with Hfr strains and generalized-transduction experiments with phage P1. Explain why each technique, by itself, is insufficient for accurate mapping.

Answer: While the interrupted-mating experiments will yield the gene order, it will be relative only to fairly distant markers. Thus, the precise location cannot be pinpointed with this technique. Generalized transduction will yield information with regard to very close markers, which makes it a poor choice for the initial experiments because of the massive amount of screening that would have to be done. Together, the two techniques allow, first, for a localization of the mutant (interrupted-mating) and, second, for precise determination of the location of the mutant (generalized transduction) within the general region.

24. In *E. coli*, four Hfr strains donate the following markers, shown in the order donated:

Strain 1:	M	Z	X	W	C
Strain 2:	L	A	N	C	W
Strain 3:	A	L	B	R	U
Strain 4:	Z	M	U	R	B

All these Hfr strains are derived from the same F⁻ strain. What is the order of these markers on the circular chromosome of the original F⁺?

Answer: This problem is analogous to forming long gene maps with a series of three-point testcrosses. Arrange the four sequences so that their regions of overlap are aligned:

$$\overline{M-Z}-X-W-C$$
$$W-C-N-A-L$$
$$A-L-B-R-U$$
$$B-R-U-\underline{M-Z}$$

The regions with the bars above or below are identical in sequence (and "close" the circular chromosome). The correct order of markers on this circular map is

—M—Z—X—W—C—N—A—L—B—R—U—

25. You are given two strains of *E. coli*. The Hfr strain is arg^+ ala^+ glu^+ pro^+ leu^+ T^s; the F^- strain is arg^- ala^- glu^- pro^- leu^- T^r. All the markers are nutritional except *T*, which determines sensitivity or resistance to phage T1. The order of entry is as given, with arg^+ entering the recipient first and T^s last. You find that the F^- strain dies when exposed to penicillin (pen^s), but the Hfr strain does not (pen^r). How would you locate the locus for *pen* on the bacterial chromosome with respect to *arg*, *ala*, *glu*, *pro*, and *leu*? Formulate your answer in logical, well-explained steps and draw explicit diagrams where possible.

Answer: First, carry out a series of crosses in which you select in a long mating each of the auxotrophic markers. Thus, select for arg^+ T^r. In each case score for penicillin resistance. Although not too informative, these crosses will give the marker that is closest to pen^r by showing which marker has the highest linkage. Then, do a second cross concentrating on the two markers on either side of the pen^r locus. Suppose that the markers are *ala* and *glu*. You can first verify the order by taking the cross in which you selected for ala^+, the first entering marker, and scoring the percentage of both pen^r and glu^+. Because of the gradient of transfer, the percentage of pen^r should be higher than the percentage of glu^+ among the selected ala^+ recombinants.

Then, take the mating in which glu^+ was the selected marker. Because this marker enters last, one can use the cross data to determine the map units by determining the percentage of colonies that are ala^+ pen^r, and by the number of ala^- pen^r colonies.

Unpacking the Problem

26. A cross is made between two *E. coli* strains: Hfr arg^+ bio^+ leu^+ × F^- arg^- bio^- leu^-. Interrupted mating studies show that arg^+ enters the recipient last, and so arg^+ recombinants are selected on a medium containing *bio* and *leu* only. These recombinants are tested for the presence of bio^+ and leu^+. The following numbers of individuals are found for each genotype:

arg_+ bio^+ leu^+	320	arg^+ bio^- leu^+	0
arg^+ bio^+ leu^-	8	arg^+ bio^- leu^-	48

a. What is the gene order?
b. What are the map distances in recombination percentages?

Answer:
a. Determine the gene order by comparing arg^+ bio^+ leu^- with arg^+ bio^- leu^+. If the order were arg leu bio, four crossovers would be required to get arg^+ leu^- bio, while only two would be required to get arg leu^+ bio^-. If the order is arg bio leu, four crossovers would be required to get arg^+ bio^- leu^+, and only two would be required to get arg^+ bio^+ leu^-. There are eight recombinants that are arg^+ bio^+ leu^- and none that are arg^+ bio^- leu^+. On the basis of the frequencies of these two classes, the gene order is arg bio leu.

b. The arg-bio distance is determined by calculating the percentage of the exconjugants that are arg^+ bio^- leu^-. These cells would have had a crossing-over event between the arg and bio genes.

$$RF = 100\%(48)/376 = 12.77\% \text{ (m.u.)}$$

Similarly, the bio-leu distance is estimated by the arg^+ $bio+$ leu^- colony type.

$$RF = 100\%(8)/376 = 2.13\% \text{ (m.u.)}$$

27. Linkage maps in an Hfr bacterial strain are calculated in units of minutes (the number of minutes between genes indicates the length of time that it takes for the second gene to follow the first in conjugation). In making such maps, microbial geneticists assume that the bacterial chromosome is transferred from Hfr to F$^-$ at a constant rate. Thus, two genes separated by 10 minutes near the origin end are assumed to be the same physical distance apart as two genes separated by 10 minutes near the F$^-$ attachment end. Suggest a critical experiment to test the validity of this assumption.

Answer: The most straightforward way would be to pick two Hfr strains that are near the genes in question but are oriented in opposite directions. Then, measure the time of transfer between two specific genes, in one case when they are transferred early and in the other when they are transferred late. For example,

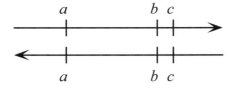

28. A particular Hfr strain normally transmits the pro^+ marker as the last one in conjugation. In a cross of this strain with an F$^-$ strain, some pro^+ recombinants are recovered early in the mating process. When these pro^+ cells are mixed with

F⁻ cells, the majority of the F⁻ cells are converted into *pro*⁺ cells that also carry the F factor. Explain these results.

Answer: The best explanation is that the integrated F factor of the Hfr looped out of the bacterial chromosome abnormally and is now an F′ that contains the *pro*⁺ gene. This F′ is rapidly transferred to F⁻ cells, converting them to *pro*⁺ (and F⁺).

29. F′ strains in *E. coli* are derived from Hfr strains. In some cases, these F′ strains show a high rate of integration back into the bacterial chromosome of a second strain. Furthermore, the site of integration is often the site occupied by the sex factor in the original Hfr strain (before production of the F′ strains). Explain these results.

Answer: The high rate of integration and the preference for the same site originally occupied by the F factor suggest that the F′ contains some homology with the original site. The source of homology could be a fragment of the F factor, or more likely, it is homology with the chromosomal copy of the bacterial gene that is also present on the F′.

30. You have two *E. coli* strains, F⁻ *str*ʳ *ala*⁻ and Hfr *str*ˢ *ala*⁺, in which the F factor is inserted close to *ala*⁺. Devise a screening test to detect strains carrying F′ *ala*⁺.

Answer: First, carry out a cross between the Hfr and F⁻, and then select for colonies that are *ala*⁺ *str*ʳ. If the Hfr donates the *ala* region late, then redo the cross but now interrupt the mating early and select for *ala*⁺. This selects for an F′, because this Hfr would not have transferred the *ala* gene early.

If the Hfr instead donates this region early, then use a Rec⁻ recipient strain that cannot incorporate a fragment of the donor chromosome by recombination. Any *ala*+ colonies from the cross should then be used in a second mating to another *ala*– strain to see whether they can donate the *ala* gene easily, which would indicate that there is F′ *ala*. (This would also require another marker to differentiate the donor and recipient strains. For example, the *ala*⁻ strain could be tetracycliner and selection would be for *ala*+ *tet*ʳ.)

31. Five Hfr strains A through E are derived from a single F⁺ strain of *E. coli*. The following chart shows the entry times of the first five markers into an F⁻ strain when each is used in an interrupted-conjugation experiment:

A		B		C		D		E	
mal⁺	(1)	*ade*⁺	(13)	*pro*⁺	(3)	*pro*⁺	(10)	*his*⁺	(7)
*str*ˢ	(11)	*his*⁺	(28)	*met*⁺	(29)	*gal*⁺	(16)	*gal*⁺	(17)

ser^+	(16)	gal^+	(38)	xyl^+	(32)	his^+	(26)	pro^+	(23)
ade^+	(36)	pro^+	(44)	mal^+	(37)	ade^+	(41)	met^+	(49)
his^+	(51)	met^+	(70)	str^s	(47)	ser^+	(61)	xyl^+	(52)

a. Draw a map of the F$^+$ strain, indicating the positions of all genes and their distances apart in minutes.

b. Show the insertion point and orientation of the F plasmid in each Hfr strain.

c. In the use of each of these Hfr strains, state which allele you would select to obtain the highest proportion of Hfr exconjugants.

Answer:

a. and b.

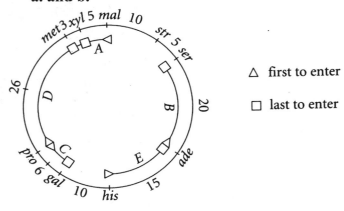

△ first to enter

☐ last to enter

c. A: Select for mal^+
B: Select for ade^+
C: Select for pro^+
D: Select for pro^+
E: Select for his^+

32. Streptococcus pneumoniae cells of genotype str^s mtl^- are transformed by donor DNA of genotype str^r mtl^+ and (in a separate experiment) by a mixture of two DNAs with genotypes str^r mtl^- and str^s mtl^+. The accompanying table shows the results:

	Percentage of cells transformed into		
Transforming DNA	str^r mtl^-	str^s mtl^+	str^r mtl^+
str^r mtl^+	4.3	0.40	0.17
str^r mtl^- + str^s mtl^+	2.8	0.85	0.0066

a. What does the first row of the table tell you? Why?

b. What does the second row of the table tell you? Why?

Answer:

a. If the two genes are far enough apart to be located on separate DNA fragments, then the frequency of double transformants should be the product of the frequency of the two single transformants, or (4.3%) × (0.40%) = 0.017%. The observed double transformant frequency is 0.17 percent, a factor of 10 greater than expected. Therefore, the two genes are located close enough together to be cotransformed at a rate of 0.17 percent.

b. Here, when the two genes must be contained on separate pieces of DNA, the rate of cotransformation is much lower, confirming the conclusion in part (a).

33. Recall that, in Chapter 4, we considered the possibility that a crossover event may affect the likelihood of another crossover. In the bacteriophage T4, gene *a* is 1.0 m.u. from gene *b*, which is 0.2 m.u. from gene *c*. The gene order is *a, b, c*. In a recombination experiment, you recover five double crossovers between *a* and *c* from 100,000 progeny viruses. Is it correct to conclude that interference is negative? Explain your answer.

Answer: The   expected   number   of   double   recombinants   is (0.01)(0.002)(100,000) = 2. Interference = 1 − (observed DCO/expected DCO) = 1 − 5/2 = −1.5. By definition, the interference is negative.

34. You have infected *E. coli* cells with two strains of T4 virus. One strain is minute (*m*), rapid lysis (*r*), and turbid (*tu*); the other is wild type for all three markers. The lytic products of this infection are plated and classified. The resulting 10,342 plaques were distributed among eight genotypes as follows:

m r tu	3467	*m + +*	520
+ + +	3729	*+ r tu*	474
m r +	853	*+ r +*	172
m + tu	162	*+ + tu*	965

a. Determine the linkage distances between *m* and *r*, between *r* and *tu*, and between *m* and *tu*.

b. What linkage order would you suggest for the three genes?

c. What is the coefficient of coincidence (see Chapter 4) in this cross? What does it signify?

(Problem 34 is reprinted with the permission of Macmillan Publishing Co., Inc., from Monroe W. Strickberger, *Genetics.* Copyright 1968 by Monroe W. Strickberger.)

Answer:

a. The parental genotypes are + + + and *m r tu*. For determining the *m–r* distance, the recombinant progeny are

$$
\begin{array}{ll}
m + tu & 162 \\
m + + & 520 \\
+ r\ tu & 474 \\
+ r + & \underline{172} \\
& 1328
\end{array}
$$

Therefore, the map distance is 100%(1328)/10,342 = 12.8 m.u.

Using the same approach, the r–tu distance is 100%(2152)/10,342 = 20.8 m.u., and the m–tu distance is 100%(2812)/10,342 = 27.2 m.u.

b. Because genes m and tu are the farthest apart, the gene order must be $m\ r\ tu$.

c. The coefficient of coincidence (c.o.c.) compares the actual number of double crossovers to the expected number, (where c.o.c. = observed double crossovers/expected double crossovers). For these data, the expected number of double recombinants is (0.128)(0.208)(10,342) = 275. Thus, c.o.c. = (162 + 172)/275 = 1.2. This indicates that there are more double crossover events than predicted and suggests that the occurrence of one crossover makes a second crossover between the same DNA molecules more likely to occur.

35. With the use of P22 as a generalized transducing phage grown on a $pur^+\ pro^+\ his^+$ bacterial donor, a recipient strain of genotype $pur^-\ pro^-\ his^-$ is infected and incubated. Afterward, transductants for pur^+, pro^+, and his^+ are selected individually in experiments I, II, and III, respectively.

a. What media are used for these selection experiments?
b. The transductants are examined for the presence of unselected donor markers, with the following results:

I			II			III		
pro^-	his^-	87%	pur^-	his^-	43%	pur^-	pro^-	21%
pro^+	his^-	0%	pur^+	his^-	0%	pur^+	pro^-	15%
pro^-	his^+	10%	pur^-	his^+	55%	pur^-	pro^+	60%
pro^+	his^+	3%	pur^+	his^+	2%	pur^+	pro^+	4%

What is the order of the bacterial genes?
c. Which two genes are closest together?
d. On the basis of the order that you proposed in part c, explain the relative proportions of genotypes observed in experiment II.

(Problem 35 is from D. Freifelder, *Molecular Biology and Biochemistry*. Copyright 1978 by W. H. Freeman and Company, New York.)

Answer:

a. I: minimal plus proline and histidine
II: minimal plus purines and histidine
III: minimal plus purines and proline

b. The order can be deduced from cotransfer rates. It is *pur–his–pro*.

c. The closer the two genes, the higher the rate of cotransfer. *his* and *pro* are closest.

d. *pro*$^+$ transduction requires a crossover on both sides of the *pro* gene. Because *his* is closer to *pro* than *pur*, you get the following:

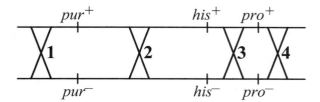

Genotypes	Frequency	Crossovers
pur$^-$ *his*$^-$	43%	4 and 3
pur$^+$ *his*$^-$	0%	4 and 3, and 2 and 1
pur$^-$ *his*$^+$	55%	4 and 2
pur$^+$ *his*$^+$	2%	4 and 1

As can be seen, a *pur*$^+$ *his*$^-$ *pro*$^+$ genotype requires four crossovers and as expected, would occur less frequently (in this example, 0%).

36. Although most λ-mediated *gal*$^+$ transductants are inducible lysogens, a small percentage of these transductants in fact are not lysogens (that is, they contain no integrated λ). Control experiments show that these transductants are not produced by mutation. What is the likely origin of these types?

Answer: In a small percentage of the cases, *gal*$^+$ transductants can arise by recombination between the *gal*$^+$ DNA of the λdgal transducing phage and the *gal*$^-$ gene on the chromosome. This will generate *gal*$^+$ transductants without phage integration.

37. An *ade*$^+$ *arg*$^+$ *cys*$^+$ *his*$^+$ *leu*$^+$ *pro*$^+$ bacterial strain is known to be lysogenic for a newly discovered phage, but the site of the prophage is not known. The bacterial map is

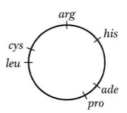

The lysogenic strain is used as a source of the phage, and the phages are added to a bacterial strain of genotype *ade⁻ arg⁻ cys⁻ his⁻ leu⁻ pro⁻*. After a short incubation, samples of these bacteria are plated on six different media, with the supplementations indicated in the following table. The table also shows whether colonies were observed on the various media.

| Medium | Nutrient supplementation in medium | | | | | | Presence of colonies |
	Ade	Arg	Cys	His	Leu	Pro	
1	−	+	+	+	+	+	N
2	+	−	+	+	+	+	N
3	+	+	−	+	+	+	C
4	+	+	+	−	+	+	N
5	+	+	+	+	−	+	C
6	+	+	+	+	+	−	N

(In this table, a plus sign indicates the presence of a nutrient supplement, a minus sign indicates that a supplement is not present, N indicates no colonies, and C indicates colonies present.)

a. What genetic process is at work here?

b. What is the approximate locus of the prophage?

Answer:

a. This appears to be specialized transduction. It is characterized by the transduction of specific markers based on the position of the integration of the prophage. Only those genes near the integration site are possible candidates for misincorporation into phage particles that then deliver this DNA to recipient bacteria.

b. The only media that supported colony growth were those lacking either cysteine or leucine. These selected for *cys⁺* or *leu⁺* transductants and indicate that the prophage is located in the *cys-leu* region.

38. In a generalized-transduction system using P1 phage, the donor is *pur⁺ nad⁺ pdx⁻* and the recipient is *pur⁻ nad⁻ pdx⁺*. The donor allele *pur⁺* is initially selected after transduction, and 50 *pur⁺* transductants are then scored for the other alleles present. Here are the results:

Genotype	Number of colonies
nad⁺ pdx⁺	3
nad⁺ pdx⁻	10
nad⁻ pdx⁺	24
nad⁻ pdx⁻	13
	50

a. What is the cotransduction frequency for *pur* and *nad*?
b. What is the cotransduction frequency for *pur* and *pdx*?
c. Which of the unselected loci is closest to *pur*?
d. Are *nad* and *pdx* on the same side or on opposite sides of *pur*? Explain.

(Draw the exchanges needed to produce the various transformant classes under either order to see which requires the minimum number to produce the results obtained.)

Answer:

a. This is simply calculated as the percentage of *pur*$^+$ colonies that are also *nad*^{d}+ = 100%(3 + 10)/50 = 26%

b. This is calculated as the percentage of *pur*$^+$ colonies that are also *pdx*$^-$ = 100%(10 + 13)/50 = 46%

c. *pdx* is closer, as determined by cotransduction rates.

d. From the cotransduction frequencies, you know that *pdx* is closer to *pur* than *nad* is, so there are two gene orders possible: *pur pdx nad* or *pdx pur nad*. Now, consider how a bacterial chromosome that is *pur*$^+$ *pdx*$^+$ *nad*$^+$ might be generated, given the two gene orders: if *pdx* is in the middle, 4 crossovers are required to get *pur*$^+$ *pdx*$^+$ *nad*$^+$; if pur is in the middle, only 2 crossovers are required (see next page). The results indicate that there are fewer *pur*$^+$ *pdx*$^+$ *nad*$^+$ transductants than any other class, suggesting that this class is "harder" to generate than the others. This implies that *pdx* is in the middle, and the gene order is *pur pdx nad*.

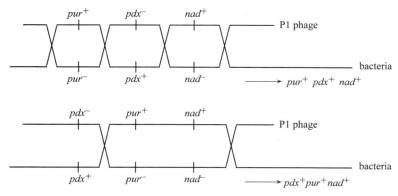

39. In a generalized-transduction experiment, phages are collected from an *E. coli* donor strain of genotype *cys*$^+$ *leu*$^+$ *thr*$^+$ and used to transduce a recipient of genotype *cys*$^-$ *leu*$^-$ *thr*$^-$. Initially, the treated recipient population is plated on a minimal medium supplemented with leucine and threonine. Many colonies are obtained.

a. What are the possible genotypes of these colonies?
b. These colonies are then replica plated onto three different media: (1) minimal plus threonine only, (2) minimal plus leucine only, and (3)

minimal. What geno types could, in theory, grow on these three media?

c. Of the original colonies, 56 percent are observed to grow on medium 1, 5 percent on medium 2, and no colonies on medium 3. What are the actual genotypes of the colonies on media 1, 2, and 3?

d. Draw a map showing the order of the three genes and which of the two outer genes is closer to the middle gene.

Answer:

a. Owing to the medium used, all colonies are cys^+ but either + or − for the other two genes.

cys^+ leu^+ thr^+
cys^+ leu^+ thr^-
cys^+ leu^- thr^+
cys^+ leu^- thr^-

b. (1) cys^+ leu^+ thr^+ and cys^+ leu^+ thr^- (supplemented with threonine)
(2) cys^+ leu^+ thr^+ and cys^+ leu^- thr^+ (supplemented with leucine)
(3) cys^+ leu^+ thr^+ (no supplements)

c. Because none grew on minimal medium, no colony was leu^+ thr^+. Therefore, medium (1) had cys^+ leu^+ thr^-, and medium (2) had cys^+ leu^- thr^+. The remaining cultures were cys^+ leu^- thr^-, and this genotype occurred in 100% − 56% − 5% = 39% of the colonies.

d. *cys* and *leu* are cotransduced 56 percent of the time, while *cys* and *thr* are cotransduced only 5 percent of the time. This indicates that *cys* is closer to *leu* than it is to *thr*. Because no leu^+ cys^+ thr^+ cotransductants are found, it indicates that *cys* is in the middle.

leu   *cys*   *thr*

40. Deduce the genotypes of the following *E. coli* strains 1 through 4:

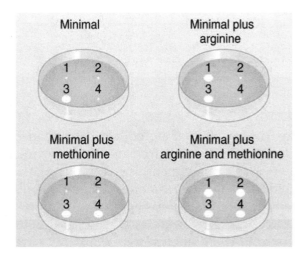

Answer: Prototrophic strains of *E. coli* will grow on minimal media, while auxotrophic strains will only grow on media supplemented with the required molecule(s). Thus, strain 3 is prototrophic (wild-type), strain 4 is *met⁻*, strain 1 is *arg⁻*, and strain 2 is *arg⁻ met⁻*.

41. In an interrupted-conjugation experiment in *E. coli*, the *pro* gene enters after the *thi* gene. A *pro⁺ thi⁺* Hfr is crossed with a *pro⁻ thi⁻* F⁻ strain, and exconjugants are plated on medium containing thiamine but no proline. A total of 360 colonies are observed, and they are isolated and cultured on fully supplemented medium. These cultures are then tested for their ability to grow on medium containing no proline or thiamine (minimal medium), and 320 of the cultures are found to be able to grow but the remainder cannot.

  a. Deduce the genotypes of the two types of cultures.
  b. Draw the crossover events required to produce these genotypes.
  c. Calculate the distance between the *pro* and *thi* genes in recombination units.

Unpacking the Problem

  1. What type of organism is *E. coli*?

     Answer: *E. coli* is a bacterium and a prokaryote.

  2. What does a culture of *E. coli* look like?

     Answer: *E. coli* can be grown in suspension or on an agar medium. The latter method allows for the identification of individual colonies, each a clone of descendants from a single cell (and visible to the naked eye when it reaches more than 107 cells).

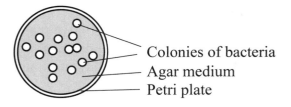

  Colonies of bacteria
  Agar medium
  Petri plate

  3. On what sort of substrates does *E. coli* generally grow in its natural habitat?

     Answer: Naturally, *E. coli* is an enteric bacterium living symbiotically within the gut of host organisms (like us).

  4. What are the minimal requirements for *E. coli* cells to divide?

Answer: Minimal medium consists of inorganic salts, a carbon source for energy, and water.

5. Define the terms *prototroph* and *auxotroph*.

Answer: Prototroph refers to the wild-type phenotype, or in other words, an organism that can grow on minimal media. Auxotroph refers to a mutant that can grow only on a medium supplemented with one or more specific nutrients not required by the wild-type strain.

6. Which cultures in this experiment are prototrophic and which are auxotrophic?

Answer: In this experiment, the Hfr and the exconjugants that can grow on minimal medium are prototrophs, whereas the recipient F^- and the exconjugants that do not grow on minimal medium are auxotrophs.

7. Given some strains of unknown genotype regarding thiamine and proline, how would you test their genotypes? Give precise experimental details, including equipment.

Answer: Unknown strains would be grown as individual colonies on medium enriched with proline and thiamine, and then cells from each colony could be picked (by a sterile toothpick, for example) and placed individually onto medium supplemented with either thiamine or proline or onto minimal medium. Proline and thiamine auxotrophs would be identified on the basis of growth patterns. For example, a *pro*⁻ strain will grow only on medium supplemented with proline.

Instead of the labor-intensive method of individually picking cells, replica plating can be used to transfer some cells of each colony from a master plate (supplemented with proline and thiamine) to plates that contain the various media described above. The physical arrangement (and positional patterns) of colonies is used to identify the various colonies as they are transferred from plate to plate.

8. What kinds of chemicals are proline and thiamine? Does it matter in this experiment?

Answer: Proline is an amino acid, and thiamine is a B1 vitamin. Their chemical nature does not matter to the experiment other than that they are necessary chemicals for cell growth that prototrophs can synthesize from ingredients in minimal medium and specific auxotrophic mutants cannot.

9.   Draw a diagram showing the full set of manipulations performed in the experiment.

Answer:

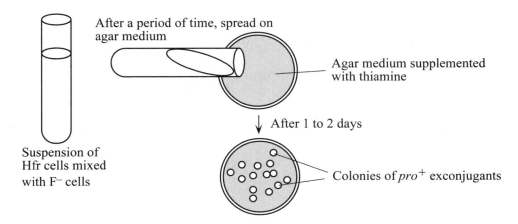

After a period of time, spread on agar medium

Agar medium supplemented with thiamine

↓ After 1 to 2 days

Suspension of Hfr cells mixed with F⁻ cells

Colonies of *pro*⁺ exconjugants

10. Why do you think the experiment was done?

Answer: Interrupted-mating experiments are used to roughly map genes onto the circular bacterial chromosome.

11.   How was it established that *pro* enters after *thi*? Give precise experimental steps.

Answer: The Hfr and F⁻ strains are mixed together in solution, and then at various times, samples are removed and put into a kitchen blender, vortexed (blender is turned on) for a few seconds to disrupt conjugation, and then plated onto a medium containing the appropriate supplements. The amount of time that has passed from the mixing of the strains to mating disruption is used as a measurement for mapping. The time of first appearance of a specific gene from the Hfr in the F⁻ cell gives the gene's relative position in minutes. Typically, the F⁻ cells are streptomycin-resistant and the Hfr cells are streptomycin-sensitive. The antibiotic is used in the various media to kill the Hfr cells (which are otherwise prototrophic) and allow only those F⁻ cells that have received the appropriate gene or genes from the Hfr to grow. In this case, it would be discovered that some of the F⁻ cells would become *thi*⁺ in samples taken earlier in the experiment than samples taken when they first become *pro*⁺.

12. In what way does an interrupted-mating experiment differ from the experiment described in this problem?

Answer: In this experiment, there is no attempt to disrupt conjugation. The two strains are mixed and at some later (unspecified) time, plated onto medium containing thiamine. This selects for strains that are pro^+, because proline is not present in this medium.

13. What is an exconjugant? How do you think that exconjugants were obtained? (It might include genes not described in this problem.)

Answer: Exconjugants are recipient cells (F^-) that now contain alleles from the donor (Hfr). Typically, the F^- cells are streptomycin-resistant and the Hfr cells are streptomycin-sensitive. The antibiotic is used in the various media to kill the Hfr cells and allow only the appropriate F^- exconjugants to grow.

14. When the *pro* gene is said to enter after *thi*, does it mean the *pro* allele, the pro^+ allele, either, or both?

Answer: The statement "*pro* enters after *thi*" is one of gene position and order relative to the transfer of the bacterial chromosome by a particular Hfr. For the Hfr in this experiment, transfer occurs such that the *pro* gene is transferred after the *thi* gene. Because this Hfr is also pro^+, it is this specific allele that is entering.

15. What is "fully supplemented medium" in the context of this question?

Answer: In this experiment, "fully supplemented" medium contains proline and thiamine.

16. Some exconjugants did not grow on minimal medium. On what medium would they grow?

Answer: All exconjugants are pro^+, because that is the way they were selected. Thus, those that do not grow on minimal medium must require thiamine.

17. State the types of crossovers that take part in Hfr × F^- recombination. How do these crossovers differ from crossovers in eukaryotes?

Answer: Genetic exchange in prokaryotes does not take place between two whole genomes as it does in eukaryotes. It takes place between one complete circular genome, the F^-, and an incomplete linear genomic fragment donated by the Hfr. In this way, exchange of genetic information is nonreciprocal (from Hfr to F^-). Only even numbers of crossovers are allowed between the two DNAs, because the circular chromosome would become linear otherwise. This results in unidirectional exchange, because part of the DNA of the recipient chromosome is replaced by the DNA of the donor, while the other product (the rest of the donor DNA now with some recombined recipient DNA) is nonviable and lost.

18. What is a recombination unit in the context of the present analysis? How does it differ from the map units used in eukaryote genetics?

Answer: In this experiment, the map distance is calculated by selecting for the last marker to enter (in this case pro^+) and then determining how often the earlier unselected marker (in this case thi^+) is also present. Look at the following diagram:

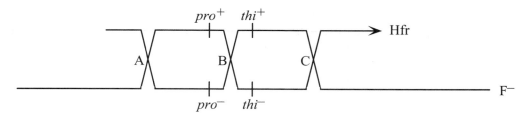

For the F$^-$ cell to become pro^+, two recombination events have to occur—one in the region to the left (marked A) and one in either region to the right (marked B or C.) Thus, the percentage of pro^+ (second recombination within either B or C) that are thi^- (second event only within region B) can be used to determine map distance where 1% = 1 map unit. (The map units calculated this way cannot be combined with other map unit calculations [from other experiments] to build a larger genomic map. The map units obtained are just giving an estimate of the relative sizes of intervals B and C as targets for crossover.)

Solution to the Problem

a. The two genotypes being cultured are pro^+ thi^- (grows only on media supplemented with thiamine) and pro^+ thi^+ (grows on minimal media).

b. Two recombination events must occur, one on either side of pro (because exconjugants were plated on medium supplemented with thiamine, only pro^+ cells would have grown). The pro^+ thi^- strains would have had recombination in regions labeled A and B, and the pro^+ thi^+ strains would have had recombination in regions labeled A and C.

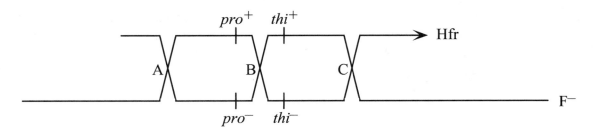

c.   The distance between *pro* and *thi* is:

$$= \frac{100\%(\text{the number of colonies that are } pro{+}\ thi^{-})}{\text{total number of } pro^{+} \text{ colonies}}$$

$$= 100\%(40)/360 = 11.1\%$$

42.   A generalized transduction experiment uses a *metE⁺ pyrD⁺* strain as donor and *metE⁻ pyrD⁻* as recipient. *metE⁺* transductants are selected and then tested for the *pyrD⁺* allele. The following numbers were obtained:

metE⁺ pyrD⁻   857
metE⁺ pyrD⁺   1

Do these results suggest that these loci are closely linked? What other explanations are there for the lone "double"?

Answer: No, closely linked loci would be expected to be cotransduced; the greater the cotransduction frequency, the closer the loci are. Because only 1 of 858 *metE⁺* was also *pyrD⁺*, the genes are not closely linked. The lone *metE⁺ pyrD⁺* could be the result of cotransduction, or it could be a spontaneous mutation of *pyrD* to *pyrD⁺*, or the result of coinfection by two separate transducing phages.

43.   An *argC⁻* strain was infected with transducing phage, and the lysate was used to transduce *metF⁻* recipients on medium containing arginine but no methionine. The *metF⁺* transductants were then tested for arginine requirement: most were *argC⁺* but a small percentage were found to be *argC⁻*. Draw diagrams to show the likely origin of the *argC⁺* and *argC⁻* strains.

Answer: The *metF⁺* colonies that are now also *argC⁻* are the result of cotransduction of the two markers from the donor strain. This will be less likely than the transduction of just the *metF⁺* allele and, in these cases, the recipient remains *argC⁺*.

The following diagram illustrates the possible recombination events that will result in *metF⁺* transductants that remain *argC⁺*:

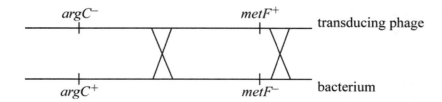

The following diagram illustrates the possible recombination events that will result in *metF⁺* transductants that are also now *argC⁻*:

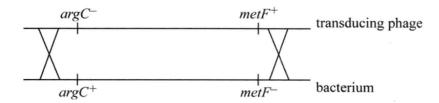

CHALLENGING PROBLEMS

44. Four *E. coli* strains of genotype $a^+ b^-$ are labeled 1, 2, 3, and 4. Four strains of genotype $a^- b^+$ are labeled 5, 6, 7, and 8. The two genotypes are mixed in all possible combinations and (after incubation) are plated to determine the frequency of $a^+ b^+$ recombinants. The following results are obtained, where M = many recombinants, L = low numbers of recombinants, and 0 = no recombinants:

	1	2	3	4
5	0	M	M	0
6	0	M	M	0
7	L	0	0	M
8	0	L	L	0

On the basis of these results, assign a sex type (either Hfr, F⁺, or F⁻) to each strain.

Answer: To interpret the data, the following results are expected:

Cross	Result
F⁺ × F⁻	(L) low number of recombinants
Hfr × F⁻	(M) many recombinants
Hfr × Hfr	(0) no recombinants
Hfr × F⁺	(0) no recombinants
F⁺ × F⁺	(0) no recombinants
F⁻ × F⁻	(0) no recombinants

The only strains that show both the (L) and the (M) result when crossed are 2, 3, and 7. These must be F⁻ because that is the only cell type that can participate in a cross and give either recombination result. Hfr strains will result in only (M) or (0), and F$^+$ will result in only (L) or (0) when crossed. Thus, strains 1 and 8 are F$^+$, and strains 4, 5, and 6 are Hfr.

45. An Hfr strain of genotype a^+ b^+ c^+ d^- str^s is mated with a female strain of genotype a^- b^- c^- d^+ str^r. At various times, the culture is shaken vigorously to separate mating pairs. The cells are then plated on agar of the following three types, where nutrient A allows the growth of a^- cells; nutrient B, of b^- cells; nutrient C, of c^- cells; and nutrient D, of d^- cells (a plus indicates the presence of streptomycin or a nutrient, and a minus indicates its absence):

Agar type	Str	A	B	C	D
1	+	+	+	−	+
2	+	−	+	+	+
3	+	+	−	+	+

a. What donor genes are being selected on each type of agar?

b. The following table shows the number of colonies on each type of agar for samples taken at various times after the strains are mixed. Use this information to determine the order of genes a, b, and c.

Time of sampling (minutes)	Number of colonies on agar of type		
	1	2	3
0	0	0	0
5	0	0	0
7.5	100	0	0
10	200	0	0
12.5	300	0	75
15	400	0	150
17.5	400	50	225
20	400	100	250
25	400	100	250

c. From each of the 25-minute plates, 100 colonies are picked and transferred to a petri dish containing agar with all the nutrients except D. The numbers of colonies that grow on this medium are 89 for the sample from agar type 1, 51 for the sample from agar type 2, and 8 for the sample from agar type 3. Using these data, fit gene d into the sequence of a, b, and c.

d. At what sampling time would you expect colonies to first appear on agar containing C and streptomycin but no A or B?

(Problem 45 is from D. Freifelder, *Molecular Biology and Biochemistry*. Copyright 1978 by W. H. Freeman and Company.)

Answer:

a.

Agar type	Selected genes
1	$c+$
2	$a+$
3	$b+$

b. The order of genes is revealed in the sequence of colony appearance. Because colonies first appear on agar type 1, which selects for c^+, c must be first. Colonies next appear on agar type 3, which selects for b^+, indicating that b follows c. Allele a^+ appears last. The gene order is $c\ b\ a$. The three genes are roughly equally spaced.

c. In this problem you are looking at the cotransfer of the selected gene with the d^- allele (both from the Hfr). Cells that are d^- do not grow because the medium is lacking D and selecting for those cells that are d^+. Therefore, the farther a gene is from gene d, the less likely cotransfer of the selected gene will occur with d^- and the more likely that colonies will grow (remain d^+). From the data, d is closest to b (only 8/100 did not cotransfer d^- with b^+.) It is also closer to a than it is to c. Thus, the gene order is $c\ b\ d\ a$ (or $a\ d\ b\ c$).

d. With no A or B in the agar, the medium selects for $a^+\ b^+$, and the first colonies should appear at about 17.5 minutes.

46. In the cross Hfr $aro^+\ arg^+\ ery^r\ str^s \times F^-\ aro^-\ arg^-\ ery^s\ str^r$, the markers are transferred in the order given (with aro^+ entering first), but the first three genes are very close together. Exconjugants are plated on a medium containing Str (streptomycin, to kill Hfr cells), Ery (erythromycin), Arg (arginine), and Aro (aromatic amino acids). The following results are obtained for 300 colonies isolated from these plates and tested for growth on various media: on Ery only, 263 strains grow; on Ery + Arg, 264 strains grow; on Ery + Aro, 290 strains grow; on Ery + Arg + Aro, 300 strains grow.

    a. Draw up a list of genotypes, and indicate the number of individuals in each genotype.
    b. Calculate the recombination frequencies.
    c. Calculate the ratio of the size of the *arg*-to-*aro* region to the size of the *ery*-to-*arg* region.

Answer:
a. To survive on the selective medium, all cultures must be *ery*r. Keep in mind that cells from all 300 colonies were each tested under four separate conditions.

263 colonies grew when only erythromycin is added, so these must be arg^+ $aro^+\ ery^r$. The remaining 37 cultures are mutant for one or both genes. One additional colony grew if arginine was also added to the medium (264 – 263

= 1). It must be *arg⁻ aro⁺ eryʳ*. A total of 290 colonies are *arg⁺* because they grew when erythromycin and aromatic amino acids were added to the medium. Of these, 27 are *aro⁻* (290 − 263 = 27). The genotypes and their frequencies are summarized below:

263	*eryʳ arg⁺ aro⁺*
27	*eryʳ arg⁺ aro⁻*
1	*eryʳ arg⁻ aro⁺*
9	*eryʳ arg⁻ aro⁻*
300	

b. Recombination in the *aro–arg* region is represented by two genotypes: *aro⁺ arg⁻* and *aro⁻ arg⁺*. The frequency of recombination is:

$$100\%(1 + 27)/300 = 9.3\% \text{ (m.u.)}$$

Recombination in the *ery–arg* region is represented by two genotypes: *aro⁺ arg⁻* and *aro⁻ arg⁻*. The frequency of recombination is:

$$100\%(1 + 9)/300 = 3.3\% \text{ (m.u.)}$$

Recombination in the *ery–aro* region is represented by three genotypes: *arg⁺ aro⁻*, *arg⁻ aro⁻*, and *arg⁻ aro⁺*. Recall that the DCO must be counted twice. The frequency of recombination is:

$$100\%(27 + 9 + 2)/300 = 12.6\% \text{ (m.u.)}$$

c. The ratio is 28:10, or 2.8:1.0

47. A transformation experiment is performed with a donor strain that is resistant to four drugs: A, B, C, and D. The recipient is sensitive to all four drugs. The treated recipient cell population is divided up and plated on media containing various combinations of the drugs. The following table shows the results.

Drugs added	Number of colonies	Drugs added	Number of colonies
None	10,000	BC	51
A	1156	BD	49
B	1148	CD	786
C	1161	BC	30
D	1139	ABD	42
AB	46	ACD	630
AC	640	BCD	36
AD	942	ABCD	30

a. One of the genes is obviously quite distant from the other three, which appear to be tightly (closely) linked. Which is the distant gene?

b. What is the probable order of the three tightly linked genes?

(Problem 47 is from Franklin Stahl, *The Mechanics of Inheritance,* 2nd ed. Copyright 1969, Prentice Hall, Englewood Cliffs, N.J. Reprinted by permission.)

Answer:

a. To determine which genes are close, compare the frequency of double transformants. Pairwise testing gives low values whenever B is involved but fairly high rates when any drug but B is involved. This suggests that the gene for B resistance is not close to the other three genes.

b. To determine the relative order of genes for resistance to A, C, and D, compare the frequency of double and triple transformants. The frequency of resistance to AC is approximately the same as resistance to ACD. This strongly suggests that D is in the middle. Also, the frequency of AD co-resistance is higher than AC (suggesting that the gene for A resistance is closer to D than to C), and the frequency of CD is higher than AC (suggesting that C is closer to D than to A).

48. You have two strains of λ that can lysogenize *E. coli*; their linkage maps are as follows:

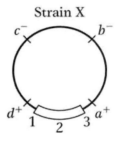
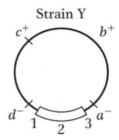

Strain X   Strain Y

The segment shown at the bottom of the chromosome, designated 1–2–3, is the region responsible for pairing and crossing over with the *E. coli* chromosome. (Keep the markers on all your drawings.)

a. Diagram the way in which λ strain X is inserted into the *E. coli* chromosome (so that the *E. coli* is lysogenized).

b. The bacteria that are lysogenic for strain X can be superinfected by using strain Y. A certain percentage of these superinfected bacteria become "doubly" lysogenic (that is, lysogenic for both strains). Diagram how it will take place. (Don't worry about how double lysogens are detected.)

c. Diagram how the two λ prophages can pair.

d. Crossover products between the two prophages can be recovered. Diagram a crossover event and the consequences.

Answer:
a. and b.

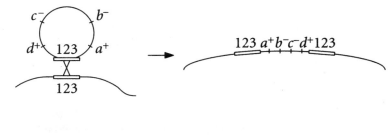

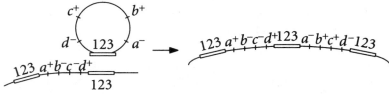

c.

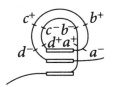

d.

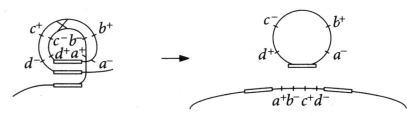

49. You have three strains of *E. coli*. Strain A is F′ *cys*⁺ *trp1*/*cys*⁺ *trp1* (that is, both the F′ and the chromosome carry *cys*⁺ and *trp1*, an allele for tryptophan requirement). Strain B is F⁻ *cys*⁻ *trp2 Z* (this strain requires cysteine for growth and carries *trp2*, another allele causing a tryptophan requirement; strain B is lysogenic for the generalized transducing phage Z). Strain C is F⁻ *cys*⁺ *trp1* (it is an F⁻ derivative of strain A that has lost the F′). How would you determine whether *trp1* and *trp2* are alleles of the same locus? (Describe the crosses and the results expected.)

Answer: If the *trp1* and *trp2* mutations map close to each other, one would cross strains A and B and select exconjugants that are *cys*⁺ and resistant to the phage Z. If *trp1* and *trp2* are alleles of the same locus, then a cross between strains A and B will not result in *trp*⁺ cells; if they are not allelic, strain B cells that have received the F′ from strain A will be *trp*⁺ (complementation). Strain C is not useful for this experiment.

If the genes are not close, then a second approach would be to try transforming B with C; select cys⁺ and see if *tryp*⁺ recombinants are ever seen even at high plating densities.

50. A generalized transducing phage is used to transduce an $a^-\ b^-\ c^-\ d^-\ e^-$ recipient strain of *E. coli* with an $a^+\ b^+\ c^+\ d^+\ e^+$ donor. The recipient culture is plated on various media with the results shown in the following table. (Note that a^- indicates a requirement for A as a nutrient, and so forth.) What can you conclude about the linkage and order of the genes?

Compounds added to minimal medium	Presence (+) or absence (–) of colonies
CDE	–
BDE	–
BCE	+
BCD	+
ADE	–
ACE	–
ACD	–
ABE	–
ABD	+
ABC	–

Answer: If a compound is not added and growth occurs, the *E. coli* recipient cell must have received the wild-type genes for production of those nutrients by transduction. Thus, the BCE culture selects for cells that are now a^+ and d^+, the BCD culture selects for cells that are a^+ and e^+, and the ABD culture selects for cells that are c^+ and e^+. These genes can be aligned (see below) to give the map order of *d a e c*. (Notice that *b* is never cotransduced and is therefore distant from this group of genes.)

$$a^+\quad d^+$$

$$e^+\quad a^+$$

$$c^+\quad e^+$$

51. In 1965, Jon Beckwith and Ethan Signer devised a method of obtaining specialized transducing phages carrying the *lac* region. They knew that the integration site, designated *att80*, for the temperate phage φ80 (a relative of phage λ) was located near *tonB*, a gene that confers resistance to the virulent phage T1:

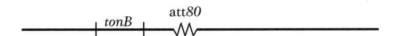

They used an F′ *lac*⁺ plasmid that could not replicate at high temperatures in a strain carrying a deletion of the *lac* genes. By forcing the cell to remain *lac*⁺ at high temperatures, the researchers could select strains in which the plasmid had integrated into the chromosome, thereby allowing the F′ *lac* to be maintained at high temperatures. By combining this selection with a simultaneous selection for resistance to T1 phage infection, they found that the only survivors were cells in which the F′ *lac* had integrated into the *tonB* locus, as shown here:

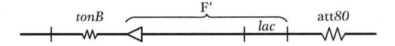

This result placed the *lac* region near the integration site for phage ϕ80. Describe the subsequent steps that the researchers must have followed to isolate the specialized transducing particles of phage ϕ80 that carried the *lac* region.

Answer: To isolate the specialized transducing particles of phage ϕ80 that carried *lac*⁺, the researchers would have had to lysogenize the strain with ϕ80, induce the phage with UV, and then use these lysates to transduce a *lac*⁻ strain to *lac*⁺. *Lac*⁺ colonies would then be used to make a new lysate, which should be highly enriched for the *lac*⁺ transducing phage.

52. Wild-type *E. coli* takes up and concentrates a certain red food dye, making the colonies blood red. Transposon mutagenesis was used, and the cells were plated on food dye. Most colonies were red, but some colonies did not take up dye and appeared white. In one white colony, the DNA surrounding the transposon insert was sequenced, with the use of a DNA replication primer identical with part of the end of the transposon sequence, and the sequence adjacent to the transposon was found to correspond to a gene of unknown function called *atoE*, spanning positions 2.322 through 2.324 Mb on the map (numbered from an arbitrary position zero). Propose a function for *atoE*. What biological process could be investigated in this way and what other types of white colonies might be expected?

Answer: *atoE* may encode a protein necessary for the transport of the food dye into the cell. All cells are surrounded by a semipermeable membrane so many substances require specific transport proteins to get into or out of the cell. Other possible white colonies expected are those in which other genes impacting uptake have been mutated or spontaneous mutants of the *atoE* gene.

6

Gene Interaction

WORKING WITH THE FIGURES

1. In Figure 6-1,

    a. what do the yellow stars represent?
    b. explain in your own words why the heterozygote is functionally wild type.

    Answer:
    a. Yellow stars represent mutations in the gene.
    b. Heterozygous is a functional genotype because there is a sufficient amount of a gene product from the second (wild type) gene on the second chromosome, which does not have a mutation.

2. In Figure 6-2, explain how the mutant polypeptide acts as a spoiler and what its net effect on phenotype is.

    Answer: Consider the lower-right-hand panel in this figure: In the heterozygote, the mutant protein imposes its shape on the wild-type protein resulting in nonfunction.

3. In Figure 6-6, assess the allele V^f: is it dominant? recessive? codominant? incompletely dominant?

    Answer: Allele V^f which produces solid white patterns in a clover leaf, appears codominant with V^{bv} and V^h, but it is dominant to the other alleles.

4. In Figure 6-11,

    a. in view of the position of HPA oxidase earlier in the pathway compared to that of HA oxidase, would you expect people with tyrosinosis to show symptoms of alkaptonuria?
    b. if a double mutant could be found, would you expect tyrosinosis to be epistatic to alkaptonuria?

Answer:

a. No, people with tyrosinosis would not accumulate homogentisic acid, since they do not have an enzyme (HPA oxidase) which could convert HPA to HA.

b. Yes, if there would be a double mutant, lacking both enzymes, an HPA oxidase mutation would be epistatic to alkaptonuria.

5. In Figure 6-12,

a. what do the dollar, pound, and yen symbols represent?
b. why can't the left-hand F_1 heterozygote synthesize blue pigment?

Answer:

a. Dollar, pound and yen symbols represent different null mutations in the genes $w1$ and $w2$, involved in flower pigment production.

b. The left hand mutations are two different types of mutants within the same $w1$ gene. They could not complement each other and therefore the phenotype would lack pigment (a precursor could not be converted to the functional next step of the metabolic pathway).

6. In Figure 6-13, explain at the protein level why this heterokaryon can grow on minimal medium.

Answer: In the figure two different arginine mutants (arg-1 and arg-2) were combined, producing a mixed heterokaryon (mixed nuclei cells). Such combination would result in fungal growth on the minimal medium, since they could now complete metabolic pathway to synthesize arginine. In another words, arg-1 has a functional gene for enzyme missing in arg-2 and vice versa.

7. In Figure 6-14, write possible genotypes for each of the four snakes illustrated.

Answer: A) b+/- o+/- (dominant for both alleles; 9/16 expected in F_2)
       B) b+/- o/o   (dominant for black only; 3/16 expected)
       C) b/b o+/- (dominant for orange only; 3/16 expected)
       D) b/b o/o   (recessive for both; 1/16 expected)

8. In Figure 6-15,

a. which panel represents the double mutant?
b. state the purpose of the regulatory gene.
c. in the situation in panel b, would protein from the active protein gene be

made?

Answer:
a. D is a double mutant.
b. Regulatory gene controls expression of a structural gene.
c. No, since this type of regulatory gene controls transcription and if the regulatory protein is absent, transcription could not start.

9.   In Figure 6-16, if you selfed 10 different F_2 pink plants, would you expect to find any white-flowered plants among the offspring? Any blue-flowered plants?

Answer: No, blue flowers would require both dominant alleles. If you cross:

w+/- m/m   ×   w+/- m/m

The expected offspring would be ¾ pink and ¼ white if selfed parent was a heterozygous (w+/w m/m), but all pink if selfed parent was a homozygous pink parent (w+/w+ m/m).

10.   In Figure 6-19,

a.   what do the square/triangular pegs and holes represent?
b.   is the suppressor mutation alone wild type in phenotype?

Answer:
a.   The square and triangular pegs, as well as the corresponding holes, represent the recognition site of a protein, whose activation depends on proper alignment of the two products.
b.   No, because if there is only a suppressor mutation in a gene, the active binding site of such a protein product would be different, and it would not be able to bind to a functional gene product of the gene *m*. Only when both genes '*m*' and '*s*' are mutant, wild type could occur at the phenotypic level.

11.   In Figure 6-21, propose a specific genetic explanation for individual Q (give a possible genotype, defining the alleles).

Answer: This pedigree shows an incomplete penetrance of a dominant gene (A). Individual Q has a genotype A/a (if it is a rare dominant allele, there is a very low likelihood that another parent would harbor the same mutation, especially since it is a dominant trait).
Two of the children in the last generation have complete penetrance of the dominant allele (with genotypes A/a), while female R might have the allele A which is not showing in this case. Probability that R female is (A/a) is ½ and that she is a/a is ½.

Basic Problems

12. In humans, the disease galactosemia causes mental retardation at an early age. Lactose (milk sugar) is broken down to galactose plus glucose. Normally, galactose is broken down further by the enzyme galactose-1-phosphate uridyltransferase (GALT). However, in patients with galactosemia, GALT is inactive, leading to a buildup of high levels of galactose, which, in the brain, causes mental retardation. How would you provide a secondary cure for galactosemia? Would you expect this disease phenotype to be dominant or recessive?

Answer: Lactose is composed of one molecule of galactose and one molecule of glucose. A secondary cure would result if all galactose and lactose were removed from the diet. The disorder would be expected not to be dominant, because one good copy of the gene should allow for at least some, if not all, breakdown of galactose. In fact, galactosemia is a recessive human disease.

13. In humans, PKU (phenylketonuria) is a disease caused by an enzyme inefficiency at step A in the following simplified reaction sequence, and AKU (alkaptonuria) is due to an enzyme inefficiency in one of the steps summarized as step B here:

$$\text{phenylalanine} \xrightarrow{\text{ A }} \text{tyrosine} \xrightarrow{\text{ B }} CO_2 + H_2O$$

A person with PKU marries a person with AKU. What phenotypes do you expect for their children? All normal, all having PKU only, all having AKU only, all having both PKU and AKU, or some having AKU and some having PKU?

Answer: Assuming homozygosity for the normal gene, the mating $A/A \cdot b/b \times a/a \cdot B/B$ would only result in $A/a \cdot B/b$ heterozygous offspring. All of their children would be normal (see Problem 23).

14. In *Drosophila*, the autosomal recessive *bw* causes a dark brown eye, and the unlinked autosomal recessive *st* causes a bright scarlet eye. A homozygote for both genes has a white eye. Thus, we have the following correspondences between genotypes and phenotypes:

$st^+/ st^+ ; bw^+/bw^+$ = red eye (wild type)
$st^+/ st^+ ; bw/bw$ = brown eye
$st/st ; bw^+/bw^+$ = scarlet eye
$st/st ; bw/bw$ = white eye

Construct a hypothetical biosynthetic pathway showing how the gene products interact and why the different mutant combinations have different phenotypes.

Answer:

bw^+

No pigment (white) ———→ Scarlet

No pigment (white) ———→ Brown

st^+

Scarlet plus brown pigments in *Drosophila* result in a wild-type (red) eye color.

15. Several mutants are isolated, all of which require compound G for growth. The compounds (A to E) in the biosynthetic pathway to G are known, but their order in the pathway is not known. Each compound is tested for its ability to support the growth of each mutant (1 to 5). In the following table, a plus sign indicates growth and a minus sign indicates no growth.

	Compound tested					
	A	B	C	D	E	G
Mutant 1	–	–	–	+	–	+
2	–	+	–	+	–	+
3	–	–	–	–	–	+
4	–	+	+	+	–	+
5	+	+	+	+	–	+

a. What is the order of compounds A to E in the pathway?
b. At which point in the pathway is each mutant blocked?
c. Would a heterokaryon composed of double mutants 1,3 and 2,4 grow on a minimal medium? Would 1,3 and 3,4? Would 1,2 and 2,4 and 1,4?

Answer: Growth will be supported by a particular compound if it is later in the pathway than the enzymatic step blocked in the mutant. Restated, the more mutants a compound supports, the later in the pathway growth must be. In this example, compound G supports growth of all mutants and can be considered the end product of the pathway. Alternatively, compound E does not support the growth of any mutant and can be considered the starting substrate for the pathway. The data indicate the following:

a. and b.

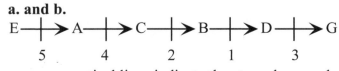

vertical lines indicate the step where each mutant is blocked.

c. A heterokaryon of double mutants 1, 3 and 2, 4 would grow as the first would supply functional 2 and 4, and the second would supply functional 1 and 3.

A heterokaryon of the double mutants 1, 3 and 3, 4 would not grow as both are mutant for 3.

A heterokaryon of the double mutants 1, 2 and 2, 4 and 1, 4 would grow as the first would supply functional 4, the second would supply functional 1, and the last would supply functional 2.

16. In a certain plant, the flower petals are normally purple. Two recessive mutations arise in separate plants and are found to be on different chromosomes. Mutation 1 (m_1) gives blue petals when homozygous (m_1/m_1). Mutation 2 (m_2) gives red petals when homozygous (m_2/m_2). Biochemists working on the synthesis of flower pigments in this species have already described the following pathway:

a. Which mutant would you expect to be deficient in enzyme A activity?
b. A plant has the genotype M_1/m_1 ; M_2/m_2. What would you expect its phenotype to be?
c. If the plant in part *b* is selfed, what colors of progeny would you expect and in what proportions?
d. Why are these mutants recessive?

Answer:
a. If enzyme A was defective or missing (m_2/m_2), red pigment would still be made, and the petals would be red.

b. Purple, because it has a wild-type allele for each gene, and you are told that the mutations are recessive.

c.
9	$M_1/-$; $M_2/-$	purple
3	m_1/m_1 ; $M_2/-$	blue
3	$M_1/-$; m_2/m_2	red
1	m_1/m_1 ; m_2/m_2	white

d. The mutant alleles do not produce functional enzyme. However, enough functional enzyme must be produced by the single wild-type allele of each gene to synthesize normal levels of pigment.

17. In sweet peas, the synthesis of purple anthocyanin pigment in the petals is controlled by two genes, B and D. The pathway is

$$\underset{\substack{\text{white} \\ \text{intermediate}}}{} \xrightarrow[\text{enzyme}]{\text{gene } B} \underset{\substack{\text{blue} \\ \text{intermediate}}}{} \xrightarrow[\text{enzyme}]{\text{gene } D} \underset{\substack{\text{anthocyanin} \\ \text{(purple)}}}{}$$

a. What color petals would you expect in a pure-breeding plant unable to catalyze the first reaction?

b. What color petals would you expect in a pure-breeding plant unable to catalyze the second reaction?

c. If the plants in parts a and b are crossed, what color petals will the F_1 plants have?

d. What ratio of purple : blue : white plants would you expect in the F_2?

Answer:

a. If enzyme B is missing, a white intermediate will accumulate and the petals will be white.

b. If enzyme D is missing, a blue intermediate will accumulate and the petals will be blue.

c. P b/b ; D/D × B/B ; d/d
     F_1 B/b ; D/d purple

d. P b/b ; D/D × B/B ; d/d
     F_1 B/b ; D/d × B/b ; D/d
     F_2 9 $B/-$; $D/-$ purple
         3 b/b ; $D/-$ white
         3 $B/-$; d/d blue
         1 b/b ; d/d white

The ratio of purple : blue : white would be 9:3:4.

18. If a man of blood-group AB marries a woman of blood-group A whose father was of blood-group O, to what different blood groups can this man and woman expect their children to belong?

Answer: The woman must be I^A/i, so the mating is $I^A/i \times I^A/I^B$. Their children will be:

Genotype	Phenotype
$1/4\ I^A/I^A$	A
$1/4\ I^A/IB$	AB
$1/4\ I^A/i$	A
$1/4\ I^B/i$	B

19. Most of the feathers of erminette fowl are light-colored, with an occasional black one, giving a flecked appearance. A cross of two erminettes produced a total of 48 progeny, consisting of 22 erminettes, 14 blacks, and 12 pure whites.

What genetic basis of the erminette pattern is suggested? How would you test your hypotheses?

Answer: You are told that the cross of two erminette fowls results in 22 erminette, 14 black, and 12 pure white. Two facts are important: (1) the parents consist of only one phenotype, yet the offspring have three phenotypes, and (2) the progeny appear in an approximate ratio of 1:2:1. These facts should tell you immediately that you are dealing with a heterozygous × heterozygous cross involving one gene and that the erminette phenotype must be the heterozygous phenotype.

When the heterozygote shows a different phenotype from either of the two homozygotes, the heterozygous phenotype results from incomplete dominance or codominance. Because two of the three phenotypes contain black, either fully or in an occasional feather, you might classify erminette as an instance of incomplete dominance because it is intermediate between fully black and fully white. Alternatively, because erminette has both black and white feathers, you might classify the phenotype as codominant. Your decision will rest on whether you look at the whole animal (incomplete dominance) or at individual feathers (codominance). This is yet another instance where what you conclude is determined by how you observe.

To test the hypothesis that the erminette phenotype is a heterozygous phenotype, you could cross an erminette with either, or both, of the homozygotes. You should observe a 1:1 ratio in the progeny of both crosses.

20. Radishes may be long, round, or oval, and they may be red, white, or purple. You cross a long, white variety with a round, red one and obtain an oval, purple F_1. The F_2 shows nine phenotypic classes as follows: 9 long, red; 15 long, purple; 19 oval, red; 32 oval, purple; 8 long, white; 16 round, purple; 8 round, white; 16 oval, white; and 9 round, red.

a. Provide a genetic explanation of these results. Be sure to define the genotypes and show the constitution of the parents, the F_1, and the F_2.

b. Predict the genotypic and phenotypic proportions in the progeny of a cross between a long, purple radish and an oval, purple one.

Answer:

a. The original cross and results were:

P   long, white × round, red

F_1 oval, purple

F_2 9 long, red   19 oval, red   8 round, white
        15 long, purple   32 oval, purple   16 round, purple

8 long, white	16 oval, white	9 round, red
32 long	67 oval	33 round

The data show that, when the results are rearranged by shape, a 1:2:1 ratio is observed for color within each shape category. Likewise, when the data are rearranged by color, a 1:2:1 ratio is observed for shape within each color category.

9 long, red	15 long, purple	8 round, white
19 oval, red	32 oval, purple	16 oval, white
9 round, red	16 round, purple	8 long, white
37 red	63 purple	32 white

A 1:2:1 ratio is observed when there is a heterozygous × heterozygous cross. Therefore, the original cross was a dihybrid cross. Both oval and purple must represent an incomplete dominant phenotype.

Let L = long, L' = round, R = red and R' = white. The cross becomes

P $L/L\ ;\ R'/R' \times L'/L'\ ;\ R/R$

F_1 $L/L'\ ;\ R/R' \times L/L'\ ;\ R/R'$

F_2

	1/4 R/R =	1/16	long, red
1/4 L/L ×	1/2 R/R' =	1/8	long, purple
	1/4 R'/R' =	1/16	long, white
	1/4 R/R =	1/8	oval, red
1/2 L/L' ×	1/2 R/R' =	1/4	oval, purple
	1/4 R'/R' =	1/8	oval, white
	1/4 R/R =	1/16	round, red
1/4 L'/L' ×	1/2 R/R' =	1/8	round, purple
	1/4 R'/R' =	1/16	round, white

b. A long, purple × oval, purple cross is as follows:

P $L/L\ ;\ R/R' \times L/L'\ ;\ R/R''$

F_1

	1/4 R/R =	1/8 long, red
1/2 L/L ×	1/2 R/R' =	1/4 long, purple
	1/4 R'/R' =	1/8 long, white
	1/4 R/R =	1/8 oval, red
1/2 L/L' ×	1/2 R/R' =	1/4 oval, purple
	1/4 R'/R' =	1/8 oval, white

21. In the multiple-allele series that determines coat color in rabbits, c^+ encodes agouti, cch encodes chinchilla (a beige coat color), and ch encodes Himalayan. Dominance is in the order $c^+ > c^{ch} > c^h$. In a cross of $c^+/cch \times c^{ch}/c^h$, what proportion of progeny will be chinchilla?

Answer: From the cross $c+/c^{ch} \times c^{ch}/c^h$ the progeny are:

1/4 c^+/c^{ch}	full color
1/4 c^+/c^h	full color
1/4 c^{ch}/c^{ch}	chinchilla
1/4 c^h/c^h	chinchilla

Thus, 50 percent of the progeny will be chinchilla.

22. Black, sepia, cream, and albino are coat colors of guinea pigs. Individual animals (not necessarily from pure lines) showing these colors were intercrossed; the results are tabulated as follows, where the abbreviations A (albino), B (black), C (cream), and S (sepia) represent the phenotypes:

		Phenotypes of progeny			
Cross	Parental phenotypes	B	S	C	A
1	B × B	22	0	0	7
2	B × A	10	9	0	0
3	C × C	0	0	34	11
4	S × C	0	24	11	12
5	B × A	13	0	12	0
6	B × C	19	20	0	0
7	B × S	18	20	0	0
8	B × S	14	8	6	0
9	S × S	0	26	9	0
10	C × A	0	0	15	17

a. Deduce the inheritance of these coat colors and use gene symbols of your own choosing. Show all parent and progeny genotypes.

b. If the black animals in crosses 7 and 8 are crossed, what progeny proportions can you predict by using your model?

Answer:

a. The data indicate that there is a single gene with multiple alleles. All the ratios produced are 3:1 (complete dominance), 1:2:1 (incomplete codominance), or 1:1 (testcross). The order of dominance is:

black (C^b)> sepia (C^s) > cream (C^c) > albino (C^a)

Cross	Parents	Progeny	Conclusion
Cross 1:	$C^b/C^a \times C^b/C^a$	$3\ C^b/- : 1\ C^a/C^a$	C^b is dominant to C^a.
Cross 2:	$C^b/C^s \times C^a/C^a$	$1\ C^b/C^a : 1\ C^s/C^a$	C^b is dominant to C^s; C^s is dominant to C^a.
Cross 3:	$C^c/C^a \times C^c/C^a$	$3\ C^c/- : 1\ C^a/C^a$	C^c is dominant to C^a.
Cross 4:	$C^s/C^a \times C^c/C^a$	$1\ C^c/C^a : 2\ C^s/- : 1\ C^a/C^a$	C^s is dominant to C^c.
Cross 5:	$C^b/C^c \times C^a/C^a$	$1\ C^b/C^a : 1\ C^c/C^a$	C^b is dominant to C^c.
Cross 6:	$C^b/C^s \times C^c/-$	$1\ C^b/- : 1\ C^s/-$	"–" can be C^c or C^a.
Cross 7:	$C^b/C^s \times C^s/-$	$1\ C^b/C^s : 1\ C^s/-$	"–" can be C^c or C^a.
or	$C^b/- \times C^s/C^s$	$1\ C^b/C^s : 1\ C^s/-$	"–" can be C^s, C^c or C^a.
Cross 8:	$C^b/C^c \times C^s/C^c$	$1\ C^s/C^c : 2\ C^b/- : 1\ C^c/C^c$	
Cross 9:	$C^s/C^c \times C^s/C^c$	$3\ C^s/- : 1\ C^c/C^c$	
Cross 10:	$C^c/C^a \times C^a/C^a$	$1\ C^c/C^a : 1\ C^a/C^a$	

b. The progeny of the cross $C^b/C^s \times C^b/C^c$ will be 3/4 black (1/4 C^b/C^b, 1/4 C^b/C^c, 1/4 C^b/C^s) : 1/4 sepia (C^s/C^c). The progeny of the cross C^b/C^a (or C^b/C^c) × C^b/C^c will be 3/4 black: 1/4 cream.

23. In a maternity ward, four babies become accidentally mixed up. The ABO types of the four babies are known to be O, A, B, and AB. The ABO types of the four sets of parents are determined. Indicate which baby belongs to each set of parents: (a) AB × O, (b) A × O, (c) A × AB, (d) O × O.

Answer: Both codominance (=) and classical dominance (>) are present in the multiple allelic series for blood type: $A = B$, $A > O$, $B > O$.

Parents' phenotype	Parents' possible genotypes	Parents' possible children
a. AB × O	$A/B \times O/O$	$A/O, B/O$
b. A × O	A/A or $A/O \times O/O$	$A/O, O/O$
c. A × AB	A/A or $A/O \times A/B$	$A/A, A/B, A/O, B/O$
d. O × O	$O/O \times O/O$	O/O

The possible genotypes of the children are

Phenotype	Possible genotypes
O	O/O
A	$A/A, A/O$
B	$B/B, B/O$
AB	A/B

Using the assumption that each set of parents had one child, the following combinations are the only ones that will work as a solution.

Parents	Child
a. AB × O	B

b. A × O   A
c. A × AB   AB
d. O × O   O

24. Consider two blood polymorphisms that humans have in addition to the ABO system. Two alleles LM and LN determine the M, N, and MN blood groups. The dominant allele *R* of a different gene causes a person to have the Rh$^+$ (rhesus positive) phenotype, whereas the homozygote for *r* is Rh$^-$ (rhesus negative). Two men took a paternity dispute to court, each claiming three children to be his own. The blood groups of the men, the children, and their mother were as follows:

Person	Blood group		
husband	O	M	Rh$^+$
wife's lover	AB	MN	Rh$^-$
wife	A	N	Rh$^+$
child 1	O	MN	Rh$^+$
child 2	A	N	Rh$^+$
child 3	A	MN	Rh$^-$

From this evidence, can the paternity of the children be established?
Answer: *M* and *N* are codominant alleles. The rhesus group is determined by classically dominant alleles. The *ABO* alleles are mixed codominance and classical dominance (see Problem 6).

Person	Blood type			Possible paternal contribution		
husband	M	Rh$^+$		*O*	*M*	*R* or *r*
wife's lover	AB	MN	Rh$^-$	*A* or *B*	*M* or *N*	*r*
wife	A	N	Rh$^+$	*A* or *O*	*N*	*R* or *r*
child 1	O	MN	Rh$^+$	*O*	*M*	*R* or *r*
child 2	A	N	Rh$^+$	*A* or *O*	*N*	*R* or *r*
child 3	A	MN	Rh$^-$	*A* or *O*	*M*	*r*

The wife must be *A/O* ; *N/N* ; *R/r*. (She has a child with type O blood and another child that is Rh$^-$ so she must carry both of these recessive alleles.) Only the husband could donate O to child 1. Only the lover could donate A and N to child 2. Both the husband and the lover could have donated the necessary alleles to child 3.

25. On a fox ranch in Wisconsin, a mutation arose that gave a "platinum" coat color. The platinum color proved very popular with buyers of fox coats, but the breeders could not develop a pure-breeding platinum strain. Every time two platinums were crossed, some normal foxes appeared in the progeny. For example, the repeated matings of the same pair of platinums produced 82

platinum and 38 normal progeny. All other such matings gave similar progeny ratios. State a concise genetic hypothesis that accounts for these results.

Answer: The key to solving this problem is in the statement that breeders cannot develop a pure–breeding stock and that a cross of two platinum foxes results in some normal progeny. Platinum must be dominant to normal color and heterozygous (*A/a*). An 82:38 ratio is very close to 2:1. Because a 1:2:1 ratio is expected in a heterozygous cross, one genotype is nonviable. It must be the *A/A*, homozygous platinum, genotype that is nonviable, because the homozygous recessive genotype is normal color (*a/a*). Therefore, the platinum allele is a pleiotropic allele that governs coat color in the heterozygous state and is lethal when homozygous.

26. For a period of several years, Hans Nachtsheim investigated an inherited anomaly of the white blood cells of rabbits. This anomaly, termed the Pelger anomaly, is the arrest of the segmentation of the nuclei of certain white cells. This anomaly does not appear to seriously inconvenience the rabbits.

   a. When rabbits showing the typical Pelger anomaly were mated with rabbits from a true-breeding normal stock, Nachtsheim counted 217 offspring showing the Pelger anomaly and 237 normal progeny. What appears to be the genetic basis of the Pelger anomaly?

   b. When rabbits with the Pelger anomaly were mated with each other, Nachtsheim found 223 normal progeny, 439 showing the Pelger anomaly, and 39 extremely abnormal progeny. These very abnormal progeny not only had defective white blood cells, but also showed severe deformities of the skeletal system; almost all of them died soon after birth. In genetic terms, what do you suppose these extremely defective rabbits represented? Why do you suppose there were only 39 of them?

   c. What additional experimental evidence might you collect to support or disprove your answers to part *b*?

   d. In Berlin, about 1 human in 1000 shows a Pelger anomaly of white blood cells very similar to that described for rabbits. The anomaly is inherited as a simple dominant, but the homozygous type has not been observed in humans. Can you suggest why, if you are permitted an analogy with the condition in rabbits?

   e. Again by analogy with rabbits, what phenotypes and genotypes might be expected among the children of a man and woman who both show the Pelger anomaly?
   (Problem 26 is from A. M. Srb, R. D. Owen, and R. S. Edgar, *General Genetics,* 2nd ed. W. H. Freeman and Company, 1965.)

Answer:

a. Because Pelger crossed with normal stock results in two phenotypes in a 1:1 ratio, either Pelger or normal is heterozygous (*A/a*), and the other is homozygous (*a/a*) recessive. The problem states that normal is true-breeding, or *a/a*. Pelger must be *A/a*.

b. The cross of two Pelger rabbits results in three phenotypes. This means that the Pelger anomaly is dominant to normal. This cross is *A/a* × *A/a*, with an expected ratio of 1:2:1. Because the normal must be *a/a*, the extremely abnormal progeny must be *A/A*. There were only 39 extremely abnormal progeny because the others died before birth.

c. The Pelger allele is pleiotropic. In the heterozygous state, it is dominant for nuclear segmentation of white blood cells. In the homozygous state, it is lethal.

You could look for the nonsurviving fetuses in utero. Because the hypothesis of embryonic death when the Pelger allele is homozygous predicts a one-fourth reduction in litter size, you could also do an extensive statistical analysis of litter size, comparing normal × normal with Pelger × Pelger.

d. By analogy with rabbits, the absence of a homozygous Pelger anomaly in humans can be explained as recessive lethality. Also, because one in 1000 people have the Pelger anomaly, a heterozygous × heterozygous mating would be expected in only one of 1 million.

(1/1000 × 1/1000) random matings, and then only one in four of the progeny would be expected to be homozygous. Thus, the homozygous Pelger anomaly is expected in only 1 of 4 million births. This is extremely rare and might not be recognized.

e. By analogy with rabbits, among the children of a man and a woman with the Pelger anomaly, two-thirds of the surviving progeny would be expected to show the Pelger anomaly and one-third would be expected to be normal. The developing fetus that is homozygous for the Pelger allele would not be expected to survive until birth.

27. Two normal-looking fruit flies were crossed, and, in the progeny, there were 202 females and 98 males.

    a. What is unusual about this result?
    b. Provide a genetic explanation for this anomaly.
    c. Provide a test of your hypothesis.

Answer:
a. The expected sex ratio is 1:1, yet in this case seems to be 2:1.

b. The female parent was heterozygous for an X-linked recessive lethal allele. This would result in 50 percent fewer males than females.

c. Half of the female progeny should be heterozygous for the lethal allele and half should be homozygous for the nonlethal allele. Individually mate the F_1 females and determine the sex ratio of their progeny.

28. You have been given a virgin *Drosophila* female. You notice that the bristles on her thorax are much shorter than normal. You mate her with a normal male (with long bristles) and obtain the following F_1 progeny: 1/3 short-bristled females, 1/3 long-bristled females, and 1/3 long-bristled males. A cross of the F1 long-bristled females with their brothers gives only long-bristled F_2. A cross of short-bristled females with their brothers gives 1/3 short-bristled females, 1/3 long-bristled females, and 1/3 long-bristled males. Provide a genetic hypothesis to account for all these results, showing genotypes in every cross.

Answer: Note that a cross of the short-bristled female with a normal male results in two phenotypes with regard to bristles and an abnormal sex ratio of two females : one male. Furthermore, all the males are normal, while the females are normal and short in equal numbers. Whenever the sexes differ with respect to phenotype among the progeny, an X-linked gene is implicated. Because only the normal phenotype is observed in males, the short-bristled phenotype must be heterozygous, and the allele must be a recessive lethal. Thus, the first cross was $A/a \times a/Y$.

Long-bristled females (a/a) were crossed with long-bristled males (a/Y). All their progeny would be expected to be long-bristled (a/a or a/Y). Short-bristled females (A/a) were crossed with long-bristled males (a/Y). The progeny expected are

1/4 A/a	short-bristled females
1/4 a/a	long-bristled females
1/4 a/Y	long-bristled males
1/4 A/Y	nonviable

29. A dominant allele H reduces the number of body bristles that *Drosophila* flies have, giving rise to a "hairless" phenotype. In the homozygous condition, H is lethal. An independently assorting dominant allele S has no effect on bristle number except in the presence of H, in which case a single dose of S suppresses the hairless phenotype, thus restoring the hairy phenotype. However, S also is lethal in the homozygous (S/S) condition.

a. What ratio of hairy to hairless flies would you find in the live progeny of a cross between two hairy flies both carrying H in the suppressed condition?

b. When the hairless progeny are backcrossed with a parental hairy fly, what phenotypic ratio would you expect to find among their live progeny?

Answer: In order to do this problem, you should first restate the information provided. The following two genes are independently assorting

h/h = hairy	*s/s* = no effect
H/h = hairless	*S/s* suppresses *H/h*, giving hairy
H/H = lethal	*S/S* = lethal

a. The cross is *H/h* ; *S/s* × *H/h* ; *S/s*. Because this is a typical dihybrid cross, the expected ratio is 9:3:3:1. However, the problem cannot be worked in this simple fashion because of the epistatic relationship of these two genes. Therefore, the following approach should be used.

For the *H* gene, you expect 1/4 *H/H*:1/2 *H/h*:1/4 *h/h*. For the *S* gene, you expect 1/4 *S/S*:1/2 *S/s*:1/4 *s/s*. To get the final ratios, multiply the frequency of the first genotype by the frequency of the second genotype.

1/4 *H/H*			all progeny die regardless of the *S* gene	
		1/4 *S/S* =	1/8 *H/h* ; *S/S*	die
1/2 *H/h*	×	1/2 *S/s* =	1/4 *H/h* ; *S/s*	hairy
		1/4 *s/s* =	1/8 *H/h* ; *s/s*	hairless
		1/4 *S/S* =	1/16 *h/h* ; *S/S*	die
1/4 *h/h*	×	1/2 *S/s* =	1/8 *h/h* ; *S/s*	hairy
		1/4 *s/s* =	1/16 *h/h* ; *s/s*	hairy

Of the 9/16 living progeny, the ratio of hairy to hairless is 7:2.

b. This cross is *H/h* ; *s/s* × *H/h* ; *S/s*. A 1:2:1 ratio is expected for the *H* gene and a 1:1 ratio is expected for the *S* gene.

1/4 *H/H*			all progeny die regardless of the *S* gene	
1/2 *H/h*	×	1/2 *S/s* =	1/4 *H/h* ; *S/s*	hairy
		1/2 *s/s* =	1/4 *H/h* ; *s/s*	hairless
1/4 *h/h*	×	1/2 *S/s* =	1/8 *h/h* ; *S/s*	hairy
		1/2 *s/s* =	1/8 *h/h* ; *s/s*	hairy

Of the 3/4 living progeny, the ratio of hairy to hairless is 2:1.

30. After irradiating wild-type cells of *Neurospora* (a haploid fungus), a geneticist finds two leucine-requiring auxotrophic mutants. He combines the two mutants in a heterokaryon and discovers that the heterokaryon is prototrophic.

a. Were the mutations in the two auxotrophs in the same gene in the pathway for synthesizing leucine or in two different genes in that pathway? Explain.
b. Write the genotype of the two strains according to your model.
c. What progeny and in what proportions would you predict from crossing the two auxotrophic mutants? (Assume independent assortment.)

Answer:

a. The mutations are in two different genes as the heterokaryon is prototrophic (the two mutations complemented each other).

b. $leu1^+$; $leu\ 2^-$ and $leu1^-$; $leu2^+$

c. With independent assortment, expect

$$1/4 \quad leu1^+ ; leu\ 2^-$$
$$1/4 \quad leu1^- ; leu2^+$$
$$1/4 \quad leu1^- ; leu\ 2^-$$
$$1/4 \quad leu1^+ ; leu\ 2^+$$

31. A yeast geneticist irradiates haploid cells of a strain that is an adenine-requiring auxotrophic mutant, caused by mutation of the gene *ade1*. Millions of the irradiated cells are plated on minimal medium, and a small number of cells divide and produce prototrophic colonies. These colonies are crossed individually with a wild-type strain. Two types of results are obtained:

(1) prototroph × wild type : progeny all prototrophic
(2) prototroph × wild type : progeny 75% prototrophic, 25% adenine-requiring auxotrophs

a. Explain the difference between these two types of results.
b. Write the genotypes of the prototrophs in each case.
c. What progeny phenotypes and ratios do you predict from crossing a prototroph of type 2 by the original *ade1* auxotroph?

Answer:

a. The first type of prototroph is due to reversion of the original *ade1* mutant to wild type. The second type of prototroph is due to a new mutation (call it sup^m) in an unlinked gene that suppresses the $ade1^-$ phenotype. For the results of crossing type 2 prototrophs to wild type
$ade1^-$; sup^m × $ade1^+$; sup
and the results would be

1/4	$ade1^+$; sup	prototroph
1/4	$ade1^+$; sup^m	prototroph
1/4	$ade1^-$; sup^m	prototroph
1/4	$ade1^-$; sup	auxotroph

b. type 1 would be *ade1*⁺ ; *sup*
type 2 would be *ade1*⁻ ; *sup*ᵐ

c. For the results of crossing type 2 prototrophs to the original mutant

$ade1^0 ; sup^m \times ade1^- ; sup$

and the results would be

1/2 *ade1*⁻ ; *sup*   auxotroph
1/2 *ade1*⁻ ; *sup*ᵐ   prototroph

32. In roses, the synthesis of red pigment is by two steps in a pathway, as follows:

$$\text{colorless intermediate} \xrightarrow{\text{gene } P}$$
$$\text{magenta intermediate} \xrightarrow{\text{gene } Q} \text{red pigment}$$

a. What would the phenotype be of a plant homozygous for a null mutation of gene *P*?
b. What would the phenotype be of a plant homozygous for a null mutation of gene *Q*?
c. What would the phenotype be of a plant homozygous for null mutations of genes *P* and *Q*?
d. Write the genotypes of the three strains in parts *a, b,* and *c.*
e. What F_2 ratio is expected from crossing plants from parts *a* and *b*? (Assume independent assortment.)

Answer:
a. colorless
b. magenta
c. colorless
d. *p/p* ; *Q/Q*
P/P ; *q/q*
p/p ; *q/q*
e. 9 red : 4 colorless : 3 magenta

33. Because snapdragons (*Antirrhinum*) possess the pigment anthocyanin, they have reddish purple petals. Two pure anthocyaninless lines of *Antirrhinum* were developed, one in California and one in Holland. They looked identical in having no red pigment at all, manifested as white (albino) flowers. However, when petals from the two lines were ground up together in buffer in the same test tube, the solution, which appeared colorless at first, gradually turned red.

a. What control experiments should an investigator conduct before proceeding with further analysis?

b. What could account for the production of the red color in the test tube?

c. According to your explanation for part *b*, what would be the genotypes of the two lines?

d. If the two white lines were crossed, what would you predict the phenotypes of the F_1 and F_2 to be?

Answer: The suggestion from the data is that the two albino lines in the snapdragons had mutations in two different genes. When the extracts from the two lines were placed in the same test tube, they were capable of producing color because the gene product of one line was capable of compensating for the absence of a gene product from the second line.

a. Grind a sample of each specimen separately (negative control) to ensure that the grinding process does not activate or release an enzyme from some compartment that allows the color change. Another control is to cross the two pure-breeding lines. The cross would be A/A ; b/b × a/a ; B/B. The progeny will be A/a ; B/b, and all should be reddish purple.

b. The most likely explanation is that the red pigment is produced by the action of at least two different gene products. When petals of the two plants were ground together, the different defective enzyme of each plant was complemented by the normal enzyme of the other.

c. The genotypes of the two lines would be A/A ; b/b and a/a ; B/B.

d. The F_1 would all be pigmented, A/a ; B/b. This is an example of complementation. The mutants are defective for different genes. The F_2 would be

9	$A/-$; $B/-$	pigmented
3	a/a ; $B/-$	white
3	$A/-$; b/b	white
1	a/a ; b/b	white

34. The frizzle fowl is much admired by poultry fanciers. It gets its name from the unusual way that its feathers curl up, giving the impression that it has been (in the memorable words of animal geneticist F. B. Hutt) "pulled backwards through a knothole." Unfortunately, frizzle fowl do not breed true: when two frizzles are intercrossed, they always produce 50 percent frizzles, 25 percent normal, and 25 percent with peculiar woolly feathers that soon fall out, leaving the birds naked.

a. Give a genetic explanation for these results, showing genotypes of all phenotypes, and provide a statement of how your explanation works.

b. If you wanted to mass-produce frizzle fowl for sale, which types would be best to use as a breeding pair?

Answer:

a. This is an example where one phenotype in the parents gives rise to three phenotypes in the offspring. The "frizzle" is the heterozygous phenotype and shows incomplete dominance.

P A/a (frizzle) × A/a (frizzle)

F_1 1 A/A (normal) : 2 A/a (frizzle) : 1 a/a (woolly)

b. If A/A (normal) is crossed to a/a (woolly), all offspring will be A/a (frizzle).

35. The petals of the plant *Collinsia parviflora* are normally blue, giving the species its common name, blue-eyed Mary. Two pure-breeding lines were obtained from color variants found in nature; the first line had pink petals, and the second line had white petals. The following crosses were made between pure lines, with the results shown:

Parents	F_1	F_2
blue × white	blue	101 blue, 33 white
blue × pink	blue	192 blue, 63 pink
pink × white	blue	272 blue, 121 white, 89 pink

a. Explain these results genetically. Define the allele symbols that you use, and show the genetic constitution of the parents, the F_1, and the F_2 in each cross.

b. A cross between a certain blue F_2 plant and a certain white F_2 plant gave progeny of which 3/8 were blue, 1/8 were pink, and 1/2 were white. What must the genotypes of these two F_2 plants have been?

Unpacking Problem 35

1. What is the character being studied?

   Answer: The character being studied is petal color.

2. What is the wild-type phenotype?

   Answer: The wild-type phenotype is blue.

3. What is a variant?

   Answer: A variant is a phenotypic difference from wild type that is observed.

4. What are the variants in this problem?

    Answer: There are two variants: pink and white.

5. What does "in nature" mean?

    Answer: *In nature* means that the variants did not appear in laboratory stock and, instead, were found growing wild.

6. In what way would the variants have been found in nature? (Describe the scene.)

    Answer: Possibly, the variants appeared as a small patch or even a single plant within a larger patch of wild type.

7. At which stages in the experiments would seeds be used?

    Answer: Seeds would be grown to check the outcome from each cross.

8. Would the way of writing a cross "blue × white," for example, mean the same as "white × blue"? Would you expect similar results? Why or why not?

    Answer: Given that no sex linkage appears to exist (sex is not specified in parents or offspring), *blue × white* means the same as *white × blue*. Similar results would be expected because the trait being studied appears to be autosomal.

9. In what way do the first two rows in the table differ from the third row?

    Answer: The first two crosses show a 3:1 ratio in the F_2, suggesting the segregation of one gene. The third cross has a 9:4:3 ratio for the F_2, suggesting that two genes are segregating.

10. Which phenotypes are dominant?

    Answer: Blue is dominant to both white and pink.

11. What is complementation?

Answer: *Complementation* refers to generation of wild-type progeny from the cross of two strains that are mutant in different genes.

12. Where does the blueness come from in the progeny of the pink × white cross?

Answer: The ability to make blue pigment requires two enzymes that are individually defective in the pink or white strains. The F_1 progeny of this cross is blue, since each has inherited one nonmutant allele for both genes and can therefore produce both functional enzymes.

13. What genetic phenomenon does the production of a blue F_1 from pink and white parents represent?

Answer: Blueness from a pink × white cross arises through complementation.

14. List any ratios that you can see.

Answer: The following ratios are observed: 3:1, 9:4:3.

15. Are there any monohybrid ratios?

Answer: There are monohybrid ratios observed in the first two crosses.

16. Are there any dihybrid ratios?

Answer: There is a modified 9:3:3:1 ratio in the third cross.

17. What does observing monohybrid and dihybrid ratios tell you?

Answer: A monohybrid ratio indicates that one gene is segregating, while a dihybrid ratio indicates that two genes are segregating.

18. List four modified Mendelian ratios that you can think of.

Answer: 15:1, 12:3:1, 9:6:1, 9:4:3, 9:7, 1:2:1, 2:1

19. Are there any modified Mendelian ratios in the problem?

Answer: There is a modified dihybrid ratio in the third cross.

20. What do modified Mendelian ratios indicate generally?

    Answer: A modified dihybrid ratio most frequently indicates the interaction of two or more genes.

21. What is indicated by the specific modified ratio or ratios in this problem?

    Answer: Recessive epistasis is indicated by the modified dihybrid ratio.

22. Draw chromosomes representing the meioses in the parents in the cross blue × white and representing meiosis in the F_1.

    Answer

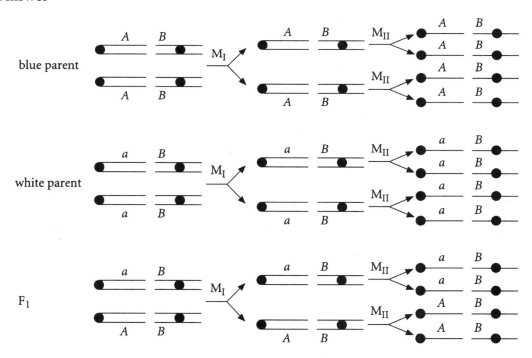

23. Repeat step 22 for the cross blue × pink.

Solution to the Problem

a. Let A = wild-type, a = white, B = wild-type, and b pink.

Cross 1: P   blue × white A/A ; B/B × a/a ; B/B
          F_1 all blue   all A/a ; B/B
          F_2 3 blue : 1 white 3 $A/-$; B/B : 1 a/a ; B/B

Cross 2: P   blue × pink A/A ; B/B × A/A ; b/b
          F_1 all blue   all A/A ; B/b
          F_2 3 blue : 1 pink   3 A/A ; $B/-$: 1 A/A ; b/b

Cross 3: P   pink × white A/A ; b/b × a/a ; B/B
          F_1 all blue   all A/a ; B/b
          F_2 9 blue   9 $A/-$; $B/-$
               4 white   3 a/a ; $B/-$: 1 a/a ; b/b
               3 pink   3 $A/-$; b/b

When the allele a is homozygous, the expression of alleles B or b is blocked or masked. The white phenotype is epistatic to the pigmented phenotypes. It is likely that the product of the A gene produces an intermediate that is then modified by the product of the B gene. If the plant is a/a, this intermediate is not made and the phenotype of the plant is the same regardless of the ability to produce functional B product.

b. The cross is

F_2 blue × white
F_3 3/8 blue
        1/8 pink
        4/8 white

Begin by writing as much of each genotype as can be assumed:

F_2 $A/-$; $B/-$ × a/a ; $-/-$
F_3 3/8 $A/-$; $B/-$
        1/8 $A/-$; b/b
        4/8 a/a ; $-/-$

Notice that both a/a and b/b appear in the F_3 progeny. In order for these homozygous recessives to occur, each parent must have at least one a and one b. Using this information, the cross becomes

F_2 A/a ; B/b × a/a ; $-/b$
F_3 3/8 A/a ; B/b
        1/8 A/a ; b/b
        4/8 a/a ; $b/-$

The only remaining question is whether the white parent was homozygous recessive, b/b, or heterozygous, B/b. If the white parent had been homozygous recessive, then the cross would have been a testcross of the blue parent, and the progeny ratio would have been

1 blue : 1 pink : 2 white, or 1 A/a ; B/b : 1 A/a ; b/b : 1 a/a ; B/b : 1 a/a ; b/b.

This was not observed. Therefore, the white parent had to have been heterozygous, and the F_2 cross was A/a ; $B/b \times a/a$; B/b.

36. A woman who owned a purebred albino poodle (an autosomal recessive phenotype) wanted white puppies; so she took the dog to a breeder, who said he would mate the female with an albino stud male, also from a pure stock. When six puppies were born, all of them were black; so the woman sued the breeder, claiming that he replaced the stud male with a black dog, giving her six unwanted puppies. You are called in as an expert witness, and the defense asks you if it is possible to produce black offspring from two pure-breeding recessive albino parents. What testimony do you give?

Answer: One would not expect black offspring from two pure-breeding recessive albino parents, yet if albinism results from mutations in two different genes this might be possible. If the cross of the two different albino parents is designated as:

$$A/A \; ; b/b \times a/a \; ; B/B$$
Albino 1   Albino 2

all offspring would be heterozygous

$$A/a \; ; B/b$$

and they would have a black phenotype because of complementation.

37. A snapdragon plant that bred true for white petals was crossed with a plant that bred true for purple petals, and all the F_1 had white petals. The F_1 was selfed. Among the F_2, three phenotypes were observed in the following numbers:

white	240
solid purple	61
spotted purple	19
Total	320

a. Propose an explanation for these results, showing genotypes of all generations (make up and explain your symbols).
b. A white F_2 plant was crossed with a solid purple F_2 plant, and the progeny were

white	50%
solid purple	25%

spotted purple   25%

What were the genotypes of the F$_2$ plants crossed?

Answer: The data indicate that white is dominant to solid purple. Note that the F$_2$ are in an approximate 12:3:1 ratio. Such a ratio indicates epistasis.

a. Because a modified 9:3:3:1 ratio was obtained in the F$_2$, the F$_1$ had to be a double heterozygote. Solid purple occurred at one-third the rate of white, which means that it will be in the form of either *D/–* ; *e/e* or *d/d* ; *E/–*. In order to achieve a double heterozygote in the F$_1$, the original white parent also has to be either *D/–* ; *e/e* or *d/d* ; *E/–*.

Arbitrarily assume that the original cross was *D/D* ; *e/e* (white) × *d/d* ; *E/E* (purple). The F$_1$ would all be *D/d* ; *E/e*. The F$_2$ would be

9 *D/–* ; *E/–*	white, by definition
3 *d/d* ; *E/–*	purple, by definition
3 *D/–* ; *e/e*	white, by deduction
1 *d/d* ; *e/e*	spotted purple, by deduction

Under these assumptions, *D* blocks the expression of both *E* and *e*. The *d* allele has no effect on the expression of *E* and *e*. *E* results in solid purple, while *e* results in spotted purple. It would also be correct, of course, to assume the opposite set of epistatic relationships (*E* blocks the expression of *D* or *d*, *D* results in solid purple, and *d* results in spotted purple).

b. The cross is white × solid purple. While the solid purple genotype must be *d/d* ; *E/–*, as defined in part *a*, the white genotype can be one of several possibilities. Note that the progeny phenotypes are in a 1:2:1 ratio and that one of the phenotypes, spotted, must be *d/d* ; *e/e*. In order to achieve such an outcome, the purple genotype must be *d/d* ; *E/e*. The white genotype of the parent must contain both a *D* and a *d* allele in order to produce both white (*D/–*) and spotted plants (*d/d*). At this point, the cross has been deduced to be *D/d* ; *–/–* (white) × *d/d* ; *E/e* (purple).

If the white plant is *E/E*, the progeny will be

1/2 *D/d* ; *E/–*	white
1/2 *d/d* ; *E/–*	solid purple

This was not observed. If the white plant is *E/e*, the progeny will be

3/8 *D/d* ; *E/–*	white
1/8 *D/d* ; *e/e*	white
3/8 *d/d* ; *E/–*	solid purple
1/8 *d/d* ; *e/e*	spotted purple

The phenotypes were observed but in a different ratio. If the white plant is *e/e*, the progeny will be

1/4 *D/d* ; *E/e*   white
1/4 *D/d* ; *e/e*   white
1/4 *d/d* ; *E/e*   solid purple
1/4 *d/d* ; *e/e*   spotted purple

This was observed in the progeny. Therefore, the parents were *D/d* ; *e/e* (white) × *d/d* ; *E/e* (purple).

38. Most flour beetles are black, but several color variants are known. Crosses of pure-breeding parents produced the following results (see table) in the F_1 generation, and intercrossing the F_1 from each cross gave the ratios shown for the F_2 generation. The phenotypes are abbreviated Bl, black; Br, brown; Y, yellow; and W, white.

Cross	Parents	F_1	F_2
1	Br × Y	Br	3 Br : 1 Y
2	Bl × Br	Bl	3 Bl : 1 Br
3	Bl × Y	Bl	3 Bl : 1 Y
4	W × Y	Bl	9 Bl : 3 Y : 4 W
5	W × Br	Bl	9 Bl : 3 Br : 4 W
6	Bl × W	Bl	9 Bl : 3 Y : 4 W

a. From these results, deduce and explain the inheritance of these colors.
b. Write the genotypes of each of the parents, the F_1, and the F_2 in all crosses.

Answer: **a. and b.** Crosses 1–3 show a 3:1 ratio, indicating that brown, black, and yellow are all alleles of one gene in flour beetles. Crosses 4–6 show a modified 9:3:3:1 ratio, indicating that at least two genes are involved. Those crosses also indicate that the presence of color is dominant to its absence. Furthermore, epistasis must be involved for there to be a modified 9:3:3:1 ratio.

By looking at the F_1 of crosses 1–3, the following allelic dominance relationships can be determined: black > brown > yellow. Arbitrarily assign the following genotypes for homozygotes: B^l/B^l = black, B^r/B^r = brown, B^y/B^y = yellow.

By looking at the F_2 of crosses 4–6, a white phenotype is composed of two categories: the double homozygote and one class of the mixed homozygote/heterozygote. Let lack of color be caused by *c/c*. Color will therefore be *C/–*.

Parents	F$_1$	F$_2$
1 *Br/Br ; C/C × By/By ; C/C*	*Br/By ; C/C*	3 *Br/– ; C/C* : 1 *By/By ; C/C*
2 *Bl/Bl ; C/C × Br/Br ; C/C*	*Bl/Br ; C/C*	3 *Bl/– ; C/C* : 1 *Br/Br ; C/C*
3 *Bl/Bl ; C/C × By/By ; C/C*	*Bl/By ; C/C*	3 *Bl/– ; C/C* : 1 *By/By ; C/C*
4 *Bl/Bl ; c/c × By/By ; C/C*	*Bl/By ; C/c*	9 *Bl/– ; C/–* : 3 *By/By ; C/–* : 3 *Bl/– ; c/c* : 1 *By/By ; c/c*
5 *Bl/Bl ; c/c × Br/Br ; C/C*	*Bl/Br ; C/c*	9 *Bl/– ; C/–* : 3 *Br/Br ; C/–* : 3 *Bl/– ; c/c* : 1 *Br/Br ; c/c*
6 *Bl/Bl ; C/C × By/By ; c/c*	*Bl/By ; C/c*	9 *Bl/– ; C/–* : 3 *By/By ; C/–* : 3 *Bl/– ; c/c* : 1 *By/By ; c/c*

39. Two albinos marry and have four normal children. How is this possible?

Answer: It is possible to produce normally pigmented offspring from albino parents if albinism results from mutations in either of two different genes. If the cross is designated

A/A · b/b × a/a · B/B

then all the offspring would be

A/a · B/b

and they would have a pigmented phenotype because of complementation.

40. Consider the production of flower color in the Japanese morning glory (*Pharbitis nil*). Dominant alleles of either of two separate genes (*A/– · b/b* or *a/a · B/–*) produce purple petals. *A/– · B/–* produces blue petals, and *a/a · b/b* produces scarlet petals. Deduce the genotypes of parents and progeny in the following crosses:

Cross	Parents	Progeny
1	blue × scarlet	1/4 blue : 1/2 purple : 1/4 scarlet
2	purple × purple	1/4 blue : 1/2 purple : 1/4 scarlet
3	blue × blue	3/4 blue : 1/4 purple
4	blue × purple	3/8 blue : 4/8 purple : 1/8 scarlet
5	purple × scarlet	1/2 purple : 1/2 scarlet

Answer: The first step in each cross is to write as much of the genotype as possible from the phenotype.

Cross 1:   *A/– ; B/– × a/a ; b/b* → 1 *A/– ; B/–* : 2 *?/a; ?/b* : 1 *a/a ; b/b*

Because the double recessive appears, the blue parent must be *A/a* ; *B/b*. The 1/2 purple then must be *A/a* ; *b/b* and *a/a* ; *B/b*.

Cross 2:   *?/?* ; *?/?* × *?/?* ; *?/?*   → 1 *A/–* ; *B/–* : 2 *?/?* ; *?/?* : 1 *a/a* ; *b/b*

The two parents must be, in either order, *A/a* ; *b/b* and *a/a* ; *B/b*. The two purple progeny must be the same. The blue progeny are *A/a* ; *B/b*.

Cross 3:   *A/–* ; *B/–* × *A/–* ; *B/–*   → 3 *A/–* ; *B/–* : 1 *?/?* ; *?/?*

The only conclusions possible here are that one parent is either *A/A* or *B/B,* and the other parent is *B/b* if the first is *A/A* or *A/a* if the first is *B/B*.

Cross 4:   *A/–* ; *B/–* × *?/?* ; *?/?*   → 3 *A/–* ; *B/–* : 4 *?/?* ; *?/?* : 1 *a/a* ; *b/b*

The purple parent can be either *A/a* ; *b/b* or *a/a* ; *B/b* for this answer. Assume the purple parent is *A/a* ; *b/b*. The blue parent must be *A/a* ; *B/b*. The progeny are

3/4 *A/–* ×	1/2 *B/b* = 3/8 *A/–* ; *B/b*	blue
	1/2 *b/b* = 3/8 *A/–* ; *b/b*	purple
1/4 *a/a* ×	1/2 *B/b* = 1/8 *a/a* ; *B/b*	purple
	1/2 *b/b* = 1/8 *a/a* ; *b/b*	scarlet

Cross 5:   *A/–* ; *b/b* × *a/a* ; *b/b*   → 1 *A/–* ; *b/b* : 1 *a/a* ; *b/b*

As written this is a testcross for gene *A*. The purple parent and progeny are *A/a* ; *b/b*. Alternatively, the purple parent and progeny could be *a/a* ; *B/b*.

41. Corn breeders obtained pure lines whose kernels turn sun red, pink, scarlet, or orange when exposed to sunlight (normal kernels remain yellow in sunlight). Some crosses between these lines produced the following results. The phenotypes are abbreviated O, orange; P, pink; Sc, scarlet; and SR, sun red.

	Phenotypes		
Cross	Parents	F$_1$	F$_2$
1	SR × P	all SR	66 SR : 20 P
2	O × SR	all SR	998 SR : 314 O
3	O × P	all O	1300 O : 429 P
4	O × Sc	all Y	182 Y : 80 O : 58 Sc

Analyze the results of each cross, and provide a unifying hypothesis to account for all the results. (Explain all symbols that you use.)

Answer: The F_1 progeny of cross 1 indicate that sun red is dominant to pink. The F_2 progeny, which are approximately in a 3:1 ratio, support this. The same pattern is seen in crosses 2 and 3, with sun red dominant to orange and orange dominant to pink. Thus, we have a multiple allelic series with sun red > orange > pink. In all three crosses, the parents must be homozygous.

If c^{sr} = sun red, c^o = orange, and c^P = pink, then the crosses and the results are

Cross	Parents	F_1	F_2
1	$c^{sr}/c^{sr} \times c^P/c^P$	c^{sr}/c^P	3 $c^{sr}/-$: 1 c^P/c^P
2	$c^o/c^o \times c^{sr}/c^{sr}$	c^{sr}/c^o	3 $c^{sr}/-$: 1 c^o/c^o
3	$c^o/c^o \times c^P/c^P$	c^o/c^{cp}	3 $c^o/-$: 1 c^P/c^P

Cross 4 presents a new situation. The color of the F_1 differs from that of either parent, suggesting that two separate genes are involved. An alternative explanation is either codominance or incomplete dominance. If either codominance or incomplete dominance is involved, then the F_2 will appear in a 1:2:1 ratio. If two genes are involved, then a 9:3:3:1 ratio, or some variant of it, will be observed. Because the wild-type phenotype appears in the F_1 and F_2, it appears that complementation is occurring. This requires two genes. The progeny actually are in a 9:4:3 ratio. This means that two genes are involved and that there is epistasis. Furthermore, for three phenotypes to be present in the F_2, the two F_1 parents must have been heterozygous.

Let a stand for the scarlet gene and A for its colorless allele, and assume that there is a dominant allele, C, that blocks the expression of the alleles that we have been studying to this point.

Cross 4:   P c^o/c^o ; $A/A \times C/C$; a/a

    F_1 C/c^o ; A/a
    F_2 9 $C/-$; $A/-$ yellow
           3 $C/-$; a/a scarlet
           3 c^o/c^o ; $A/-$ orange
           1 c^o/co ; a/a orange (epistasis, with c^o blocking the expression of a/a)

42. Many kinds of wild animals have the agouti coloring pattern, in which each hair has a yellow band around it.

   a. Black mice and other black animals do not have the yellow band; each of their hairs is all black. This absence of wild agouti pattern is called *nonagouti*. When mice of a true-breeding agouti line are crossed with nonagoutis, the F_1 is all agouti and the F_2 has a 3 : 1 ratio of agoutis to nonagoutis. Diagram this cross, letting A represent the allele responsible for the agouti phenotype and *a*, nonagouti. Show the phenotypes and genotypes of the parents, their gametes, the F_1, their gametes, and the F_2.

b. Another inherited color deviation in mice substitutes brown for the black color in the wild-type hair. Such brown-agouti mice are called cinnamons. When wild-type mice are crossed with cinnamons, all of the F_1 are wild type and the F_2 has a 3 : 1 ratio of wild type to cinnamon. Diagram this cross as in part *a*, letting B stand for the wild-type black allele and b stand for the cinnamon brown allele.

c. When mice of a true-breeding cinnamon line are crossed with mice of a true-breeding nonagouti (black) line, all of the F_1 are wild type. Use a genetic diagram to explain this result.

d. In the F_2 of the cross in part c, a fourth color called *chocolate* appears in addition to the parental cinnamon and nonagouti and the wild type of the F_1. Chocolate mice have a solid, rich brown color. What is the genetic constitution of the chocolates?

e. Assuming that the *A/a* and *B/b* allelic pairs assort independently of each other, what do you expect to be the relative frequencies of the four color types in the F_2 described in part *d*? Diagram the cross of parts c and d, showing phenotypes and genotypes (including gametes).

f. What phenotypes would be observed in what proportions in the progeny of a backcross of F_1 mice from part c with the cinnamon parental stock? With the nonagouti (black) parental stock? Diagram these backcrosses.

g. Diagram a testcross for the F_1 of part c. What colors would result and in what proportions?

h. Albino (pink-eyed white) mice are homozygous for the recessive member of an allelic pair *C/c*, which assorts independently of the *A/a* and *B/b* pairs. Suppose that you have four different highly inbred (and therefore presumably homozygous) albino lines. You cross each of these lines with a true-breeding wild-type line, and you raise a large F_2 progeny from each cross. What genotypes for the albino lines can you deduce from the following F_2 phenotypes?

	Phenotypes of progeny				
F_2 of line	Wild type	Black	Cinnamon	Chocolate	Albino
1	87	0	32	0	39
2	62	0	0	0	18
3	96	30	0	0	41
4	287	86	92	29	164

(Problem 42 is adapted from A. M. Srb, R. D. Owen, and R. S. Edgar, *General Genetics,* 2nd ed. W. H. Freeman and Company, 1965.)

Answer:

a. P   *A/A* (agouti) × *a/a* (nonagouti)
    gametes   *A* and *a*

     F_1 *A/a* (agouti)
    gametes   *A* and *a*

     F_2 1 *A/A* (agouti) : 2 *A/a* (agouti) : 1 *a/a* (nonagouti)

b. P   *B/B* (wild type) × *b/b* (cinnamon)
    gametes   *B* and *b*

     F_1 *B/b* (wild type)
    gametes   *B* and *b*

     F_2 1 *B/B* (wild type) : 2 *B/b* (wild type) : 1 *b/b* (cinnamon)

c. P   *A/A* ; *b/b* (cinnamon or brown agouti) × *a/a* ; *B/B* (black nonagouti)
    gametes   *A* ; *b* and *a* ; *B*

     F_1 *A/a* ; *B/b* (wild type, or black agouti)

d.   9 *A/–* ; *B/–*   black agouti
    3 *a/a* ; *B/–*   black nonagouti
    3 *A/–*; *b/b*   cinnamon
    1 *a/a* ; *b/b*   chocolate

e. P   *A/A* ; *b/b* (cinnamon) × *a/a* ; *B/B* (black nonagouti)
    gametes   *A* ; *b* and *a* ; *B*

     F_1 *A/a* ; *B/b* (wild type)
    gametes   *A* ; *B*, *A* ; *b*, *a* ; *B*, and *a* ; *b*

     F_2 9   *A/–* ; *B/–*   wild type
              1 *A/A* ; *B/B*
              2 *A/a* ; *B/B*
              2 *A/A* ; *B/b*
              4 *A/a* ; *B/b*
      3   *a/a* ; *B/–*   black nonagouti
              1 *a/a* ; *B/B*
              2 *a/a* ; *B/b*
      3   *A/–* ; *b/b*   cinnamon
              1 *A/A* ; *b/b*

2 A/a ; b/b

   1   a/a ; b/b chocolate

f. P A/a ; $B/b \times A/A$; b/b A/a ; $B/b \times a/a$; B/B
         (wild type) (cinnamon)   (wild type) (black nonagouti)

    F_1 1/4 A/A ; B/b wild type   1/4 A/a ; B/B wild type
         1/4 A/a ; B/b wild type   1/4 A/a ; B/b wild type
         1/4 A/A ; b/b cinnamon   1/4 a/a ; B/B black nonagouti
         1/4 A/a ; b/b cinnamon   1/4 a/a ; B/b black nonagouti

g. P A/a ; $B/b \times a/a$; b/b
         (wild type) (chocolate)

    F_1 1/4 A/a ; B/b wild type
         1/4 A/a ; b/b cinnamon
         1/4 a/a ; B/b black nonagouti
         1/4 a/a ; b/b chocolate

h. To be albino, the mice must be c/c, but the genotype with regard to the A and B genes can be determined only by looking at the F_2 progeny.

Cross 1:   P c/c ; $?/?$; $?/? \times C/C$; A/A ; B/B

        F_1 C/c ; $A/-$; $B/-$

        F_2 87 wild type $C/-$; $A/-$; $B/-$
               32 cinnamon $C/-$; $A/-$; b/b
               39 albino c/c ; $?/?$; $?/?$

For cinnamon to appear in the F_2, the F_1 parents must be B/b. Because the wild type is B/B, the albino parent must have been b/b. Now the F_1 parent can be written C/c ; $A/-$; B/b. With such a cross, one-fourth of the progeny would be expected to be albino (c/c), which is what is observed. Three-fourths of the remaining progeny would be black, either agouti or nonagouti, and one-fourth would be either cinnamon, if agouti, or chocolate, if nonagouti. Because chocolate is not observed, the F_1 parent must not carry the allele for nonagouti. Therefore, the F_1 parent is A/A and the original albino must have been c/c ; A/A ; b/b.

Cross 2:   P c/c ; $?/?$; $?/? \times C/C$; A/A ; B/B

        F_1 C/c ; $A/-$; $B/-$

        F_2 62 wild type $C/-$; $A/-$; $B/-$
               18 albino c/c ; $?/?$; $?/?$

This is a 3:1 ratio, indicating that only one gene is heterozygous in the F_1. That gene must be C/c. Therefore, the albino parent must be c/c ; A/A ; B/B.

Cross 3:   P c/c ; $?/?$; $?/? \times C/C$; A/A ; B/B

F_1 C/c ; $A/-$; $B/-$

F_2 96 wild type $C/-$; $A/-$; $B/-$
          30 black $C/-$; a/a ; $B/-$
          41 albino c/c ; $?/?$; $?/?$

For a black nonagouti phenotype to appear in the F_2, the F_1 must have been heterozygous for the A gene. Therefore, its genotype can be written C/c ; A/a; $B/-$ and the albino parent must be c/c ; a/a ; $?/?$. Among the colored F_2 a 3:1 ratio is observed, indicating that only one of the two genes is heterozygous in the F_1. Therefore, the F_1 must be C/c ; A/a ; B/B and the albino parent must be c/c ; a/a ; B/B.

Cross 4:   P c/c ; $?/?$; $?/? \times C/C$; A/A ; B/B

F_1 C/c ; $A/-$; $B/-$

F_2 287 wild type $C/-$; $A/-$; $B/-$
          86 black $C/-$; a/a ; $B/-$
          92 cinnamon $C/-$; $A/-$; b/b
          29 chocolate $C/-$; a/a ; b/b
          164 albino c/c ; $?/?$; $?/?$

To get chocolate F_2 progeny the F_1 parent must be heterozygous for all genes and the albino parent must be c/c ; a/a ; b/b.

Unpacking the Problem

43. An allele A that is not lethal when homozygous causes rats to have yellow coats. The allele R of a separate gene that assorts independently produces a black coat. Together, A and R produce a grayish coat, whereas a and r produce a white coat. A gray male is crossed with a yellow female, and the F_1 is 3/8 yellow, 3/8 gray, 1/8 black, and 1/8 white. Determine the genotypes of the parents.
Answer: To solve this problem, first restate the information.

     $A/-$ yellow $A/-$; $R/-$ gray
     $R/-$ black a/a ; r/r white

The cross is gray × yellow, or $A/-$; $R/- \times A/-$; r/r. The F_1 progeny are

     3/8 yellow   1/8 black

3/8 gray   1/8 white

For white progeny, both parents must carry an *r* and an *a* allele. Now the cross can be rewritten as: *A/a* ; *R/r* × *A/a* ; *r/r*

44. The genotype *r/r* ; *p/p* gives fowl a single comb, *R/–* ; *P/–* gives a walnut comb, *r/r* ; *P/–* gives a pea comb, and *R/–* ; *p/p* gives a rose comb (see the illustrations). Assume independent assortment.

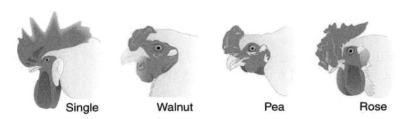

Single   Walnut   Pea   Rose

a. What comb types will appear in the F₁ and in the F₂ and in what proportions if single-combed birds are crossed with birds of a true-breeding walnut strain?

b. What are the genotypes of the parents in a walnut × rose mating from which the progeny are 3/8 rose, 3/8 walnut, 1/8 pea, and 1/8 single?

c. What are the genotypes of the parents in a walnut × rose mating from which all the progeny are walnut?

d. How many genotypes produce a walnut phenotype? Write them out.

Answer:

a. The stated cross is

P   single-combed (*r/r* ; *p/p*) × walnut-combed (*R/R* ; *P/P*)

   F₁   *R/r* ; *P/p*   walnut

   F₂   9 *R/–* ; *P/–*   walnut
           3 *r/r* ; *P/–*   pea
           3 *R/–* ; *p/p*   rose
           1 *r/r* ; *p/p*   single

b. The stated cross is

P   Walnut-combed × rose-combed

and the F₁ progeny are

Phenotypes		Possible genotypes
3/8	rose	*R/–* ; *p/p*
3/8	walnut	*R/–* ; *P/–*
1/8	pea	*r/r* ; *P/–*

$^1/_8$ single r/r ; p/p

The 3 $R/-$: 1 r/r ratio indicates that the parents were heterozygous for the R gene. The 1 $P/-$: 1 p/p ratio indicates a testcross for this gene. Therefore, the parents were R/r ; P/p and R/r ; p/p.

c. The stated cross is

P   walnut-combed × rose-combed

F_1 walnut ($R/-$; P/p)

To get this result, one of the parents must be homozygous R, but both need not be, and the walnut parent must be homozygous P/P.

d. The following genotypes produce the walnut phenotype:

R/R ; P/P, R/r ; P/P, R/R ; P/p, R/r ; P/p

45. The production of eye-color pigment in *Drosophila* requires the dominant allele A. The dominant allele P of a second independent gene turns the pigment to purple, but its recessive allele leaves it red. A fly producing no pigment has white eyes. Two pure lines were crossed with the following results:

P   red-eyed female × white-eyed male
↓

F_1 purple-eyed females
                red-eyed males
            F_1 × F_1
↓

F_2 both males and females: $\frac{3}{8}$ purple eyed
                    $\frac{3}{8}$ red eyed
                    $\frac{2}{8}$ white eyed

Explain this mode of inheritance and show the genotypes of the parents, the F_1, and the F_2.

Answer: Notice that the F_1 shows a difference in phenotype correlated with sex. At least one of the two genes is X-linked. The F_2 ratio suggests independent assortment between the two genes. Because purple is present in the F_1, the parental white-eyed male must have at least one P allele. The presence of white eyes in the F_2 suggests that the F_1 was heterozygous for pigment production, which means that the male also must carry the a allele. A start on the parental genotypes can now be made

P A/A ; $P/P \times a/-$; $P/-$,   where – could be either a Y chromosome or a second allele.

The question now is, which gene is X-linked? If the A gene is X-linked, the cross is

P A/A ; $p/p \times a/Y$; P/P

F$_1$ A/a ; $P/p \times A/Y$; P/p

All F$_2$ females will inherit the A allele from their father. Under this circumstance, no white-eyed females would be observed. Therefore, the A gene cannot be X-linked. The cross is

P A/A ; $p/p \times a/a$; P/Y

F$_1$ A/a ; P/p purple-eyed females
          A/a ; p/Y red-eyed males

F$_2$ Females   Males
          3/8 $A/-$; P/p purple   3/8 $A/-$; P/Y purple
          3/8 $A/-$; p/p red   3/8 $A/-$; p/Y red
          1/8 a/a ; P/p white   1/8 a/a ; P/Y white
          1/8 a/a ; p/p white   1/8 a/a ; p/Y white

46. When true-breeding brown dogs are mated with certain true-breeding white dogs, all the F$_1$ pups are white. The F$_2$ progeny from some F$_1 \times$ F$_1$ crosses were 118 white, 32 black, and 10 brown pups. What is the genetic basis for these results?

Answer: The results indicate that two independently assorting genes are involved (modified 9:3:3:1 ratio), with white blocking the expression of color by the other gene. The ratio of white : color pups is 3:1, indicating that the F$_1$ is heterozygous (W/w). Among colored dogs, the ratio is 3 black : 1 brown, indicating that black is dominant to brown, and the F$_1$ is heterozygous (B/b). The original brown dog is w/w ; b/b, and the original white dog is W/W ; B/B. The F$_1$ progeny are W/w ; B/b, and the F$_2$ progeny are: 12: 3: 1.

    9 $W/-$; $B/-$ white

    3 $W/-$; b/b white
    3 w/w ; $B/-$ black
    1 w/w ; b/b brown

47. Wild-type strains of the haploid fungus *Neurospora* can make their own tryptophan. An abnormal allele *td* renders the fungus incapable of making its own tryptophan. An individual of genotype *td* grows only when its medium supplies tryptophan. The allele *su* assorts independently of *td*; its only known effect is to suppress the *td* phenotype. Therefore, strains carrying both *td* and *su* do not require tryptophan for growth.

a. If a *td* ; *su* strain is crossed with a genotypically wild-type strain, what genotypes are expected in the progeny and in what proportions?

b. What will be the ratio of tryptophan-dependent to tryptophan-independent progeny in the cross of part *a*?

Answer:

a. The cross is

P   *td* ; *su* (wild type) × *td+* ; *su+* (wild type)

F_1 1 *td* ; *su*   wild type
       1 *td* ; *su+*   requires tryptophan
       1 *td+* ; *su+*   wild type
       1 *td+* ; *su*   wild type

b. 1 tryptophan–dependent : 3 tryptophan–independent

48. Mice of the genotypes *A/A* ; *B/B* ; *C/C* ; *D/D* ; *S/S* and *a/a* ; *b/b* ; *c/c* ; *d/d* ; *s/s* are crossed. The progeny are intercrossed. What phenotypes will be produced in the F_2 and in what proportions? (The allele symbols stand for the following: *A* = agouti, *a* = solid (nonagouti); *B* = black pigment, *b* = brown; *C* = pigmented, *c* = albino; *D* = nondilution, *d* = dilution (milky color); *S* = unspotted, *s* = pigmented spots on white background.)

Answer:
P   *A/A* ; *B/B* ; *C/C* ; *D/D* ; *S/S* × *a/a* ; *b/b* ; *c/c* ; *d/d* ; *s/s*

F_1 *A/a* ; *B/b* ; *C/c* ; *D/d* ; *S/s*

F_2 *A/a* ; *B/b* ; *C/c* ; *D/d* ; *S/s* × *A/a* ; *B/b* ; *C/c* ; *D/d* ; *S/s*

(3/4 *A/–*)(3/4 *B/–*)(3/4 *C/–*)(3/4 *D/–*)(3/4 *S/–*) = 243/1024	agouti
(3/4 *A/–*)(3/4 *B/–*)(3/4 *C/–*)(3/4 *D/–*)(1/4 *s/s*) = 81/1024	spotted agouti
(3/4 *A/–*)(3/4 *B/–*)(3/4 *C/–*)(1/4 *d/d*)(3/4 *S/–*) = 81/1024	dilute agouti
(3/4 *A/–*)(3/4 *B/–*)(3/4 *C/–*)(1/4 *d/d*)(1/4 *s/s*) = 27/1024	dilute spotted agouti
(3/4 *A/–*)(3/4 *B/–*)(1/4 *c/c*)(3/4 *D/–*)(3/4 *S/–*) = 81/1024	albino
(3/4 *A/–*)(3/4 *B/–*)(1/4 *c/c*)(3/4 *D/–*)(1/4 *s/s*) = 27/1024	albino
(3/4 *A/–*)(3/4 *B/–*)(1/4 *c/c*)(1/4 *d/d*)(3/4 *S/–*) = 27/1024	albino
(3/4 *A/–*)(3/4 *B/–*)(1/4 *c/c*)(1/4 *d/d*)(1/4 *s/s*) = 9/1024	albino
(3/4 *A/–*)(1/4 *b/b*)(3/4 *C/–*)(3/4 *D/–*)(3/4 *S/–*) = 81/1024	agouti brown

$(3/4 \ A/-)(1/4 \ b/b)(3/4 \ C/-)(3/4 \ D/-)(1/4 \ s/s) = 27/1024$ agouti brown spotted
$(3/4 \ A/-)(1/4 \ b/b)(3/4 \ C/-)(1/4 \ d/d)(3/4 \ S/-) = 27/1024$ dilute agouti brown
$(3/4 \ A/-)(1/4 \ b/b)(3/4 \ C/-)(1/4 \ d/d)(1/4 \ s/s) = 9/1024$ dilute spotted agouti
   brown

$(3/4 \ A/-)(1/4 \ b/b)(1/4 \ c/c)(3/4 \ D/-)(3/4 \ S/-) = 27/1024$ albino
$(3/4 \ A/-)(1/4 \ b/b)(1/4 \ c/c)(3/4 \ D/-)(1/4 \ s/s) = 9/1024$ albino
$(3/4 \ A/-)(1/4 \ b/b)(1/4 \ c/c)(1/4 \ d/d)(3/4 \ S/-) = 9/1024$ albino
$(3/4 \ A/-)(1/4 \ b/b)(1/4 \ c/c)(1/4 \ d/d)(1/4 \ s/s) = 3/1024$ albino
$(1/4 \ a/a)(3/4 \ B/-)(3/4 \ C/-)(3/4 \ D/-)(3/4 \ S/-) = 81/1024$ black
$(1/4 \ a/a)(3/4 \ B/-)(3/4 \ C/-)(3/4 \ D/-)(1/4 \ s/s) = 27/1024$ black spotted
$(1/4 \ a/a)(3/4 \ B/-)(3/4 \ C/-)(1/4 \ d/d)(3/4 \ S/-) = 27/1024$ dilute black
$(1/4 \ a/a)(3/4 \ B/-)(3/4 \ C/-)(1/4 \ d/d)(1/4 \ s/s) = 9/1024$ dilute spotted black
$(1/4 \ a/a)(3/4 \ B/-)(1/4 \ c/c)(3/4 \ D/-)(3/4 \ S/-) = 27/1024$ albino
$(1/4 \ a/a)(3/4 \ B/-)(1/4 \ c/c)(3/4 \ D/-)(1/4 \ s/s) = 9/1024$ albino
$(1/4 \ a/a)(3/4 \ B/-)(1/4 \ c/c)(1/4 \ d/d)(3/4 \ S/-) = 9/1024$ albino
$(1/4 \ a/a)(3/4 \ B/-)(1/4 \ c/c)(1/4 \ d/d)(1/4 \ s/s) = 3/1024$ albino
$(1/4 \ a/a)(1/4 \ b/b)(3/4 \ C/-)(3/4 \ D/-)(3/4 \ S/-) = 27/1024$ brown
$(1/4 \ a/a)(1/4 \ b/b)(3/4 \ C/-)(3/4 \ D/-)(1/4 \ s/s) = 9/1024$ brown spotted
$(1/4 \ a/a)(1/4 \ b/b)(3/4 \ C/-)(1/4 \ d/d)(3/4 \ S/-) = 9/1024$ dilute brown
$(1/4 \ a/a)(1/4 \ b/b)(3/4 \ C/-)(1/4 \ d/d)(1/4 \ s/s) = 3/1024$ dilute spotted brown
$(1/4 \ a/a)(1/4 \ b/b)(1/4 \ c/c)(3/4 \ D/-)(3/4 \ S/-) = 9/1024$ albino
$(1/4 \ a/a)(1/4 \ b/b)(1/4 \ c/c)(3/4 \ D/-)(1/4 \ s/s) = 3/1024$ albino
$(1/4 \ a/a)(1/4 \ b/b)(1/4 \ c/c)(1/4 \ d/d)(3/4 \ S/-) = 3/1024$ albino
$(1/4 \ a/a)(1/4 \ b/b)(1/4 \ c/c)(1/4 \ d/d)(1/4 \ s/s) = 1/1024$ albino

49. Consider the genotypes of two lines of chickens: the pure-line mottled Honduran is i/i ; D/D ; M/M ; W/W, and the pure-line leghorn is I/I ; d/d ; m/m ; w/w, where

I = white feathers, i = colored feathers
D = duplex comb, d = simplex comb
M = bearded, m = beardless
W = white skin, w = yellow skin

These four genes assort independently. Starting with these two pure lines, what is the fastest and most convenient way of generating a pure line that has colored feathers, has a simplex comb, is beardless, and has yellow skin? Make sure that you show

a. the breeding pedigree.
b. the genotype of each animal represented.
c. how many eggs to hatch in each cross, and why this number.
d. why your scheme is the fastest and the most convenient.

Answer:

a. and b. The two starting lines are *i/i* ; *D/D* ; *M/M* ; *W/W* and *I/I* ; *d/d* ; *m/m* ; *w/w*, and you are seeking *i/i* ; *d/d* ; *m/m* ; *w/w*. There are many ways to proceed, one of which follows below:

I: *i/i* ; *D/D* ; *M/M* ; *W/W* × *I/I* ; *d/d* ; *m/m* ; *w/w*

II: *I/i* ; *D/d* ; *M/m* ; *W/w* × *I/i* ; *D/d* ; *M/m* ; *W/w*

II: Select *i/i* ; *d/d* ; *m/m* ; *w/w*, which has a probability of $(1/4)4 = 1/256$.

c. and d. In the first cross, all progeny chickens will be of the desired genotype to proceed to the second cross. Therefore, the only problems are to be sure that progeny of both sexes are obtained for the second cross, which is relatively easy, and that enough females are obtained to make the time required for the desired genotype to appear feasible. Because chickens lay one to two eggs a day, the more females who are egg–laying, the faster the desired genotype will be obtained. In addition, it will be necessary to obtain a male and a female of the desired genotype in order to establish a pure breeding line.

Assume that each female will lay two eggs a day, that money is no problem, and that excess males cause no problems. By hatching 200 eggs from the first cross, approximately 100 females will be available for the second cross. These 100 females will produce 200 eggs each day. Thus, in one week a total of 1,400 eggs will be produced. Of these 1,400 eggs, there will be approximately 700 of each sex. At a probability of 1/256, the desired genotype should be achieved 2.7 times for each sex within that first week.

50.   The following pedigree is for a dominant phenotype governed by an autosomal allele. What does this pedigree suggest about the phenotype, and what can you deduce about the genotype of individual A?

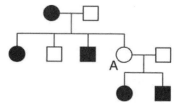

Answer: Pedigrees like this are quite common. They indicate lack of penetrance due to epistasis or environmental effects. Individual A must have the dominant autosomal gene, even though she does not express the trait, as both her children are affected.

51.   Petal coloration in foxgloves is determined by three genes. *M* encodes an enzyme that synthesizes anthocyanin, the purple pigment seen in these petals;

m/m produces no pigment, resulting in the phenotype albino with yellowish spots. *D* is an enhancer of anthocyanin, resulting in a darker pigment; *d/d* does not enhance. At the third locus, *w/w* allows pigment deposition in petals, but *W* prevents pigment deposition except in the spots and so results in the white, spotted phenotype. Consider the following two crosses:

Cross	Parents	Progeny
1	dark × white with purple yellowish spots	1/2 dark purple : 1/2 light purple
2	white with × light yellowish purple spots	1/2 white with purple spots : 1/ 4 dark purple : 1/4 light purple

In each case, give the genotypes of parents and progeny with respect to the three genes.

Answer: In cross 1, the following can be written immediately

P   *M/– ; D/– ; w/w* (dark purple) × *m/m ; ?/? ; ?/?* (white with yellowish spots)

F$_1$ 1/2 *M/– ; D/– ; w/w*   dark purple
         1/2 *M/– ; d/d ; w/w*   light purple

All progeny are colored, indicating that no *W* allele is present in the parents. Because the progeny are in a 1:1 ratio, only one of the genes in the parents is heterozygous. Also, the light purple progeny, *d/d*, indicates which gene that is. Therefore, the genotypes must be

P   *M/M ; D/d ; w/w* × *m/m ; d/d ; w/w*

F$_1$ 1/2 *M/m ; D/d ; w/w*
         1/2 *M/m ; d/d ; w/w*

In cross 2, the following can be written immediately

P   *m/m ; ?/? ; ?/?* (white with yellowish spots) × *M/– ; d/d ; w/w* (light purple)

F$_1$ 1/2 *M/– ; ?/? ; W/–*   white with purple spots
         1/4 *M/– ; D/– ; w/w*   dark purple
         1/4 *M/– ; d/d ; w/w*   light purple

For light and dark purple to appear in a 1:1 ratio among the colored plants, one of the parents must be heterozygous *D/d*. The ratio of white to colored is 1:1, a testcross, so one of the parents is heterozygous *W/w*. All plants are purple, indicating that one parent is homozygous *M/M*. Therefore, the genotypes are:

P m/m ; D/d ; W/w × M/M ; d/d ; w/w

F$_1$ 1/2 M/m ; $-/d$; W/w
1/4 M/m ; D/d ; w/w
1/4 M/m ; d/d ; w/w

52. In one species of *Drosophila,* the wings are normally round in shape, but you have obtained two pure lines, one of which has oval wings and the other sickle-shaped wings. Crosses between pure lines reveal the following results:

Parents		F$_1$	
Female	Male	Female	Male
sickle	round	sickle	sickle
round	sickle	sickle	round
sickle	oval	oval	sickle

a. Provide a genetic explanation of these results, defining all allele symbols.
b. If the F$_1$ oval females from cross 3 are crossed with the F$_1$ round males from cross 2, what phenotypic proportions are expected for each sex in the progeny?

Answer:
a. Note that the first two crosses are reciprocal and that the male offspring differ in phenotype between the two crosses. This indicates that the gene is on the X chromosome.

Also note that the F$_1$ females in the first two crosses are sickle. This indicates that sickle is dominant to round. The third cross also indicates that oval is dominant to sickle. Therefore, this is a multiple allelic series with oval > sickle > round.

Let W^o = oval, W^s = sickle, and W^r = round. The three crosses are:

Cross 1: W^s/W^s × W^r/Y → W^s/W^r and W^s/Y

Cross 2: W^r/W^r × W^s/Y → W^s/W^r and W^r/Y

Cross 3: W^s/W^s × W^o/Y → W^o/W^s and W^s/Y

b. W^o/W^s × W^r/Y

1/4 W^o/W^r female oval
1/4 W^s/W^r female sickle
1/4 W^o/Y male oval

$1/4 \; W^s/Y$ male sickle

53. Mice normally have one yellow band on each hair, but variants with two or three bands are known. A female mouse having one band was crossed with a male having three bands. (Neither animal was from a pure line.) The progeny were

Females   1/2 one band   Males   1/2 one band
             1/2 three bands   1/2 two bands

a. Provide a clear explanation of the inheritance of these phenotypes.
b. In accord with your model, what would be the outcome of a cross between a three-banded daughter and a one-banded son?

Answer:
a. First, note that there is a phenotypic difference between males and females, indicating X linkage. This means that the male progeny express both alleles of the female parent. A beginning statement of the genotypes could be as follows, where H indicates the gene and the numbers 1, 2, and 3 indicate the variants.

P $H^1/H^2 \times H^3/Y$

F_1 $1/4 \; H^1/H^3$ one-banded female
      $1/4 \; H^3/H^2$ three-banded female
      $1/4 \; H^1/Y$ one-banded male
      $1/4 \; H^2/Y$ two-banded male

Because both female F_1 progeny obtain allele H^3 from their father, yet only one has a three-banded pattern, there is obviously a dominance relationship among the alleles. The mother indicates that H^1 is dominant to H^2. The female progeny must be H^1/H^3 and H^3/H^2, and they have a one-banded and three-banded pattern, respectively. H^3 must be dominant to H^2 because the daughter with that combination of alleles has to be the three-banded daughter. In other words, there is a multiple–allelic series with $H^1 > H^3 > H^2$.

b. The cross is:

P $H^3/H^2 \times H^1/Y$

F_1 $1/4 \; H^1/H^3$ one-banded female
      $1/4 \; H^1/H^2$ one-banded female
      $1/4 \; H^2/Y$ two-banded male
      $1/4 \; H^3/Y$ three-banded male

54. In minks, wild types have an almost black coat. Breeders have developed many pure lines of color variants for the mink-coat industry. Two such pure lines are platinum (blue gray) and aleutian (steel gray). These lines were used in crosses, with the following results:

Cross	Parents	F₁	F₂
1	wild × platinum	wild	18 wild, 5 platinum
2	wild × aleutian	wild	27 wild, 10 aleutian
3	platinum × aleutian	wild	133 wild 41 platinum 46 aleutian 17 sapphire (new)

a. Devise a genetic explanation of these three crosses. Show genotypes for the parents, the F₁, and the F₂ in the three crosses, and make sure that you show the alleles of each gene that you hypothesize for every mink.

b. Predict the F₁ and F₂ phenotypic ratios from crossing sapphire with platinum and with aleutian pure lines.

Answer:

a. The first two crosses indicate that wild type is dominant to both platinum and aleutian. The third cross indicates that two genes are involved rather than one gene with multiple alleles because a 9:3:3:1 ratio is observed.

Let platinum be *a*, aleutian be *b*, and wild type be *A/–* ; *B/–*.

Cross 1: P   *A/A* ; *B/B* × *a/a* ; *B/B*   wild type × platinum
          F₁ *A/a* ; *B/B*   all wild type
          F₂ 3 *A/–* ; *B/B* : 1 *a/a* ; *B/B*   3 wild type : 1 platinum

Cross 2: P   *A/A* ; *B/B* × *A/A* ; *b/b*   wild type × aleutian
          F₁ *A/A* ; *B/b*   all wild type
          F₂ 3 *A/A* ; *B/–* : 1 *A/A* ; *b/b*   3 wild type : 1 aleutian

Cross 3: P   *a/a* ; *B/B* × *A/A* ; *b/b*   platinum × aleutian
          F₁ *A/a* ; *B/b*   all wild type
          F₂ 9 *A/–* ; *B/–*   wild type
              3 *A/–* ; *b/b*   aleutian
              3 *a/a* ; *B/–*   platinum
              1 *a/a* ; *b/b*   sapphire

b. sapphire × platinum   sapphire × aleutian
    P   *a/a* ; *b/b* × *a/a* ; *B/B*   *a/a* ; *b/b* × *A/A* ; *b/b*
    F₁   *a/a* ; *B/b*   platinum   *A/a* ; *b/b*   aleutian
    F₂   3 *a/a* ; *B/–* platinum   3 *A/–* ; *b/b* aleutian
         1 *a/a* ; *b/b* sapphire   1 *a/a* ; *b/b* sapphire

55. In *Drosophila,* an autosomal gene determines the shape of the hair, with *B* giving straight and *b* giving bent hairs. On another autosome, there is a gene of which a dominant allele *I* inhibits hair formation so that the fly is hairless (*i* has no known phenotypic effect).

   a. If a straight-haired fly from a pure line is crossed with a fly from a pure-breeding hairless line known to be an inhibited bent genotype, what will the genotypes and phenotypes of the F_1 and the F_2 be?

   b. What cross would give the ratio 4 hairless : 3 straight : 1 bent?

Answer:

a. The genotypes of such fruit flies are:

   P B/B ; i/i × b/b ; I/I

   F_1 B/b ; I/i hairless

   F_2 9 $B/–$; $I/–$ hairless
           3 $B/–$; i/i straight
           3 b/b ; $I/–$ hairless
           1 b/b ; i/i bent

b. In order to solve this problem, first write as much as you can of the progeny genotypes.

   4 hairless $–/–$; $I/–$
   3 straight $B/–$; i/i
   1 bent b/b ; i/i

Each parent must have a *b* allele and an *i* allele. The partial genotypes of the parents are:

$–/b$; $–/i$ × $–/b$; $–/i$

At least one parent carries the *B* allele, and at least one parent carries the *I* allele. Assume for a moment that the first parent carries both. The partial genotypes become:

B/b ; I/i × $–/b$; $–/i$

Note that 1/2 the progeny are hairless. This must come from a I/i × i/i cross. Of those progeny with hair, the ratio is 3:1, which must come from a B/b × B/b cross. The final genotypes are therefore:

B/b ; I/i × B/b ; i/i

56.   The following pedigree concerns eye phenotypes in *Tribolium* beetles. The solid symbols represent black eyes, the open symbols represent brown eyes, and the cross symbols (X) represent the "eyeless" phenotype, in which eyes are totally absent.

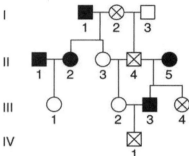

   a.   From these data, deduce the mode of inheritance of these three phenotypes.
   b.   Using defined gene symbols, show the genotype of beetle II-3.

Answer:

   a.   The first question is whether there are two genes or one gene with three alleles. Note that a black × eyeless cross produces black and brown progeny in one instance but black and eyeless progeny in the second instance. Further note that a black × black cross produces brown, a brown × eyeless cross produces brown, and a brown × black cross produces eyeless. The results are confusing enough that the best way to proceed is by trial and error.

   There are two possibilities: one gene or two genes.

   One gene. Assume one gene for a moment. Let the gene be *E* and assume the following designations

   E1 = black
   E2 = brown
   E3 = eyeless

   If you next assume, based on the various crosses, that black > brown > eyeless, genotypes in the pedigree become

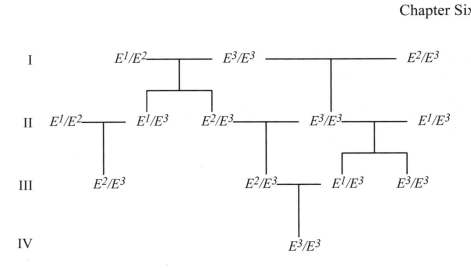

b. The genotype of individual II-3 is E^2/E^3.

Two genes. If two genes are assumed, then questions arise regarding whether both are autosomal or if one might be X-linked. Eye color appears to be autosomal. The presence or absence of eyes appears to be X-linked. Let

B = black $\qquad$ X^E = normal eyes
b = brown $\qquad$ X^e = eyeless

The pedigree then is

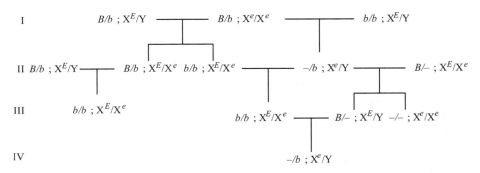

With this interpretation of the pedigree, individual II–3 is b/b ; X^E/X^e.

Without further data, it is impossible to choose scientifically between the two possible explanations.

57. A plant believed to be heterozygous for a pair of alleles B/b (where B encodes yellow and b encodes bronze) was selfed, and, in the progeny, there were 280 yellow and 120 bronze plants. Do these results support the hypothesis that the plant is B/b?

Answer: For the cross $B/b \times B/b$, you expect 3/4 $B/-$ and 1/4 b/b or for 400 progeny, 300 and 100, respectively. This should be compared to the actual data of 280 and 120. Alternatively, if the B/B genotype was actually lethal, then you would expect 2/3 B/b and 1/3 b/b or for 400 progeny, 267 and 133, respectively. This compares more favorably to the actual data. (The χ^2 value for the 3:1 hypothesis is 5.33 (reject). χ^2 value for the 2:1 hypothesis is 2.253 (accept).)

58. A plant thought to be heterozygous for two independently assorting genes (P/p ; Q/q) was selfed, and the progeny were

88 $P/-$; $Q/-$ 25 p/p ; $Q/-$
32 $P/-$; q/q 14 p/p ; q/q

Do these results support the hypothesis that the original plant was P/p ; Q/q?

Answer: There are a total of 159 progeny that should be distributed in 9:3:3:1 ratio if the two genes are assorting independently. You can see that

Observed	Expected
88 $P/-$; $Q/-$	90
32 $P/-$; q/q	30
25 p/p ; $Q/-$	30
14 p/p ; q/q	10

(For the 9:3:3:1 hypothesis, the χ^2 value is 2.61, accept.)

59. A plant of phenotype 1 was selfed, and, in the progeny, there were 100 plants of phenotype 1 and 60 plants of an alternative phenotype 2. Are these numbers compatible with expected ratios of 9 : 7, 13 : 3, and 3 : 1? Formulate a genetic hypothesis on the basis of your calculations.

Answer: For 160 progeny, there should be 90:70, 130:30, or 120:40 of phenotype 1 : phenotype 2 progeny for the various expected ratios. Only the 9:7 ratio seems feasible and that would indicate that two genes are assorting independently where

$A/-$; $B/-$ phenotype 1
$A/-$; b/b phenotype 2
a/a ; $B/-$ phenotype 2
a/a ; b/b phenotype 2

(χ^2 values are 2.53, accept, for 90:70 ratio; > 30, reject, for 130 to 30 ratio; and >10, reject, for 120 to 40 ratio.)

60. Four homozygous recessive mutant lines of *Drosophila melanogaster* (labeled 1 through 4) showed abnormal leg coordination, which made their walking highly erratic. These lines were intercrossed; the phenotypes of the F_1 flies are shown in the following grid, in which "+" represents wild-type walking and "–" represents abnormal walking:

	1	2	3	4
1	–	+	+	+
2	+	–	–	+
3	+	–	–	+
4	+	+	+	–

a. What type of test does this analysis represent?
b. How many different genes were mutated in creating these four lines?
c. Invent wild-type and mutant symbols and write out full genotypes for all four lines and for the F_1 flies.
d. Do these data tell us which genes are linked? If not, how could linkage be tested?
e. Do these data tell us the total number of genes taking part in leg coordination in this animal?

Answer:

a. Intercrossing mutant strains that all share a common recessive phenotype is the basis of the complementation test. This test is designed to identify the number of different genes that can mutate to a particular phenotype. If the progeny of a given cross still express the mutant phenotype, the mutations fail to complement and are considered alleles of the same gene; if the progeny are wild type, the mutations complement and the two strains carry mutant alleles of separate genes.

b. There are 3 genes represented in these crosses. All crosses except 2 × 3 (or 3 × 2) complement and indicate that the strains are mutant for separate genes. Strains 2 and 3 fail to complement and are mutant for the same gene.

c. Let A and a represent alleles of gene 1; B and b represent alleles of gene 4; and c^2, c^3, and C represent alleles of gene 3.

Line 1: a/a . B/B . C/C
Line 2: A/A . B/B . c^2/c^2
Line 3: A/A . B/B . c^3/c^3
Line 4: A/A . b/b . C/C

	Cross	Genotype	Phenotype
F_1s	1 × 2	A/a . B/B . C/c^2	wild type
	1 × 3	A/a . B/B . C/c^3	wild type

	1 × 4	A/a . B/b . C/C	wild type
	2 × 3	A/A . B/B . c^2/c^3	mutant
	2 × 4	A/A . B/b . C/c^2	wild type
	3 × 4	A/A . B/b . C/c^3	wild type

d. With the exception that strain 2 and 3 fail to complement and therefore have mutations in the same gene, this test does not give evidence of linkage. To test linkage, the F_1s should be crossed to tester strains (homozygous recessive strains) and segregation of the mutant phenotype followed. If the genes are unlinked, for example A/a ; B/b × a/a ; b/b, then 25 percent of the progeny will be wild type (A/a ; B/b) and 75 percent will be mutant (25 percent A/a ; b/b, 25 percent a/a ; B/b, and 25 percent a/a ; b/b). If the genes are linked (a B/a B × A b/A b) then only one half of the recombinants (i.e., less than 25 percent of the total progeny) will be wild type (A B/a b).

e. No. All it tells you is that among these strains, there are three genes represented. If genetic dissection of leg coordination was desired, large screens for the mutant phenotype would be executed with the attempt to "saturate" the genome with mutations in all genes involved in the process.

61. Three independently isolated tryptophan-requiring mutants of haploid yeast are called *trpB*, *trpD*, and *trpE*. Cell suspensions of each are streaked on a plate of nutritional medium supplemented with just enough tryptophan to permit weak growth for a *trp* strain. The streaks are arranged in a triangular pattern so that they do not touch one another. Luxuriant growth is noted at both ends of the *trpE* streak and at one end of the *trpD* streak (see the figure below).

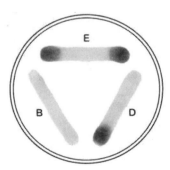

a. Do you think complementation has a role?
b. Briefly explain the pattern of luxuriant growth.
c. In what order in the tryptophan-synthesizing pathway are the enzymatic steps that are defective in mutants *trpB*, *trpD*, and *trpE*?

d. Why was it necessary to add a small amount of tryptophan to the medium to demonstrate such a growth pattern?

Answer:

a. Complementation refers to gene products within a cell, which is not what is happening here. Most likely, what is known as cross-feeding is occurring, whereby a product made by one strain diffuses to another strain and allows growth of the second strain. This is equivalent to supplementing the medium. Because cross-feeding seems to be taking place, the suggestion is that the strains are blocked at different points in the metabolic pathway.

b. For cross-feeding to occur, the growing strain must have a block that occurs earlier in the metabolic pathway than does the block in the strain from which it is obtaining the product for growth.

c. The *trpE* strain grows when cross-fed by either *trpD* or *trpB,* but the converse is not true (placing *trpE* earlier in the pathway than either *trpD* or *trpB*), and *trpD* grows when cross-fed by *trpB* (placing *trpD* prior to *trpB*). This suggests that the metabolic pathway is

$$trpE \rightarrow trpD \rightarrow trpB$$

d. Without some tryptophan, no growth at all would occur, and the cells would not have lived long enough to produce a product that could diffuse.

CHALLENGING PROBLEMS

62. A pure-breeding strain of squash that produced disk-shaped fruits (see the accompanying illustration) was crossed with a pure-breeding strain having long fruits. The F_1 had disk fruits, but the F_2 showed a new phenotype, sphere, and was composed of the following proportions:

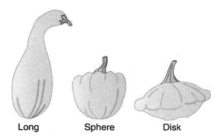

Long   Sphere   Disk

Propose an explanation for these results, and show the genotypes of the P, F_1, and F_2 generations. (Illustration from P. J. Russell, *Genetics,* 3rd ed. HarperCollins, 1992.)

Answer: Note that the F_2 are in an approximate 9:6:1 ratio. This suggests a dihybrid cross in which $A/-$; b/b has the same appearance as a/a ; $B/-$. Let the disc phenotype be the result of *A/–* ; *B/–* and the long phenotype be the result of *a/a* ; *b/b*. The crosses are:

P   *A/A* ; *B/B* (disc) × *a/a* ; *b/b* (long)

F$_1$ *A/a* ; *B/b* (disc)

F$_2$ 9   *A/–* ; *B/–* (disc)
      3   *a/a* ; *B/–* (sphere)
      3   *A/–* ; *b/b* (sphere)
      1   *a/a* ; *b/b* (long)

63. Marfan's syndrome is a disorder of the fibrous connective tissue, characterized by many symptoms, including long, thin digits; eye defects; heart disease; and long limbs. (Flo Hyman, the American volleyball star, suffered from Marfan's syndrome. She died from a ruptured aorta.)

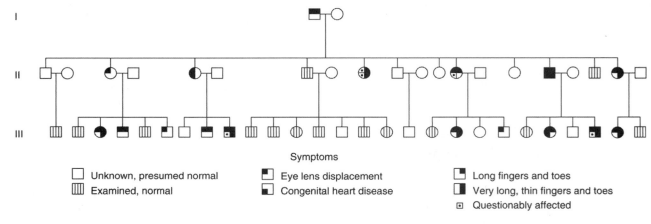

Symptoms

☐ Unknown, presumed normal	◪ Eye lens displacement	◨ Long fingers and toes
▥ Examined, normal	◪ Congenital heart disease	◨ Very long, thin fingers and toes
		⊡ Questionably affected

a. Use the pedigree above to propose a mode of inheritance for Marfan's syndrome.
b. What genetic phenomenon is shown by this pedigree?
c. Speculate on a reason for such a phenomenon.

(Illustration from J. V. Neel and W. J. Schull, *Human Heredity*. University of Chicago Press, 1954.)

Answer:
a. The best explanation is that Marfan's syndrome is inherited as a dominant autosomal trait, because roughly half of the children of all affected individuals also express the trait. If it were recessive, then all individuals marrying affected spouses would have to be heterozygous for an allele that, when homozygous, causes Marfan's.

b. The pedigree shows both pleiotropy (multiple affected traits) and variable expressivity (variable degree of expressed phenotype). Penetrance is the percentage of individuals with a specific genotype who express the associated phenotype. There is no evidence of decreased penetrance in this pedigree.

c. Pleiotropy indicates that the gene product is required in a number of different tissues, organs, or processes. When the gene is mutant, all tissues needing the gene product will be affected. Variable expressivity of a phenotype for a given genotype indicates modification by one or more other genes, random noise, or environmental effects.

64. In corn, three dominant alleles, called *A, C,* and *R*, must be present to produce colored seeds. Genotype *A/–* ; *C/–* ; *R/–* is colored; all others are colorless. A colored plant is crossed with three tester plants of known genotype. With tester *a/a* ; *c/c* ; *R/R*, the colored plant produces 50 percent colored seeds; with *a/a* ; *C/C* ; *r/r*, it produces 25 percent colored; and with *A/A* ; *c/c* ; *r/r*, it produces 50 percent colored. What is the genotype of the colored plant?

Answer:

Cross	Results	Conclusion
A/– ; *C/–* ; *R/–* × *a/a* ; *c/c* ; *R/R*	50% colored	Colored or white will depend on the *A* and *C* genes. Because half the seeds are colored, one of the two genes is heterozygous.
A/– ; *C/–* ; *R/–* × *a/a* ; *C/C* ; *r/r*	25% colored	Color depends on *A* and *R* in this cross. If only one gene were heterozygous, 50% would be colored. Therefore, both *A* and *R* are heterozygous. The colored plant is *A/a* ; *C/C*; *R/r*.
A/– ; *C/–* ; *R/–* × *A/A* ; *c/c* ; *r/r*	50% colored	This supports the above conclusion.

65. The production of pigment in the outer layer of seeds of corn requires each of the three independently assorting genes *A, C,* and *R* to be represented by at least one dominant allele, as specified in Problem 64. The dominant allele *Pr* of a fourth independently assorting gene is required to convert the biochemical precursor into a purple pigment, and its recessive allele *pr* makes the pigment red. Plants that do not produce pigment have yellow seeds. Consider a cross of a strain of genotype *A/A* ; *C/C* ; *R/R* ; *pr/pr* with a strain of genotype *a/a* ; *c/c* ; *r/r* ; *Pr/Pr*.

a. What are the phenotypes of the parents?
b. What will be the phenotype of the F_1?
c. What phenotypes, and in what proportions, will appear in the progeny of a

selfed F$_1$?

d. What progeny proportions do you predict from the testcross of an F$_1$?

Answer:

a. The *A/A* ; *C/C* ; *R/R* ; *pr/pr* parent produces pigment that is not converted to purple. The phenotype is red. The *a/a* ; *c/c* ; *r/r* ; *Pr/Pr* does not produce pigment. The phenotype is yellow.

b. The F$_1$ will be *A/a* ; *C/c* ; *R/r* ; *Pr/pr*, which will produce pigment. The pigment will be converted to purple.

c. The difficult way to determine the phenotypic ratios is to do a branch diagram, yielding the following results:

81/256 *A/–* ; *C/–* ; *R/–* ; *Pr/–*	purple	9/256 *a/a* ; *C/–* ; *R/–* ; *Pr/–* yellow
27/256 *A/–* ; *C/–* ; *R/–* ; *pr/pr*	red	9/256 *a/a* ; *C/–* ; *R/–* ; *pr/pr* yellow
27/256 *A/–* ; *C/–* ; *r/r* ; *Pr/–*	yellow	9/256 *a/a* ; *C/–* ; *r/r* ; *Pr/–* yellow
9/256 *A/–* ; *C/–* ; *r/r* ; *pr/pr*	yellow	3/256 *a/a* ; *C/–* ; *r/r* ; *pr/pr* yellow
27/256 *A/–* ; *c/c* ; *R/–* ; *Pr/–*	yellow	27/256 *a/a* ; *c/c* ; *R/–* ; *Pr/–* yellow
9/256 *A/–* ; *c/c* ; *R/–* ; *pr/pr*	yellow	3/256 *a/a* ; *c/c* ; *R/–* ; *pr/pr* yellow
9/256 *A/–* ; *c/c* ; *r/r* ; *Pr/–*	yellow	3/256 *a/a* ; *c/c* ; *r/r* ; *Pr/–* yellow
3/256 *A/–* ; *c/c* ; *r/r* ; *pr/pr*	yellow	1/256 *a/a* ; *c/c* ; *r/r* ; *pr/pr* yellow

The final phenotypic ratio is 81 purple : 27 red : 148 yellow.

The easier method of determining phenotypic ratios is to recognize that four genes are involved in a heterozygous × heterozygous cross. Purple requires a dominant allele for each gene. The probability of all dominant alleles is $(3/4)4 = 81/256$. Red results from all dominant alleles except for the *Pr* gene. The probability of that outcome is $(3/4)3(1/4) = 27/256$. The remainder of the outcomes will produce no pigment, resulting in yellow. That probability is $1 - 81/256 - 27/256 = 148/256$.

d. The cross is *A/a* ; *C/c* ; *R/r* ; *Pr/pr* × *a/a* ; *c/c* ; *r/r* ; *pr/pr*. Again, either a branch diagram or the easier method can be used. The final probabilities are

purple = $(1/2)4 = 1/16$
red = $(1/2)3(1/2) = 1/16$
yellow = $1 - 1/16 - 1/16 = 7/8$

66. The allele *B* gives mice a black coat, and *b* gives a brown one. The genotype *e/e* of another, independently assorting gene prevents the expression of *B* and *b*, making the coat color beige, whereas *E/–* permits the expression of *B* and *b*. Both genes are autosomal. In the following pedigree, black symbols indicate a black coat, pink symbols indicate brown, and white symbols indicate beige.

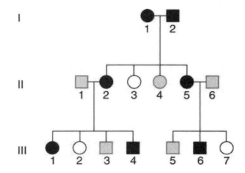

a. What is the name given to the type of gene interaction in this example?

b. What are the genotypes of the individual mice in the pedigree? (If there are alternative possibilities, state them.)

Answer:

a. This type of gene interaction is called *epistasis*. The phenotype of *e/e* is epistatic to the phenotypes of *B/–* or *b/b*.

b. The progeny of generation I have all possible phenotypes. Progeny II-3 is beige (*e/e*) so both parents must be heterozygous *E/e*. Progeny II-4 is brown (*b/b*) so both parents must also be heterozygous *B/b*. Progeny III-3 and III-5 are brown so II-2 and II-5 must be *B/b*. Progeny III-2 and III-7 are beige (*e/e*), so all their parents must be *E/e*.

The following are the inferred genotypes:

I	1 (*B/b E/e*)	2 (*B/b E/e*)			
II	1 (*b/b E/e*)	2 (*B/b E/e*)	3 (*–/– e/e*)	4 (*b/b E/–*)	5 (*B/b E/e*)
	6 (*b/b E/e*)				
III	1 (*B/b E/–*)	2 (*–/b e/e*)	3 (*b/b, E/–*)	4 (*B/b E/–*)	5 (*b/b E/–*)
	6 (*B/b E/–*)	7 (*–/b e/e*)			

67. A researcher crosses two white-flowered lines of *Antirrhinum* plants as follows and obtains the following results:

pure line 1 × pure line 2
↓
F_1 all white
F_1 × F_1
↓
F_2 131 white
       29 red

a. Deduce the inheritance of these phenotypes; use clearly defined gene symbols. Give the genotypes of the parents, F_1, and F_2.

b. Predict the outcome of crosses of the F_1 with each parental line.

Answer:

a. This is a dihybrid cross resulting in a 13 white : 3 red ratio of progeny in the F_2. This ratio of white to red indicates that the double recessive is not the red phenotype. Instead, the general formula for color is represented by *a/a* ; *B/–*.

Let line 1 be *A/A* ; *B/B* and line 2 be *a/a* ; *b/b*. The F_1 is *A/a* ; *B/b*. Assume that *A* blocks color in line 1 and *b/b* blocks color in line 2. The F_1 will be white because of the presence of *A*. The F_2 would be

9 *A/–* ; *B/–*	white
3 *A/–* ; *b/b*	white
3 *a/a* ; *B/–*	red
1 *a/a* ; *b/b*	white

b. Cross 1:   *A/A* ; *B/B* × *A/a* ; *B/b* → all *A/–* ; *B/–* white

Cross 2:   *a/a* ; *b/b* × *A/a* ; *B/b*

1/4 *A/a* ; *B/b*	white
1/4 *A/a* ; *b/b*	white
1/4 *a/a* ; *b/b*	white
1/4 *a/a* ; *B/b*	red

68. Assume that two pigments, red and blue, mix to give the normal purple color of petunia petals. Separate biochemical pathways synthesize the two pigments, as shown in the top two rows of the accompanying diagram. "White" refers to compounds that are not pigments. (Total lack of pigment results in a white petal.) Red pigment forms from a yellow intermediate that is normally at a concentration too low to color petals.

pathway I $\cdots\longrightarrow$ white$_1$ $\xrightarrow{\text{E}}$ blue

pathway II $\cdots\longrightarrow$ white$_2$ $\xrightarrow{\text{A}}$ yellow $\xrightarrow{\text{B}}$ red

C

pathway III $\cdots\longrightarrow$ white$_3$ $\xrightarrow{\text{D}}$ white$_4$

A third pathway, whose compounds do not contribute pigment to petals, normally does not affect the blue and red pathways, but, if one of its intermediates (white$_3$) should build up in concentration, it can be converted into the yellow intermediate of the red pathway.

In the diagram, the letters A through E represent enzymes; their corresponding genes, all of which are unlinked, may be symbolized by the same letters.

Assume that wild-type alleles are dominant and encode enzyme function and that recessive alleles result in a lack of enzyme function. Deduce which combinations of true-breeding parental genotypes could be crossed to produce F_2 progeny in the following ratios:

a. 9 purple : 3 green : 4 blue
b. 9 purple : 3 red : 3 blue : 1 white
c. 13 purple : 3 blue
d. 9 purple : 3 red : 3 green : 1 yellow

(**Note:** Blue mixed with yellow makes green; assume that no mutations are lethal.)

Answer:
a. Note that blue is always present, indicating E/E (blue) in both parents. Because of the ratios that are observed, neither C nor D is varying. In this case, the gene pairs that are involved are A/a and B/b. The parents are A/A ; $b/b \times a/a$; B/B or A/A ; $B/B \times a/a$; b/b.

The F_1 are A/a ; B/b and the F_2 are:

9	$A/-$; $B/-$	blue + red, or purple
3	$A/-$; b/b	blue + yellow, or green
3	a/a ; $B/-$	blue + white$_2$, or blue
1	a/a ; b/b	blue + white$_2$, or blue

b. Blue is not always present, indicating E/e in the F_1. Because green never appears, the F_1 must be B/B ; C/C ; D/D. The parents are A/A ; $e/e \times a/a$; E/E or A/A ; $E/E \times a/a$; e/e.

The F_1 are A/a ; E/e, and the F_2 are:

9	$A/-$; $E/-$	red + blue, or purple
3	$A/-$; e/e	red + white$_1$, or red
3	a/a ; $E/-$	white$_2$ + blue, or blue
1	a/a ; e/e	white$_2$ + white$_1$, or white

c. Blue is always present, indicating that the F_1 is E/E. No green appears, indicating that the F_1 is also B/B. The two genes involved are A and D. The parents are A/A ; $d/d \times a/a$; D/D or A/A ; $D/D \times a/a$; d/d.

The F_1 are A/a ; D/d and the F_2 are:

9	$A/-$; $D/-$	blue + red + white$_4$, or purple
3	$A/-$; d/d	blue + red, or purple
3	a/a ; $D/-$	blue + white$_2$ + white$_4$, or blue
1	a/a ; d/d	white$_2$ + blue + red, or purple

d. The presence of yellow indicates b/b ; e/e in the F_2. Therefore, the parents are B/B ; e/e × b/b ; E/E or B/B ; E/E × b/b ; e/e.

The F_1 are B/b ; E/e and the F_2 are

9	$B/-$; $E/-$	red + blue, or purple
3	$B/-$; e/e	red + white$_1$, or red
3	b/b ; $E/-$	yellow + blue, or green
1	b/b ; e/e	yellow + white$_1$, or yellow

e. Mutations in D suppress mutations in A.

f. Recessive alleles of A will be epistatic to mutations in B.

69. The flowers of nasturtiums (*Tropaeolum majus*) may be single (S), double (D), or superdouble (Sd). Superdoubles are female sterile; they originated from a double-flowered variety. Crosses between varieties gave the progeny listed in the following table, in which pure means "pure breeding."

Cross	Parents	Progeny
1	pure S × pure D	All S
2	cross 1 F_1 × cross 1 F_1	78 S : 27 D
3	pure D × Sd	112 Sd : 108 D
4	pure S × Sd	8 Sd : 7 S
5	pure D × cross 4 Sd progeny	18 Sd : 19 S
6	pure D × cross 4 S progeny	14 D : 16 S

Using your own genetic symbols, propose an explanation for these results, showing

a. all the genotypes in each of the six rows.
b. the proposed origin of the superdouble.

Answer:
a. Note that cross 1 suggests that one gene is involved and that single is dominant to double. Cross 2 supports this conclusion. Now, note that in crosses 3 and 4, a 1:1 ratio is seen in the progeny, suggesting that superdouble is an allele of both single and double. Superdouble must be heterozygous, however, and it must be dominant to both single and double. Because the heterozygous superdouble yields both single and double when crossed with the appropriate plant, it cannot be heterozygous for the single allele. Therefore, it must be heterozygous for the double allele. A multiple allelic series has been detected: superdouble > single > double.

For now, assume that only one gene is involved and attempt to rationalize the crosses with the assumptions made above.

Cross	Parents	Progeny	Conclusion
1	$A^S/A^S \times A^D/A^D$	A^S/A^D	A^S is dominant to A^D
2	$A^S/A^D \times A^S A^D$	$3 A^S/- : 1 A^D/A^D$	supports above conclusion
3	A^D/A^D $\times A^{Sd}/A^D$	$1 A^{Sd}/A^D : 1 A^D/A^D$	A^{Sd} is dominant to A^D
4	A^S/AS $\times A^{Sd}/A^D$	$1 A^{Sd}/A^S : 1 A^S/A^D$	A^{Sd} is dominant to A^S
5	A^D/A^D $\times A^{Sd}/A^S$	$1 A^{Sd}/A^D : 1 A^D/A^S$	supports conclusion of heterozygous superdouble
6	$A^D/A^D \times A^S/A^D$	$1 A^D/A^D : 1 A^D/A^S$	supports conclusion of heterozygous superdouble

b. While this explanation does rationalize all the crosses, it does not take into account either the female sterility or the origin of the superdouble plant from a double-flowered variety.

A number of genetic mechanisms could be proposed to explain the origin of superdouble from the double-flowered variety. Most of the mechanisms will be discussed in later chapters and so will not be mentioned here. However, it can be safely assumed at this point that, whatever the mechanism, it was aberrant enough to block the proper formation of the complex structure of the female flower. Because of female sterility, no homozygote for superdouble can be observed.

70. In a certain species of fly, the normal eye color is red (R). Four abnormal phenotypes for eye color were found: two were yellow (Y1 and Y2), one was brown (B), and one was orange (O). A pure line was established for each phenotype, and all possible combinations of the pure lines were crossed. Flies of each F_1 were intercrossed to produce an F_2. The F_1 and the F_2 flies are shown within the following square; the pure lines are given at the top and at the left-hand side.

		Y1	Y2	B	O
	F_1	all Y	all R	all R	all R
Y1	F_2	all Y	9 R	9 R	9 R
			7 Y	4 Y	4 O
				3 B	3 Y
	F_1		all Y	all R	all R
Y2	F_2		all Y	9 R	9 R
				4 Y	4 Y
				3 B	3 O

	F$_1$			all B	all R
B	F$_2$			all B	9 R
					4 O
					3 B
	F$_1$				all O
O	F$_2$				all O

a. Define your own symbols and list the genotypes of all four pure lines.

b. Show how the F$_1$ phenotypes and the F$_2$ ratios are produced.

c. Show a biochemical pathway that explains the genetic results, indicating which gene controls which enzyme.

Answer:

a. All the crosses suggest two independently assorting genes. However, that does not mean that there are a total of only two genes governing eye color. In fact, there are four genes controlling eye color that are being studied here. Let

a/a = defect in the yellow line 1
b/b = defect in the yellow line 2
d/d = defect in the brown line
e/e = defect in the orange line

The genotypes of each line are as follows:

yellow 1: a/a ; B/B ; D/D ; E/E
yellow 2: A/A ; b/b ; D/D ; E/E
brown: A/A ; B/B ; d/d ; E/E
orange: A/A ; B/B ; D/D ; e/e

b. P a/a ; B/B ; D/D ; E/E × A/A ; b/b ; D/D ; E/E yellow 1 × yellow 2

F$_1$ A/a ; B/b ; D/D ; E/E red

F$_2$ 9 $A/-$; $B/-$; D/D ; E/E red
3 a/a ; $B/-$; D/D ; E/E yellow
3 $A/-$; b/b ; D/D ; E/E yellow
1 a/a ; b/b ; D/D ; E/E yellow

P a/a ; B/B ; D/D ; E/E × A/A ; B/B ; d/d ; E/E yellow 1 × brown

F$_1$ A/a ; B/B ; D/d ; E/E red

F$_2$ 9 $A/-$; B/B ; $D/-$; E/E red
3 a/a ; B/B ; $D/-$; E/E yellow
3 $A/-$; B/B ; d/d ; E/E brown
1 a/a ; B/B ; d/d ; E/E yellow

P a/a ; B/B ; D/D ; E/E × A/A ; B/B ; D/D ; e/e yellow 1 × orange

F$_1$ A/a ; B/B ; D/D ; E/e red

F$_2$ 9 $A/-$; B/B ; D/D ; $E/-$ red
     3 a/a ; B/B ; D/D ; $E/-$ yellow
     3 $A/-$; B/B ; D/D ; e/e orange
     1 a/a ; B/B ; D/D ; e/e orange

P A/A ; b/b ; D/D ; E/E × A/A ; B/B ; d/d ; E/E yellow 2 × brown

F$_1$ A/A ; B/b ; D/d ; E/E red

F$_2$ 9 A/A ; $B/-$; $D/-$; E/E red
     3 A/A ; b/b ; $D/-$; E/E yellow
     3 A/A ; $B/-$; d/d ; E/E brown
     1 A/A ; b/b ; d/d ; E/E yellow

P A/A ; b/b ; D/D ; E/E × A/A ; B/B ; D/D ; e/e yellow 2 × orange

F$_1$ A/A ; B/b ; D/D ; E/e red

F$_2$ 9 A/A ; B/– ; D/D ; $E/-$ red
     3 A/A ; b/b ; D/D ; $E/-$ yellow
     3 A/A ; $B/-$; D/D ; e/e orange
     1 A/A ; b/b ; D/D ; e/e yellow

P A/A ; B/B ; d/d ; E/E × A/A ; B/B ; D/D ; e/e brown × orange

F$_1$ A/A ; B/B ; D/d ; E/e red

F$_2$ 9 A/A ; B/B ; $D/-$; $E/-$ red
     3 A/A ; B/B ; d/d ; $E/-$ brown
     3 A/A ; B/B ; $D/-$; e/e orange
     1 A/A ; B/B ; d/d ; e/e orange

c. When constructing a biochemical pathway, remember that the earliest gene that is defective in a pathway will determine the phenotype of a doubly defective genotype. Look at the following doubly recessive homozygotes from the crosses. Notice that the double defect d/d ; e/e has the same phenotype as the defect a/a ; e/e. This suggests that the E gene functions earlier than do the A and D genes. Using this logic, the following table can be constructed:

Genotype	Phenotype	Conclusion
a/a ; *B/B* ; *D/D* ; *e/e*	orange	*E* functions before *A*
A/A ; *B/B* ; *d/d* ; *e/e*	orange	*E* functions before *D*
a/a ; *b/b* ; *D/D* ; *E/E*	yellow	*B* functions before *A*
A/A ; *b/b* ; *D/D* ; *e/e*	yellow	*B* functions before *E*
a/a ; *B/B* ; *d/d* ; *E/E*	yellow	*A* functions before *D*
A/A ; *b/b* ; *d/d* ; *E/E*	yellow	*B* functions before *D*

The genes function in the following sequence: *B, E, A, D*. The metabolic path is:

yellow 2 $\rightarrow$ orange $\rightarrow$ yellow $\rightarrow$ brown $\rightarrow$ red

     B   *E*   *A*   *D*

71. In common wheat, *Triticum aestivum*, kernel color is determined by multiply duplicated genes, each with an *R* and an *r* allele. Any number of *R* alleles will give red, and a complete lack of *R* alleles will give the white phenotype. In one cross between a red pure line and a white pure line, the F_2 was 63/64 red and 1/64 white.

    a. How many *R* genes are segregating in this system?
    b. Show the genotypes of the parents, the F_1, and the F_2.
    c. Different F_2 plants are backcrossed with the white parent. Give examples of genotypes that would give the following progeny ratios in such backcrosses: (1) 1 red : 1 white, (2) 3 red : 1 white, (3) 7 red : 1 white.
    d. What is the formula that generally relates the number of segregating genes to the proportion of red individuals in the F_2 in such systems?

Answer:
    a. A trihybrid cross would give a 63:1 ratio. Therefore, there are three *R* loci segregating in this cross.

    b. P R_1/R_1 ; R_2/R_2 ; R_3/R_3 $\times$ r_1/r_1 ; r_2/r_2 ; r_3/r_3

        F_1 R_1/r_1 ; R_2/r_2 ; R_3/r_3

F_2	27	$R_1/-$; $R_2/-$; $R_3/-$	red
	9	$R_1/-$; $R_2/-$; r_3/r_3	red
	9	$R_1/-$; r_2/r_2 ; $R_3/-$	red
	9	r_1/r_1 ; $R_2/-$; $R_3/-$	red
	3	$R_1/-$; r_2/r_2 ; r_3/r_3	red
	3	$r1/r_1$; $R_2/-$; r_3/r_3	red
	3	r_1/r_1 ; r_2/r_2 ; $R_3/-$	red
	1	r_1/r_1 ; r_2/r_2 ; r_3/r_3	white

c. (1) In order to obtain a 1:1 ratio, only one of the genes can be heterozygous. A representative cross would be $R_1/r_1 ; r_2/r_2 ; r_3/r_3 \times r_1/r_1 ; r_2/r_2 ; r_3/r_3$.

(2) In order to obtain a 3 red : 1 white ratio, two alleles must be segregating, and they cannot be within the same gene. A representative cross would be $R_1/r_1 ; R_2/r_2 ; r_3/r_3 \times r_1/r_1 ; r_2/r_2 ; r_3/r_3$.

(3) In order to obtain a 7 red : 1 white ratio, three alleles must be segregating, and they cannot be within the same gene. The cross would be $R_1/r_1 ; R_2/r_2 ; R_3/r_3 \times r_1/r_1 ; r_2/r_2 ; r_3/r_3$.

d. The formula is $1 - (1/4)n$, where n = the number of loci that are segregating in the representative crosses above.

72. The following pedigree shows the inheritance of deaf-mutism.

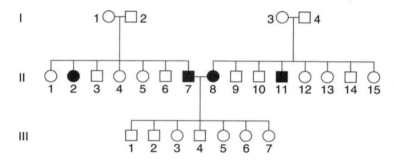

a. Provide an explanation for the inheritance of this rare condition in the two families in generations I and II, showing the genotypes of as many persons as possible; use symbols of your own choosing.
b. Provide an explanation for the production of only normal persons in generation III, making sure that your explanation is compatible with the answer to part *a*.

Answer:
a. The trait is recessive (parents without the trait have children with the trait) and autosomal (daughters can inherit the trait from unaffected fathers). Looking to generation III, there is also evidence that there are two different genes that, when defective, result in deaf-mutism.

Assuming that one gene has alleles A and a, and the other has B and b, the following genotypes can be inferred:

I-1 and I-2	*A/a ; B/B*	I-3 and I-4	*A/A ; B/b*
II-(1, 3, 4, 5, 6)	*A/– ; B/B*	II- (9, 10, 12, 13, 14, 15)	*A/A ; B/–*
II-2 and II-7	*a/a ; B/B*	I-8 and II-11	*A/A ; b/b*

b. Generation III shows complementation. All are *A/a ; B/b*.

73. The pedigree below is for blue sclera (bluish thin outer wall of the eye) and brittle bones.

  a. Are these two abnormalities caused by the same gene or by separate genes? State your reasons clearly.
  b. Is the gene (or genes) autosomal or sex-linked?
  c. Does the pedigree show any evidence of incomplete penetrance or expressivity? If so, make the best calculations that you can of these measures.

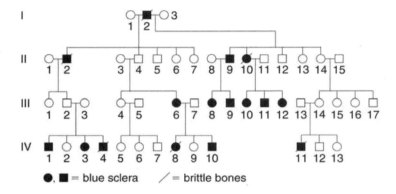

●, ■ = blue sclera   ╱ = brittle bones

Answer:

a. The first impression from the pedigree is that the gene causing blue sclera and brittle bones is pleiotropic with variable expressivity. If two genes were involved, it would be highly unlikely that all people with brittle bones also had blue sclera.

b. Sons and daughters inherit from affected fathers so the allele appears to be autosomal.

c. The trait appears to be inherited as a dominant but with incomplete penetrance. For the trait to be recessive, many of the non-related individuals marrying into the pedigree would have to be heterozygous (e.g., I-1, I-3, II-8, II-11). Individuals II-4, II-14, III-2, and III-14 have descendants with the disorder although they do not themselves express the disorder. Therefore, 4/20 people that can be inferred to carry the gene, do not express the trait. That is 80 percent penetrance. (Penetrance could be significantly less than that since many possible carriers have no shown progeny.) The pedigree also exhibits variable expressivity. Of the 16 individuals who have blue sclera, 10 do not have brittle bones. Usually, expressivity is put in terms of none, variable, and highly variable, rather than expressed as percentages.

74. Workers of the honeybee line known as Brown (nothing to do with color) show what is called "hygienic behavior"; that is, they uncap hive compartments containing dead pupae and then remove the dead pupae. This behavior prevents the spread of infectious bacteria through the colony. Workers of the Van Scoy line, however, do not perform these actions, and therefore this line is said to be "nonhygienic." When a queen from the Brown line was mated with Van Scoy drones, all the F_1 were nonhygienic. When drones from this F_1 inseminated a queen from the Brown line, the progeny behaviors were as follows:

1/4 hygienic
1/4 uncapping but no removing of pupae
1/2 nonhygienic

However, when the compartment of dead pupae was uncapped by the beekeeper and the nonhygienic honeybees were examined further, about half the bees were found to remove the dead pupae, but the other half did not.

a. Propose a genetic hypothesis to explain these behavioral patterns.
b. Discuss the data in relation to epistasis, dominance, and environmental interaction.

(**Note:** Workers are sterile, and all bees from one line carry the same alleles.)

Answer:
a. and b. Assuming that both the *Brown* and the *Van Scoy* lines were homozygous, the parental cross suggests that nonhygienic behavior is dominant to hygienic. A consideration of the specific behavior of the $F_1 \times$ *Brown* progeny, however, suggests that the behavior is separable into two processes; one involves uncapping and one involves removal of dead pupae. If this is true, uncapping and removal of dead pupae—two behaviors that normally go together in the *Brown* line—have been separated in the F_2 progeny, suggesting that they are controlled by different and unlinked genes. Those bees that lack uncapping behavior are still able to express removal of dead pupae if environmental conditions are such that they do not need to uncap a compartment first. Also, the uncapping behavior is epistatic to removal of dead pupae.

Let U = no uncapping, u = uncapping, R = no removal, r = removal

P u/u ; r/r × U/U ; R/R
      (*Brown*) (*Van Scoy*)

      U/u ; R/r × u/u ; r/r (F_1 cross *Brown*)

Progeny   1/4 U/u ; R/r nonhygienic
           1/4 U/u ; r/r nonhygienic (no uncapping but removal if uncapped)
           1/4 u/u ; R/r uncapping but no removal

$$1/4 \; u/u \; ; \; r/r \qquad \text{hygienic}$$

75. The normal color of snapdragons is red. Some pure lines showing variations of flower color have been found. When these pure lines were crossed, they gave the following results (see the table):

Cross	Parents	F_1	F_2
1	orange × yellow	orange	3 orange : 1 yellow
2	red × orange	red	3 red : 1 orange
3	red × yellow	red	3 red : 1 yellow
4	red × white	red	3 red : 1 white
5	yellow × white	red	9 red : 3 yellow : 4 white
6	orange × white	red	9 red : 3 orange : 4 white
7	red × white	red	9 red : 3 yellow : 4 white

a. Explain the inheritance of these colors.
b. Write the genotypes of the parents, the F_1, and the F_2.

Answer:
a. and b. Cross 1 indicates that orange is dominant to yellow. Crosses 2–4 indicate that red is dominant to orange, yellow, and white. Crosses 5–7 indicate that there are two genes involved in the production of color. Cross 5 indicates that yellow and white are different genes. Cross 6 indicates that orange and white are different genes.

In other words, epistasis is involved and the homozygous recessive white genotype seems to block production of color by a second gene.

Begin by explaining the crosses in the simplest manner possible until such time as it becomes necessary to add complexity. Therefore, assume that orange and yellow are alleles of gene A with orange dominant.

Cross 1:   P $A/A \times a/a$
       F_1 A/a
       F_2 $3 \; A/- : 1 \; a/a$

Immediately, there is trouble in trying to write the genotypes for cross 2, unless red is a third allele of the same gene. Assume the following dominance relationships: red > orange > yellow. Let the alleles be designated as follows:

red A^R
orange A^O
yellow A^Y

Crosses 1-3 now become:

P $A^O/A^O \times A^Y/A^Y$ $A^R/A^R \times A^O/A^O$ $A^R/A^R \times A^Y/A^Y$

F$_1$ A^O/A^Y A^R/A^O A^R/A^Y

F$_2$ $3\ A^O/- : 1\ A^Y/A^Y$ $3\ A^R/- : 1\ A^O/A^O$ $3\ A^R/- : 1\ A^Y/A^Y$

Cross 4: To do this cross you must add a second gene. You must also rewrite the above crosses to include the second gene. Let B allow color and b block color expression, producing white. The first three crosses become

P $A^O/A^O ; B/B \times A^Y/A^Y ; B/B$ $A^R/A^R ; B/B \times A^O/A^O ; B/B$ $A^R/A^R ; B/B \times A^Y/A^Y ; B/B$

F$_1$ $A^O/A^Y ; B/B$ $A^R/A^O ; B/B$ $A^R/A^Y ; B/B$

F$_2$ $3\ A^O/- ; B/B : 1\ A^Y/A^Y ; B/B$ $3\ A^R/- ; B/B : 1\ A^O/A^O ; B/B$ $3\ A^R/- : 1\ A^Y/A^Y ; B/B$

The fourth cross is:

P $A^R/A^R ; B/B \times A^R/A^R ; b/b$

F$_1$ $A^R/A^R ; B/b$

F$_2$ $3\ A^R/A^R ; B/- : 1\ A^R/A^R ; b/b$

Cross 5: To do this cross, note that there is no orange appearing. Therefore, the two parents must carry the alleles for red and yellow, and the expression of red must be blocked.

P $A^Y/A^Y ; B/B \times A^R/A^R ; b/b$

F$_1$ $A^R/A^Y ; B/b$

F$_2$ $9\ A^R/- ; B/-$ red
     $3\ A^R/- ; b/b$ white
     $3\ A^Y/A^Y; B/-$ yellow
     $1\ A^Y/A^Y ; b/b$ white

Cross 6: This cross is identical with cross 5 except that orange replaces yellow.

P $A^O/A^O ; B/B \times A^R/A^R ; b/b$

F$_1$ $A^R/A^O ; B/b$

F$_2$ $9\ A^R/- ; B/-$ red
     $3\ A^R/- ; b/b$ white
     $3\ A^O/A^O ; B/-$ orange

$1\ A^O/A^O\ ;\ b/b$ white

Cross 7: In this cross, yellow is suppressed by b/b.

P $A^R/A^R\ ;\ B/B\ \times\ A^Y/A^Y\ ;\ b/b$

F$_1$ $A^R/A^Y\ ;\ B/b$

F$_2$ $9\ A^R/-\ ;\ B/-$ red
          $3\ A^R/-\ ;\ b/b$ white
          $3\ A^Y/A^Y\ ;\ B/-$ yellow
          $1\ A^Y/A^Y\ ;\ b/b$ white

76. Consider the following F$_1$ individuals in different species and the F$_2$ ratios produced by selfing:

F$_1$	Phenotypic ratio in the F$_2$		
1 cream	12/16 cream	3/16 black	1/16 gray
2 orange	9/16 orange	7/16 yellow	
3 black	3/16 black	3/16 white	
4 solid red	9/16 solid red	3/16 mottled red	4/16 small red dots

If each F$_1$ were testcrossed, what phenotypic ratios would result in the progeny of the testcross?

Answer:
The ratios observed in the F$_2$ of these crosses are all variations of 9:3:3:1, indicating that two genes are assorting and epistasis is occurring.

(1)   F$_1$ $A/a\ ;\ B/b\ \times\ A/a\ ;\ B/b$

F$_2$ $9\ A/-\ ;\ B/-$ cream
          $3\ A/-\ ;\ b/b$ cream
          $3\ a/a\ ;\ B/-$ black
          $1\ a/a\ ;\ b/b$ gray

A testcross of the F$_1$ would be $A/a\ ;\ B/b\ \times\ a/a\ ;\ b/b$ and the offspring:

$1/4\ A/a\ ;\ B/b$ cream
$1/4\ a/a\ ;\ B/b$ black
$1/4\ A/a\ ;\ b/b$ cream
$1/4\ a/a\ ;\ b/b$ gray

(2)   F$_1$ $A/a\ ;\ B/b\ \times\ A/a\ ;\ B/b$

F$_2$ 9 *A/–* ; *B/–*   orange
    3 *A/–* ; *b/b*   yellow
    3 *a/a* ; *B/–*   yellow
    1 *a/a* ; *b/b*   yellow

A testcross of the F$_1$ would be *A/a* ; *B/b* × *a/a* ; *b/b* and the offspring:

1/4 *A/a* ; *B/b*   orange
1/4 *a/a* ; *B/b*   yellow
1/4 *A/a* ; *b/b*   yellow
1/4 *a/a* ; *b/b*   yellow

(3)   F$_1$ *A/a* ; *B/b* × *A/a* ; *B/b*

F$_2$ 9 *A/–* ; *B/–*   black
    3 *A/–* ; *b/b*   white
    3 *a/a* ; *B/–*   black
    1 *a/a* ; *b/b*   black

A testcross of the F$_1$ would be *A/a* ; *B/b* × *a/a* ; *b/b* and the offspring:

1/4 *A/a* ; *B/b*   black
1/4 *a/a* ; *B/b*   black
1/4 *A/a* ; *b/b*   white
1/4 *a/a* ; *b/b*   black

(4)   F$_1$ *A/a* ; *B/b* × *A/a* ; *B/b*

F$_2$ 9 *A/–* ; *B/–*   solid red
    3 *A/–* ; *b/b*   mottled red
    3 *a/a* ; *B/–*   small red dots
    1 *a/a* ; *b/b*   small red dots

A testcross of the F$_1$ would be *A/a* ; *B/b* × *a/a* ; *b/b* and the offspring:

1/4 *A/a* ; *B/b*   solid red
1/4 *a/a* ; *B/b*   small red dots
1/4 *A/a* ; *b/b*   mottled red
1/4 *a/a* ; *b/b*   small red dots

77. To understand the genetic basis of locomotion in the diploid nematode *Caenorhabditis elegans*, recessive mutations were obtained, all making the worm "wiggle" ineffectually instead of moving with its usual smooth gliding motion. These mutations presumably affect the nervous or muscle systems. Twelve homozygous mutants were intercrossed, and the F$_1$ hybrids were examined to see if they wiggled. The results were as follows, where a plus sign

means that the F_1 hybrid was wild type (gliding) and "w" means that the hybrid wiggled.

	1	2	3	4	5	6	7	8	9	10	11	12
1	w	+	+	+	w	+	+	+	+	+	+	+
2		w	+	+	+	w	+	w	+	w	+	+
3			w	w	+	+	+	+	+	+	+	+
4				w	+	+	+	+	+	+	+	+
5					w	+	+	+	+	+	+	+
6						w	+	w	+	w	+	+
7							w	+	+	+	w	w
8								w	+	w	+	+
9									w	+	+	+
10										w	+	+
11											w	w
12												w

a. Explain what this experiment was designed to test.
b. Use this reasoning to assign genotypes to all 12 mutants.
c. Explain why the phenotype of the F_1 hybrids between mutants 1 and 2 differed from that of the hybrids between mutants 1 and 5.

Answer:

a. Intercrossing mutant strains that all share a common recessive phenotype is the basis of the complementation test. This test is designed to identify the number of different genes that can mutate to a particular phenotype. In this problem, if the progeny of a given cross still express the wiggle phenotype, the mutations fail to complement and are considered alleles of the same gene; if the progeny are wild type, the mutations complement and the two strains carry mutant alleles of separate genes.

b. From the data,

1 and 5	fail to complement	gene A
2, 6, 8, and 10	fail to complement	gene B
3 and 4	fail to complement	gene C
7, 11, and 12	fail to complement	gene D
9	complements all others	gene E

There are five complementation groups (genes) identified by this data.

c. mutant 1: $a^1/a^1 \cdot b^+/b^+ \cdot c^+/c^+ \cdot d^+/d^+ \cdot e^+/e^+$ (although, only the mutant alleles are usually listed)
   mutant 2: $a^+/a^+ \cdot b^2/b^2 \cdot c^+/c^+ \cdot d^+/d^+ \cdot e^+/e^+$
   mutant 5: $a^5/a^5 \cdot b^+/b^+ \cdot c^+/c^+ \cdot d^+/d^+ \cdot e^+/e^+$

   1/5 hybrid: $a^1/a^5 \cdot b^+/b^+ \cdot c^+/c^+ \cdot d^+/d^+ \cdot e^+/e^+$ phenotype: wiggles

   Conclusion: 1 and 5 are both mutant for gene A.

2/5 hybrid: $a^+/a^5 ,\cdot b^+/b^2 \cdot c^+/c^+ \cdot d^+/d^+ \cdot e^+/e^+$ phenotype: wild type

Conclusion: 2 and 5 are mutant for different genes.

78. A geneticist working on a haploid fungus makes a cross between two slow-growing mutants called mossy and spider (referring to the abnormal appearance of the colonies). Tetrads from the cross are of three types (A, B, C), but two of them contain spores that do not germinate.

Spore	A	B	C
1	wild type	wild type	spider
2	wild type	spider	spider
3	no germination	mossy	mossy
4	no germination	no germination	mossy

Devise a model to explain these genetic results, and propose a molecular basis for your model.

Answer: Assume that the two genes are unlinked.   Mossy would be genotypically m^- ; s^+ and spider would be m^+ ; s^-. After mating and sporulation, independent assortment would give three types of tetrads. The doubly mutant spores do not germinate. Because either single mutant is slow-growing, the additive effect of both mutant genes is lethal.

A		B		C	
$m+$; $s+$	wild type	$m+$; $s+$	wild type	$m+$; $s-$	spider
$m+$; $s+$	wild type	$m+$; $s-$	spider	$m+$; $s-$	spider
$m-$; $s-$	dead	$m-$; $s+$	mossy	$m-$; $s+$	mossy
$m-$; $s-$	dead	$m-$; $s-$	dead	$m-$; $s+$	mossy

7

DNA: Structure and Replication

WORKING WITH THE FIGURES

1. In Table 7-1, why are there no entries for the first four tissue sources? For the last three entries, what is the most likely explanation for the slight differences in the composition of human DNA from the three tissue sources?

    Answer: The first four are single-celled organisms; therefore, a tissue category does not apply. The differences in the values are most likely attributable to experimental error.

2. In Figure 7-7, do you recognize any of the components used to make Watson and Crick's DNA model? Where have you seen them before?
    Answer: Watson and Crick made use of molecular models, ring stands, and clamps, typically found in a chemistry lab. The vertically oriented pentagons are the deoxyribose components of DNA. These are components of the two sugar-phosphate backbones of the DNA structure, one behind James Watson and the other just to the left of Francis Crick. Nitrogenous bases oriented horizontally are held in place in between the two backbones by clamps.

3. Referring to Figure 7-20, answer the following questions:
    a. What is the DNA polymerase I enzyme doing?
    b. What other proteins are required for the DNA polymerase III on the left to continue synthesizing DNA?
    c. What other proteins are required for the DNA polymerase III on the right to continue synthesizing DNA?

    Answer:
    a. PolI is removing ribonucleotide primers and filling gaps between Okazaki fragments.
    b. β-clamp and helicase
    c. β-clamp, helicase, primase, and ssb

4.   What is different about the reaction catalyzed by the green helicase in Figure 7-20 and the yellow gyrase in Figure 7-21?

Answer: Gyrase breaks the phosphodiester linkages in the DNA backbone, while helicase disrupts the hydrogen bonds between the paired bases of antiparallel strands.

5.   In Figure 7-24(a), label all the leading and lagging strands.

Answer: Each fork has both a leading and a lagging strand, as labeled below:

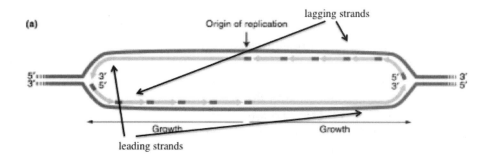

BASIC PROBLEMS

6.   Describe the types of chemical bonds in the DNA double helix.

Answer: The DNA double helix is held together by two types of bonds, covalent and hydrogen. Covalent bonds occur within each linear strand and strongly bond the bases, sugars, and phosphate groups (both within each component and between components). Hydrogen bonds occur between the two strands and involve a base from one strand with a base from the second in complementary pairing. These hydrogen bonds are individually weak but collectively quite strong.

7.   Explain what is meant by the terms *conservative* and *semiconservative* *replication*.

Answer: Conservative replication is a hypothetical form of DNA synthesis in which the two template strands remain together but dictate the synthesis of two new DNA strands, which then form a second DNA helix. The end point is two double helices, one containing only old DNA and one containing only new DNA. This hypothesis was found to be not correct. Semiconservative replication is a form of DNA synthesis in which the two template strands separate and each dictates the synthesis of a new strand. The end point is two double helices, both containing one new and one old strand of DNA. This hypothesis was found to be correct.

8. What is meant by a *primer,* and why are primers necessary for DNA replication?

Answer: A primer is a short segment of RNA that is synthesized by primase using DNA as a template during DNA replication. Once the primer is synthesized, DNA polymerase then adds DNA to the 3′ end of the RNA. Primers are required because the major DNA polymerase involved with DNA replication is unable to initiate DNA synthesis and, rather, requires a 3′ end. (It is the 3′-OH group that is required to create the next phosphodiester bond.) The RNA is subsequently removed and replaced with DNA so that no gaps exist in the final product.

9. What are helicases and topoisomerases?

Answer: Helicases are enzymes that disrupt the hydrogen bonds that hold the two DNA strands together in a double helix. This breakage is required for both RNA and DNA synthesis. Topoisomerases are enzymes that create and relax supercoiling in the DNA double helix. The supercoiling itself is a result of the twisting of the DNA helix that occurs when the two strands separate.

10. Why is DNA synthesis continuous on one strand and discontinuous on the opposite strand?

Answer: Because the DNA polymerase is capable of adding new nucleotides only at the 3′ end of a DNA strand, and because the two strands are antiparallel, at least two molecules of DNA polymerase must be involved in the replication of any specific region of DNA. When a region becomes single-stranded, the two strands have an opposite orientation. Imagine a single-stranded region that runs from right to left. The 5′ end is at the right, with the 3′ end pointing to the left; synthesis can initiate and continue uninterrupted toward the right end of this strand. Remember: new nucleotides are added in a 5′→ 3′ direction, so the template must be copied from its 3′ end. The other strand has a 5′ end at the left with the 3′ end pointing right. Thus, the two strands are oriented in opposite directions (antiparallel), and synthesis (which is 5′→ 3′) must proceed in opposite directions. For the leading strand (say, the top strand), replication is to the right, following the replication fork. It is continuous and may be thought of as moving "downstream." Replication on the bottom strand cannot move in the direction of the fork (to the right) because, for this strand, that would mean adding nucleotides to its 5′ end. Therefore, this strand must replicate discontinuously: as the fork creates a new single-stranded stretch of DNA, this is replicated to the left (away from the direction of fork movement). For this lagging strand, the replication fork is always opening new single-stranded DNA for replication upstream of the previously replicated stretch, and a new fragment of DNA is replicated back to the previously created fragment. Thus, one

(Okazaki) fragment follows the other in the direction of the replication fork, but each fragment is created in the opposite direction.

11. If the four deoxynucleotides showed nonspecific base pairing (A to C, A to G, T to G, and so on), would the unique information contained in a gene be maintained through round after round of replication? Explain.
Answer: No. The information of DNA is dependent on a faithful copying mechanism. The strict rules of complementarity ensure that replication and transcription are reproducible.

12. If the helicases were missing during replication, what would happen to the replication process?

Answer: Helicases are enzymes that disrupt the hydrogen bonds that hold the two DNA strands together in a double helix. This breakage exposes lengths of single-stranded DNA that will act as the template and are required for DNA replication. Therefore, the absence of helicases would prevent the replication process.

13. Both strands of a DNA molecule are replicated simultaneously in a continuous fashion on one strand and a discontinuous one on the other. Why can't one strand be replicated in its entirety (from end to end) before replication of the other is initiated?

Answer: Theoretically, DNA could be replicated this way but not with the replisome, which is organized to replicate both stands simultaneously. Further, this would leave one strand of the DNA single-stranded where mutagenic events would be more likely, and it would certainly take longer.

14. What would happen if, in the course of replication, the topoisomerases were unable to reattach the DNA fragments of each strand after unwinding (relaxing) the DNA molecule?

Answer: The chromosome would become hopelessly fragmented.

15. Which of the following would happen if DNA synthesis were discontinuous on both strands?
a. The DNA fragments from the two new strands could become mixed, producing possible mutations.
b. DNA synthesis would not take place, because the appropriate enzymes to carry out discontinuous replication on both strands would not be present.

c. DNA synthesis might take longer, but otherwise there would be no noticeable difference.

d. DNA synthesis would not take place, because the entire length of the chromosome would have to be unwound before both strands could be replicated in a discontinuous fashion.

Answer: **c.** DNA synthesis might take longer.

16. Which of the following is *not* a key property of hereditary material?
a. It must be capable of being copied accurately.
b. It must encode the information necessary to form proteins and complex structures.
c. It must occasionally mutate.
d. It must be able to adapt itself to each of the body's tissues.

Answer: **d.** It does not have to adapt to each of the body's tissues.

17. It is essential that RNA primers at the ends of Okazaki fragments be removed and replaced by DNA because otherwise which of the following events would result?
a. The RNA might not be accurately read during transcription, thus interfering with protein synthesis.
b. The RNA would be more likely to contain errors because primase lacks a proofreading function.
c. The stretches of RNA would destabilize and begin to break up into ribonucleotides, thus creating gaps in the sequence.
d. The RNA primers would be likely to hydrogen bond to each other, forming complex structures that might interfere with the proper formation of the DNA helix.

Answer: **b.** The RNA would be more likely to contain errors.

18. Polymerases usually add only about 10 nucleotides to a DNA strand before dissociating. However, during replication, DNA pol III can add tens of thousands of nucleotides at a moving fork. How is this addition accomplished?

Answer: Part of the replisome is the sliding clamp, which encircles the DNA and keeps pol III attached to the DNA molecule. Thus, pol III is transformed into a processive enzyme capable of adding tens of thousands of nucleotides.

19. At each origin of replication, DNA synthesis proceeds bidirectionally from two replication forks. Which of the following would happen if a mutant arose having only one functional fork per replication bubble? (See diagram.)
a. No change at all in replication.

b. Replication would take place only on one half of the chromosome.
c. Replication would be complete only on the leading strand.
d. Replication would take twice as long.

Normal   Mutant

Answer: **d.** Replication would take twice as long.

20. In a diploid cell in which $2n = 14$, how many telomeres are there in each of the following phases of the cell cycle?
    (a) G1   **(c)** mitotic prophase
    (b) G2   **(d)** mitotic telophase

    Answer:
    a. Prior to the S phase, each chromosome has two telomeres, so in the case of $2n = 14$, there are 14 chromosomes and 28 telomeres.

    b. After S, each chromosome consists of two chromatids, each with two telomeres, for a total of four telomeres per chromosome. So, for 14 chromosomes, there would be $14 \times 4 = 56$ telomeres.

    c. At prophase, the chromosomes still consist of two chromatids each, so there would be $14 \times 4 = 56$ telomeres.

    d. At telophase, there would be 28 telomeres in each of the soon-to-be daughter cells.

21. If thymine makes up 15 percent of the bases in a specific DNA molecule, what percentage of the bases is cytosine?

    Answer: If the DNA is double stranded, $A = T$ and $G = C$ and $A + T + C + G = 100\%$. If $T = 15\%$, then $C = [100 - 15(2)]/2 = 35\%$.

22. If the GC content of a DNA molecule is 48 percent, what are the percentages of the four bases (A, T, G, and C) in this molecule?

    Answer: If the DNA is double stranded, $G = C = 24\%$ and $A = T = 26\%$.

23. Bacteria called extremophiles are able to grow in hot springs such as Old Faithful at Yellowstone National Park in Wyoming. Do you think that the DNA

of extremophiles would have a higher content of GC or AT base pairs? Justify your answer.

Answer: Higher GC content. The increased number of hydrogen bonds would help to counteract the destabilizing effect of increased heat.

24. Assume that a certain bacterial chromosome has one origin of replication. Under some conditions of rapid cell division, replication could start from the origin before the preceding replication cycle is complete. How many replication forks would be present under these conditions?
Answer: Six. The first replication start would have two replication forks proceeding to completion, and the now replicated origins would each start replication again. Each would have two more replication forks, for a total of six.

25. A molecule of composition

5′-AAAAAAAAAAA-3′
3′-TTTTTTTTTTTT-5′

is replicated in a solution of adenine nucleoside triphosphate with all its phosphorus atoms in the form of the radioactive isotope 32P. Will both daughter molecules be radioactive? Explain. Then repeat the question for the molecule

5′-ATATATATATATAT-3′
3′-TATATATATATATA-5′

Answer: Only the DNA molecule that used the poly-T strand as a template would be radioactive. The other daughter molecule would not be radioactive because it would not have required any dATP for its replication.
Because each strand of the second molecule contains T, both daughter molecules would require dATP for replication, so each would be radioactive.

26. Would the Meselson and Stahl experiment have worked if diploid eukaryotic cells had been used instead?

Answer: Yes. DNA replication is also semi-conservative in diploid eukaryotes.

27. Consider the following segment of DNA, which is part of a much longer molecule constituting a chromosome:

5′....ATTCGTACGATCGACTGACTGACAGTC....3′

3′....TAAGCATGCTAGCTGACTGACTGTCAG....5′

If the DNA polymerase starts replicating this segment from the right,

a. which will be the template for the leading strand?

b. Draw the molecule when the DNA polymerase is halfway along this segment.

c. Draw the two complete daughter molecules.

d. Is your diagram in part *b* compatible with bidirectional replication from a single origin, the usual mode of replication?

Answer:

a. If replication is proceeding such that the DNA on the right is replicated first, then the top strand is the template for the leading strand.

b.

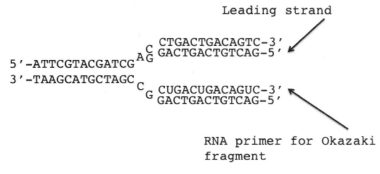

Leading strand

```
                                    CTGACTGACAGTC-3'
                                  C GACTGACTGTCAG-5'
                               A G
        5'-ATTCGTACGATCG
        3'-TAAGCATGCTAGC
                              C    CUGACUGACAGUC-3'
                               G   GACTGACTGTCAG-5'
```

RNA primer for Okazaki fragment

c.

```
....5'-ATTCGTACGATCGACTGACTGACAGTC-3'....
....3'-TAAGCATGCTAGCTGACTGACTGTCAG-5'....

....5'-ATTCGTACGATCGACTGACTGACAGTC-3'....
....3'-TAAGCATGCTAGCTGACTGACTGTCAG-5'....
```

d. Yes, it simply represents replication at one of the forks.

28. The DNA polymerases are positioned over the following DNA segment (which is part of a much larger molecule) and moving from right to left. If we assume that an Okazaki fragment is made from this segment, what will be the fragment's sequence? Label its 5′ and 3′ ends.

5′....CCTTAAGACTAACTACTTACTGGGATC....3′

3′....GGAATTCTGATTGATGAATGACCCTAG....5′

Answer: The bottom strand will serve as the template for the Okazaki fragment so its sequence will be:

5′ ....CCTTAAGACTAACTACTTACTGGGATC....3′

29. *E. coli* chromosomes in which every nitrogen atom is labeled (that is, every nitrogen atom is the heavy isotope 15N instead of the normal isotope 14N) are allowed to replicate in an environment in which all the nitrogen is 14N. Using a solid line to represent a heavy polynucleotide chain and a dashed line for a light chain, sketch each of the following descriptions:

a. The heavy parental chromosome and the products of the first replication after transfer to a 14N medium, assuming that the chromosome is one DNA double helix and that replication is semiconservative.

b. Repeat part *a,* but now assume that replication is conservative.

c. Repeat part *a,* but assume that the chromosome is in fact two side-by-side double helices, each of which replicates semiconservatively.

d. Repeat part *c,* but assume that each side-by-side double helix replicates conservatively and that the overall *chromosome* replication is semiconservative.

e. If the daughter chromosomes from the first division in 14N are spun in a cesium chloride density gradient and a single band is obtained, which of the possibilities in parts *a* through *d* can be ruled out? Reconsider the Meselson and Stahl experiment: What does it *prove*?

Answer: Model (b) is ruled out by the experiment. The results are compatible with semiconservative replication, but the exact structure could not be predicted. The other models would all give one band of intermediate density.

CHALLENGING PROBLEMS

30. If a mutation that inactivated telomerase occurred in a cell (telomerase activity in the cell = zero), what do you expect the outcome to be?

Answer: Without functional telomerase, the telomeres would shorten at each replication cycle, leading to eventual loss of essential coding information and death. In fact, there are some current observations that decline or loss of telomerase activity plays a role in the mechanism of aging in humans.

31. On the planet Rama, the DNA is of six nucleotide types: A, B, C, D, E, and F. Types A and B are called *marzines,* C and D are *orsines,* and E and F are *pirines.* The following rules are valid in all Raman DNAs:

> Total marzines = total orsines = total pirines
> $$A = C = E$$
> $$B = D = F$$

a. Prepare a model for the structure of Raman DNA.
b. On Rama, mitosis produces three daughter cells. Bearing this fact in mind, propose a replication pattern for your DNA model.
c. Consider the process of meiosis on Rama. What comments or conclusions can you suggest?

Answer:
a. A very plausible model is of a triple helix, which would look like a braid, with each strand interacting by hydrogen bonding to the other two.
b. Replication would have to be terti-conservative. The three strands would separate, and each strand would dictate the synthesis of the other two strands.
c. The reductional division would have to result in three daughter cells, and the equational would have to result in two daughter cells, in either order. Thus, meiosis would yield six gametes

32. If you extract the DNA of the coliphage φX174, you will find that its composition is 25 percent A, 33 percent T, 24 percent G, and 18 percent C. Does this composition make sense in regard to Chargaff's rules? How would you interpret this result? How might such a phage replicate its DNA?

Answer: Chargaff's rules are that A = T and G = C. Because this is not observed, the most likely interpretation is that the DNA is single-stranded. The phage would first have to synthesize a complementary strand before it could begin to make multiple copies of itself.

8

RNA: Transcription and Processing

WORKING WITH THE FIGURES

1. In Figure 8-3, why are the arrows for genes 1 and 2 pointing in opposite directions?

    Answer: The arrows for genes 1 and 2 indicate the direction of transcription, which is always 5′ to 3′. The two genes are transcribed from opposite DNA strands, which are antiparallel, so the genes must be transcribed in opposite directions to maintain the 5′ to 3′ direction of transcription.

2. In Figure 8-5, draw the "one gene" at much higher resolution with the following components: DNA, RNA polymerase(s), RNA(s).

    Answer: At the higher resolution, the feathery structures become RNA transcripts, with the longer transcripts occurring nearer the termination of the gene. The RNA in this drawing has been straightened out to illustrate the progressively longer transcripts.

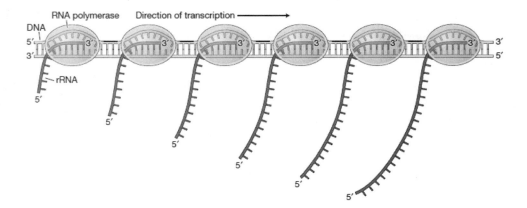

3. In Figure 8-6, describe where the gene promoter is located.

Answer: The promoter is located to the left (upstream) of the 3′ end of the template strand. From this sequence it cannot be determined how far the promoter would be from the 5′ end of the mRNA.

4.   In Figure 8-9b, write a sequence that could form the hairpin loop structure.

Answer: Any sequence that contains inverted complementary regions separated by a noncomplementary one would form a hairpin. One sequence would be:
ACG**CAAGCUUAC**CGAUUAUU**GUAAGCUUG**AAG
The two bold-faced sequences would pair and form a hairpin. The intervening non-bold sequence would be the loop.

5.   How do you know that the events in Figure 8-13 are occurring in the nucleus?

Answer: The figure shows a double-stranded DNA molecule from which RNA is being transcribed. This process only occurs in the nucleus.

6.   In Figure 8-15, what do you think would be the effect of a G to A mutation in the first G residue of the intron?

Answer: A mutation of G to A would alter the U1 SNP binding site and prevent formation of the spliceosome. This would prevent splicing of the intron.

7.   In Figure 8-23, show how the double-stranded RNA is able to silence the transgene. What would have to happen for the transgene to also silence the flanking cellular gene (in yellow)?

Answer:
a. The double-stranded RNA formed from the sense and antisense transgene transcripts would be processed by Dicer, then bind to RISC. RISC would separate the dsRNA to produce an antisense RNA/RISC complex. The RISC/RNA complex would bind to the transgene mRNA and deactivate it.

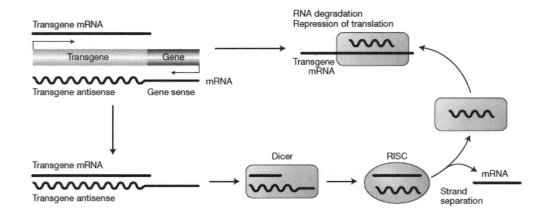

b. The flanking gene would be silenced if transcription of the transgene continued into the flanking gene. This would produce an antisense transcript of the flanking gene, leading to the formation of dsRNA and activation of Dicer and RISC.

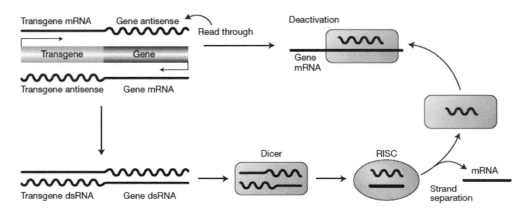

BASIC PROBLEMS

8. The two strands of λ phage DNA differ from each other in their GC content. Owing to this property, they can be separated in an alkaline cesium chloride gradient (the alkalinity denatures the double helix). When RNA synthesized by l phage is isolated from infected cells, it is found to form DNA–RNA hybrids with both strands of λ DNA. What does this finding tell you? Formulate some testable predictions.

    Answer: Because RNA can hybridize to both strands, the RNA must be transcribed from both strands. This does not mean, however, that both strands are used as a template *within each gene*. The expectation is that only one strand is used within a gene but that different genes are transcribed in different directions along the DNA. The most direct test would be to purify a specific RNA coding for a specific protein and then hybridize it to the λ genome. Only one strand should hybridize to the purified RNA.

9.   In both prokaryotes and eukaryotes, describe what else is happening to the RNA while RNA polymerase is synthesizing a transcript from the DNA template.

Answer: In prokaryotes, translation is beginning at the 5′ end while the 3′ end is still being transcribed. In eukaryotes, processing (capping, splicing) is occurring at the 5′ end while the 3′ end is still being transcribed.

10.   List three examples of proteins that act on nucleic acids.

Answer: There are many examples of proteins that act on nucleic acids, but some mentioned in this chapter are RNA polymerase, GTFs (general transcription factors), σ (sigma factor), rho, TBP (TATA binding protein), and snRNPs (a combination of proteins and snRNAs).

11.   What is the primary function of the sigma factor? Is there a protein in eukaryotes analogous to the sigma factor?

Answer: Sigma factor, as part of the RNA polyermase holoenzyme, recognizes and binds to the −35 and −10 regions of bacterial promoters. It positions the holoenzyme to correctly initiate transcription at the start site. In eukaryotes, TBP (TATA binding protein) and other GTFs (general transcription factors) have an analagous function.

12.   You have identified a mutation in yeast, a unicellular eukaryote, that prevents the capping of the 5′ end of the RNA transcript. However, much to your surprise, all the enzymes required for capping are normal. You determine that the mutation is, instead, in one of the subunits of RNA polymerase II. Which subunit is mutant and how does this mutation result in failure to add a cap to yeast RNA?

Answer: The CTD (carboxy tail domain) of the ß subunit of RNA polymerase II contains binding sites for enzymes and other proteins that are required for RNA processing and is located near the site where nascent RNA emerges. If mutations in this subunit prevent the correct binding and/or localization of the proteins necessary for capping, then this modification will not occur even though all the required enzymes are normal.

13.   Why is RNA produced only from the template DNA strand and not from both strands?

Answer: For a given gene, only one strand of DNA is transcribed. This strand (called the template) will be complementary to the RNA and also to the other strand (called the nontemplate or coding strand). Consequently, the nucleotide sequence of the RNA must be the same as that of the nontemplate strand of the DNA (except that the Ts are instead Us). Ultimately, it is the nucleotide sequence that gives the RNA its function. Transcription of both strands would give two complementary RNAs that would code for completely different polypeptides. Also, double-stranded RNA initiates cellular processes that lead to its degradation.

14. A linear plasmid contains only two genes, which are transcribed in opposite directions, each one from the end, toward the center of the plasmid. Draw diagrams of

   a. the plasmid DNA, showing the 5′ and 3′ ends of the nucleotide strands.
   b. the template strand for each gene.
   c. the positions of the transcription-initiation site.
   d. the transcripts, showing the 5′ and 3′ ends.

   Answer:
   a., b., c., and d.

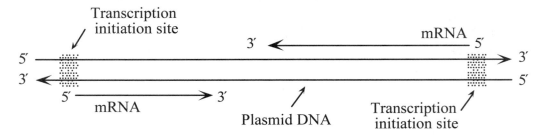

15. Are there similarities between the DNA replication bubbles and the transcription bubbles found in eukaryotes? Explain.

   Answer: Yes. Both replication and transcription is performed by large, multi-subunit molecular machines (the replisome and RNA polymerase II, respectively) and both require helicase activity at the fork of the bubble. However, transcription proceeds in only one direction and only one DNA strand is copied.

16. Which of the following statements are true about eukaryotic mRNA?

   a. The sigma factor is essential for the correct initiation of transcription.
   b. Processing of the nascent mRNA may begin before its transcription is complete.

c. Processing takes place in the cytoplasm.
d. Termination is accomplished by the use of a hairpin loop or the use of the rho factor.
e. Many RNAs can be transcribed simultaneously from one DNA template.

Answer:
a. False. Sigma factor is required in prokaryotes, not eukaryotes.
b. True. Processing begins at the 5′ end, while the 3′ end is still being synthesized.
c. False. Processing occurs in the nucleus and only mature RNA is transported out to the cytoplasm.
d. False. Hairpin loops or rho factor (in conjunction with the *rut* site) is used to terminate transcription in prokaryotes. In eukaryotes the conserved sequences AAUAAA or AUUAAA, near the 3′ end of the transcript, are recognized by an enzyme that cuts off the end of the RNA approximately 20 bases downstream.
e. True. Multiple RNA polymerases may transcribe the same template simultaneously.

17. A researcher was mutating prokaryotic cells by inserting segments of DNA. In this way, she made the following mutation:

Original TTGACAT 15 to 17 bp TATAAT
Mutant   TATAAT 15 to 17 bp TTGACAT

a. What does this sequence represent?
b. What do you predict will be the effect of such a mutation? Explain.

Answer:
a. The original sequence represents the –35 and –10 consensus sequences (with the correct number of intervening spaces) of a bacterial promoter. Sigma factor, as part of the RNA polymerase holoenzyme, recognizes and binds to these sequences.
b. The mutated (transposed) sequences would not be a binding site for sigma factor. The two regions are not in the correct orientation to each other and therefore would not be recognized as a promoter.

18. You will learn more about genetic engineering in Chapter 10, but for now, put on your genetic engineer's cap and try to solve this problem. E. coli is widely used in laboratories to produce proteins from other organisms.

a. You have isolated a yeast gene that encodes a metabolic enzyme and want to produce this enzyme in *E. coli*. You suspect that the yeast promoter will not work in *E. coli*. Why?

b. After replacing the yeast promoter with an *E. coli* promoter, you are pleased to detect RNA from the yeast gene but are confused because the RNA is almost twice the length of the mRNA from this gene isolated from yeast. Explain why this result might have occurred.

Answer:

a. The promoters of eukaryotes and prokaryotes do not have the same conserved sequences. In yeast, the promoter would have the required TATA box located about –30, whereas bacteria would have conserved sequences at –35 and –10 that would interact with sigma factor as part of the RNA polymerase holoenzyme.

b. There are two possible reasons that the mRNA is longer than expected. First, many eukaryotic genes contain introns, and bacteria would not have the splicing machinery necessary for their removal. Second, termination of transcription is not the same in bacteria and yeast; the sequences necessary for correct termination in *E. coli* would not be expected in the yeast gene.

19. Draw a prokaryotic gene and its RNA product. Be sure to include the promoter, transcription start site, transcription termination site, untranslated regions, and labeled 5′ and 3′ ends.

Answer:

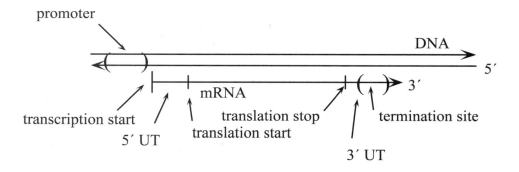

20. Draw a two-intron eukaryotic gene and its pre-mRNA and mRNA products. Be sure to include all the features of the prokaryotic gene included in your answer to Problem 19, plus the processing events required to produce the mRNA.

Answer:

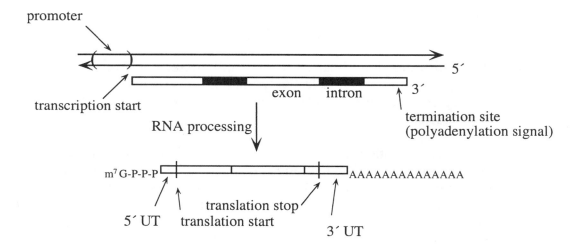

21. A certain *Drosophila* protein-encoding gene has one intron. If a large sample of null alleles of this gene is examined, will any of the mutant sites be expected

    a. in the exons?
    b. in the intron?
    c. in the promoter?
    d. in the intron–exon boundary?

Answer:
    a. Yes. The exons encode the protein, so null mutations would be expected to map within exons.
    b. Possibly. There are sequences near the boundaries of and within introns that are necessary for correct splicing. If these are altered by mutation, correct splicing will be disrupted. Although transcribed, it is likely that translation will not occur.
    c. Yes. If the promoter is deleted or altered such that GTFs cannot bind, transcription will be disrupted.
    d. Yes. There are sequences near the boundaries of and within introns that are necessary for correct splicing.

22. What are self-splicing introns and why does their existence support the theory that RNA evolved before protein?

Answer: Self-splicing introns are capable of excising themselves from a primary transcript without the need of additional enzymes or energy source. They are one of many examples of RNA molecules that are catalytic, and for this property, they are also known as ribozymes. With this additional function, RNA is the only known biological molecule to encode genetic information and catalyze biological reactions. In simplest terms, it is possible that life began with an RNA molecule, or group of molecules, that evolved the ability to self-replicate.

23. Antibiotics are drugs that selectively kill bacteria without harming animals. Many antibiotics act by selectively binding to certain proteins that are critical for bacterial function. Explain why some of the most successful antibiotics target bacterial RNA polymerase.

Answer: Antibiotics need to selectively target bacterial structures and functions that are essential for life but unique or sufficiently different from the equivalent structure and functions of their animal hosts. Bacterial RNA polymerase fits these criteria as its function is obviously essential, yet its structure is sufficiently different from the several eukaryotic RNA polymerases. These differences make it possible to develop drugs that specifically bind bacterial RNA polymerase but have little or no affinity for eukaryotic RNA polymerases.

CHALLENGING PROBLEMS

24. The following data represent the base compositions of double-stranded DNA from two different bacterial species and their RNA products obtained in experiments conducted in vitro:

Species	$(A + T)/(G + C)$	$(A + U)/(G + C)$	$(A + G)/(U + C)$
Bacillus subtilis	1.36	1.30	1.02
E. coli	1.00	0.98	0.80

a. From these data, can you determine whether the RNA of these species is copied from a single strand or from both strands of the DNA? How? Drawing a diagram will make it easier to solve this problem.
b. Explain how you can tell whether the RNA itself is single stranded or double stranded.

(Problem 24 is reprinted with the permission of Macmillan Publishing Co., Inc., from M. Strickberger, *Genetics*. Copyright 1968, Monroe W. Strickberger.)

Answer:
a. The data cannot indicate whether one or both strands are used for transcription. You do not know how much of the DNA is transcribed nor which regions of DNA are transcribed. Only when the purine/pyrimidine ratio is not unity can you deduce that only one strand is used as template.

b. If the RNA is double-stranded, the percentage of purines $(A + G)$ would equal the percentage of pyrimidines $(U + C)$ and the $(A + G)/(U + C)$ ratio would be 1.0. This is clearly not the case for *E. coli*, which has a ratio of

0.80. The ratio for *B. subtilis* is 1.02. This is consistent with the RNA being double-stranded but does not rule out single-stranded if there are an equal number of purines and pyrimidines in the strand.

25. A human gene was initially identified as having three exons and two introns. The exons are 456, 224, and 524 bp, whereas the introns are 2.3 kb and 4.6 kb.

    a. Draw this gene, showing the promoter, introns, exons, and transcription start and stop sites.

    b. Surprisingly, this gene is found to encode not one but two mRNAs that have only 224 nucleotides in common. The original mRNA is 1204 nucleotides, and the new mRNA is 2524 nucleotides. Use your drawing to show how this one region of DNA can encode these two transcripts.

Answer:
a.

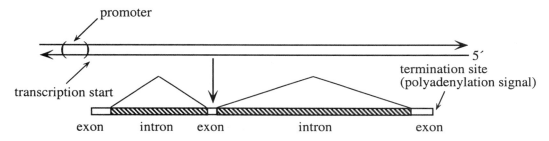

b. Alternative splicing of the primary transcript would result in mRNAs that were only partially identical. In this case, the two transcripts share 224 nucleotides in common. As this is the exact length of the second exon, one possible solution to this problem is that this exon is shared by the two alternatively-spliced mRNAs. The second transcript also contains 2.3 kb of sequence not found in the first. Perhaps what was considered the first intron is actually also part of the second transcript as that would result in the 2524 nucleotides stated as this transcript's length. Of course, other combinations of alternative splicing would also fit the data.

26. While working in your laboratory, you isolate an mRNA from *C. elegans* that you suspect is essential for embryos to develop successfully. With the assumption that you are able to turn mRNA into double-stranded RNA, design an experiment to test your hypothesis.

Answer: Double-stranded RNA, composed of a sense strand and a complementary antisense strand, can be used in *C. elegans* (and likely all organisms) to selectively prevent the synthesis of the encoded gene product (a discovery awarded the 2006 Nobel Prize in Medicine). This process, called gene silencing, blocks the synthesis of the encoded protein from the endogenous gene and is thus equivalent to "knocking out" the gene. To test whether a specific mRNA encodes an essential embryonic protein, eggs or very early embryos should be injected with the double-stranded RNA produced from your mRNA, thus activating the RNAi pathway. The effects of knocking out the specific gene product can then be followed by observing what happens in these, versus control, embryos. If the encoded protein is essential, embryonic development should be perturbed when your gene is silenced.

27. Glyphosate is an herbicide used to kill weeds. It is the main component of a product made by the Monsanto Company called Roundup. Glyphosate kills plants by inhibiting an enzyme in the shikimate pathway called EPSPS. This herbicide is considered safe because animals do not have the shikimate pathway. To sell even more of their herbicide, Monsanto commissioned its plant geneticists to engineer several crop plants, including corn, to be resistant to glyphosate. To do so, the scientists had to introduce an EPSPS enzyme that was resistant to inhibition by glyphosate into crop plants and then test the transformed plants for resistance to the herbicide.

Imagine that you are one of these scientists and that you have managed to successfully introduce the resistant EPSPS gene into the corn chromosomes. You find that some of the transgenic plants are resistant to the herbicide, whereas others are not. Your supervisor is very upset and demands an explanation of why some of the plants are not resistant even though they have the transgene in their chromosomes. Draw a picture to help him understand.

Answer: Transgene silencing is a common phenomenon in plants. Silencing may occur at the transcriptional or post-transcriptional level. Since you cannot control where transgenes insert, some may insert into transcriptionally inactive parts of the genome. Post-transcriptional silencing may be the result of activation of the RNAi pathway due to the misexpression of both strands of your transgene. See Figure 8-22 in the companion text as one example of how this can happen. In these cases, the RNAi pathway will be activated and the EPSPS gene product will be silenced.

28. Many human cancers result when a normal gene mutates and leads to uncontrolled growth (a tumor). Genes that cause cancer when they mutate are called oncogenes. Chemotherapy is effective against many tumors because it targets rapidly dividing cells and kills them. Unfortunately, chemotherapy has many side effects, such as hair loss or nausea, because it also kills many of our normal cells that are rapidly dividing, such as those in the hair follicles or stomach lining.

Many scientists and large pharmaceutical companies are excited about the prospects of exploiting the RNAi pathway to selectively inhibit oncogenes in life-threatening tumors. Explain in very general terms how gene-silencing therapy might work to treat cancer and why this type of therapy would have fewer side effects than chemotherapy.

Answer: RNAi has the potential to selectively prevent protein production from any targeted gene. Oncogenes are mutant versions of "normal" genes (called proto-oncogenes) and their altered gene products are partly or wholly responsible for causing cancer. In theory, it may be possible to design appropriate siRNA molecules (small interfering RNAs) that specifically silence the mutant oncogene product but do not silence the closely related proto-oncogene product. The latter is necessary to prevent serious side effects as the products of proto-oncogenes are essential for normal cellular function.

9

Proteins and Their Synthesis

WORKING WITH THE FIGURES

1. The primary protein structure is shown in Figure 9-3(a). Where in the mRNA (near the 5′ or 3′ end) would a mutation in R_2 be encoded?

   Answer: As mRNA is translated 5′ to 3′, polypeptides are assembled from amino end to carboxyl end (the carboxyl end of the growing polypeptide chain is bound to the amino end of the incoming amino acid) Since R_2 is near the amino end of the protein, a mutation for R_2 would be near the 5′ end of the mRNA.

2. In this chapter you were introduced to nonsense suppressor mutations in tRNA genes. However, suppressor mutations also occur in protein-coding genes. Using the tertiary structure of the β subunit of hemoglobin shown in Figure 9-3(c), explain in structural terms how a mutation could cause the loss of globin protein function. Now explain how a mutation at a second site in the same protein could suppress this mutation and lead to a normal or near-normal protein.

   Answer: Proper folding is dependent on amino acid sequence and necessary for protein function. An amino acid replacement that disrupted folding of the β subunit would cause a loss of function for the protein because the correct tertiary structure could not form. If a subsequent mutation in another region of β complemented the first mutation by at least partially reestablishing the normal folding pattern, adequate tertiary structure could form and the first mutation would be suppressed. For example, if bonding between points A and B resulted in proper folding, a mutation that changed A would disrupt folding and cause loss of function. A subsequent mutation that changed B so it could bond with mutant A would reestablish the normal folding and normal function.

3. Using the quarternary structure of hemoglobin shown in Figure 9-3(d), explain in structural terms how a mutation in the β subunit protein could be suppressed by a mutation in the α subunit gene.

Answer: A mutation in the β subunit that prevented bonding of α and β subunits would prevent formation of the quaternary structure and block protein function. A complementary mutation in the α subunit that established the capacity to bond with the mutant β subunit would allow formation of a normal quaternary structure and effectively suppress the mutation in β.

4. Transfer RNAs (tRNAs) are examples of RNA molecules that do not encode protein. Based on Figures 9-6 and 9-8, what is the significance of the sequence of tRNA molecules? What do you predict would be the impact on translation of a mutation in one of the bases of one of the stems in the tRNA structure? On the mutant organism?

    Answer: The sequence of the tRNA molecules determines the three dimensional structure, producing the characteristic L shape. The conservation of the L shape among the different tRNAs implies an important function. If one of the bases in one of the stems were mutant, the formation of the L shape would likely be impaired, reducing or eliminating the capacity of the tRNA to act as an adapter molecule. For example, a mutant tRNA might not be able to bind with the synthetase molecule to become charged, or it might not form a sufficiently stable complex with the ribosome during translation. In the first case, the mutant tRNA would not participate in translation at all; in the second, it could disrupt translation when it was inserted. The overall effect on the mutant organism would probably be minimal because there are several copies of tRNA genes and the normal function would be served by the remaining normal copies of the gene.

5. Ribosomal RNAs (rRNAs) are another example of a functional RNA molecule. Based on Figure 9-11, what do you think is the significance of the secondary structure of rRNA?

    Answer: The amount of double stranded pairing in the rRNA indicates that a large part of the molecule will have a double-helical structure. The double-helical regions could potentially interact with ribosomal proteins via their major grooves. rRNA could also interact with other RNAs. The size of the rRNA indicates a complex function. The 16S rRNA contains the Shine–Dalgarno sequence, which directs the 30S subunit to the start codon, and it could also stabilize bonding between the 30S subunit and mRNA, and between the 30S and 50S subunits.

6. The components of prokaryotic and eukaryotic ribosomes are shown in Figure 9-10. Based on this figure, do you think that the large prokaryotic ribosomal RNA (23S rRNA) would be able to substitute for the eukaryotic 28S rRNA? Justify your answer.

Answer: It would be unlikely that the 23S rRNA of prokaryotes would be able to substitute for the 28S rRNA of eukaryotes. Its different size implies a different folding pattern which would affect its affinity for protein components of the 60S subunit. Also, the 60S subunit of eukaryotes contains 49 proteins, whereas the 50S of prokaryotes has 31 proteins. Thus, the 23S rRNA would have to be able to target a related but different set of proteins to function effectively. In both prokaryotes and eukaryotes, rRNA and ribosomal proteins have coevolved as a unit and interact closely during the formation of the ribosome and translation. Although specifics about the differences in the rRNA and proteins cannot be determined from the figure, the differences in rRNA sizes and number of protein components between prokaryotes and eukaryotes indicate evolutionary divergence in this intricate interaction. It would not be expected that a bacterial rRNA would interact functionally with the protein components of a eukaryotic ribosome.

7.   In Figure 9-12, is the terminal amino acid emerging from the ribosome encoded by the 5′ or 3′ end of the mRNA?

Answer: Figure 9-12(b) shows a growing polypeptide chain attached to the P site. The terminal amino acid shown emerging from the ribosome is at the amino end of the polypeptide and so was the first one added to the polypeptide chain. The amino end of the growing polypeptide chain will be the initial amino acid because the tRNA binds to the carboxyl end, leaving the amino end free. Since translation begins at the 5′ end of the mRNA and terminates at the 3′ end, this initial amino acid would be encoded at the 5′ end.

8.   In Figure 9-12(b), what do you think happens to the tRNA that is released from the E site?

Answer: Once the tRNA is released from the E site, it returns to the cytoplasm where it will be recharged with another amino acid. In this way, tRNAs are recycled through the translational machinery.

9.   In Figure 9-17, what do you think happens next to the ribosomal subunits after they are finished translating that mRNA?

Answer: Once translation is terminated, the ribosomal subunits are released from the mRNA and the 30S subunit is free to form a new initiation complex. Both subunits will ultimately participate in the translation of another mRNA, but it will not necessarily be the same mRNA.

10.   Based on Figure 9-19, can you predict the position of a mutation that would affect the synthesis of one isoform but not the other?

Answer: Exon 8 (in blue) is present in one isoform only, exon 9 (in green) is present in the other isoform only. A mutation in the 5′ GU in exon 8 would prevent splicing and block synthesis of the blue isoform. This mutation would not affect the green isoform because splicing for the green isoform does not involve the 5′ GU in exon 8. Likewise, a mutation in the 5′ GU in exon 9 would block synthesis of the green isoform only. Additionally, a nonsense mutation in either of these exons would block synthesis of that isoform but not the other.

11. Based on Figure 9-24, can you predict the position of a mutation that would produce an active protein that was not directed to the correct location?

Answer: A mutation in the signal sequence could prevent transfer of the protein to the lumen of the ER, thus preventing it from reaching its proper destination.

BASIC PROBLEMS

Unpacking Problem 12
  a. Use the codon dictionary in Figure 9-5 to complete the following table. Assume that reading is from left to right and that the columns represent transcriptional and translational alignments.

C												DNA double helix
						T	G	A				
	C	A			U							mRNA transcribed
									G	C	A	Appropriate tRNA anticodon
			Trp	Trp								Amino acids incorporated into protein

  b. Label the 5′ and 3′ ends of DNA and RNA, as well as the amino and carboxyl ends of the protein.

Answer:
3′ CGT ACC ACT GCA 5′   DNA double helix (transcribed strand)
5′ GCA TGG TGA CGT 3′   DNA double helix
5′ GCA UGG UGA CGU 3′   mRNA transcribed
3′ CGU ACC ACU GCA 5′   appropriate tRNA anticodon
NH₃ – Ala – Trp – (stop) –COOH   amino acids incorporated

13. Consider the following segment of DNA:

5′ GCTTCCCAA 3′
3′ CGAAGGGTT 5′

Assume that the top strand is the template strand used by RNA polymerase.

a. Draw the RNA transcribed.
b. Label its 5′ and 3′ ends.
c. Draw the corresponding amino acid chain.
d. Label its amino and carboxyl ends.

Repeat parts *a* through *d*, assuming the bottom strand to be the template strand.

Answer:
a. and b.   5′ UUG GGA AGC 3′

c. and d.   Assuming the reading frame starts at the first base:

NH3 – Leu – Gly – Ser - COOH

For the bottom strand, the mRNA is 5′ GCU UCC CAA 3′ and assuming the reading frame starts at the first base, the corresponding amino acid chain is NH3 - Ala - Ser - Gln - COOH.

14.   A mutational event inserts an extra nucleotide pair into DNA. Which of the following outcomes do you expect? (1) No protein at all; (2) a protein in which one amino acid is changed; (3) a protein in which three amino acids are changed; (4) a protein in which two amino acids are changed; (5) a protein in which most amino acids after the site of the insertion are changed.

Answer: (5) With an insertion, the reading frame is disrupted. This will result in a drastically altered protein from the insertion to the end of the protein (which may be much shorter or longer than wild type because of the location of stop signals in the altered reading frame).

15.   Before the true nature of the genetic coding process was fully understood, it was proposed that the message might be read in overlapping triplets. For example, the sequence GCAUC might be read as GCA CAU AUC:

Devise an experimental test of this idea.

Answer: A single nucleotide change should result in three adjacent amino acid changes in a protein. One and two adjacent amino acid changes would be expected to be much rarer than the three changes. This is directly the opposite of what is observed in proteins. Also, given any triplet coding for an amino acid, the next triplet could only be one of four. For example, if the first is GGG, then the next must be GGN (N = any base). This puts severe limits on which amino acids could be adjacent to each other. You could check amino acid sequences of various proteins to show that this is not the case.

16. In protein-synthesizing systems in vitro, the addition of a specific human mRNA to the *E. coli* translational apparatus (ribosomes, tRNA, and so forth) stimulates the synthesis of a protein very much like that specified by the mRNA. What does this result show?

Answer: It suggests very little evolutionary change between *E. coli* and humans with regard to the translational apparatus. The code is universal, the ribosomes are interchangeable, the tRNAs are interchangeable, and the enzymes involved are interchangeable. (Initiation of translation in prokaryotes *in vivo* requires specific base-pairing between the 3′ end of the 16s rRNA and a Shine–Dalgarno sequence found in the 5′ untranslated region of the mRNA. A Shine–Dalgarno sequence would not be expected (unless by chance) in a eukaryotic mRNA and therefore initiation of translation might not occur.)

17. Which anticodon would you predict for a tRNA species carrying isoleucine? Is there more than one possible answer? If so, state any alternative answers.

Answer: There are three codons for isoleucine: 5′ AUU 3′, 5′ AUC 3′, and 5′ AUA 3′. Possible anticodons are 3′ UAA 5′ (complementary), 3′ UAG 5′ (complementary), and 3′ UAI 5′ (wobble). 5′ UAU 3′, although complementary, would also base-pair with 5′ AUG 3′ (methionine) due to wobble and therefore would not be an acceptable alternative.

18. a. In how many cases in the genetic code would you fail to know the amino acid specified by a codon if you knew only the first two nucleotides of the codon?
    b. In how many cases would you fail to know the first two nucleotides of the codon if you knew which amino acid is specified by it?

Answer:
a. By studying the genetic code table provided in the textbook, you will discover that there are 28 codons that do not specify a particular amino acid with the first two positions (32 if you count Tyr and the stop codons starting with UA).

b. If you knew the amino acid, you would not know the first two nucleotides in the cases of Arg, Ser, and Leu.

19. Deduce what the six wild-type codons may have been in the mutants that led Brenner to infer the nature of the amber codon UAG.

Answer: The codon for amber is UAG. Listed below are the amino acids that would have been needed to be inserted to continue the wild-type chain and their codons:

glutamine	CAA, CAG*
lysine	AAA, AAG*
glutamic acid	GAA, GAG*
tyrosine	UAU*, UAC*
tryptophan	UGG*
serine	AGU, AGC, UCU, UCC, UCA, UCG*

In each case, the codon marked by an asterisk would require a single base change to become UAG.

20. If a polyribonucleotide contains equal amounts of randomly positioned adenine and uracil bases, what proportion of its triplets will encode (a) phenylalanine, (b) isoleucine, (c) leucine, (d) tyrosine?

Answer:
a. The codons for phenylalanine are UUU and UUC. Only the UUU codon can exist with randomly positioned A and U. Therefore, the chance of UUU is $(1/2)(1/2)(1/2) = 1/8$.

b. The codons for isoleucine are AUU, AUC, and AUA. AUC cannot exist. The probability of AUU is $(1/2)(1/2)(1/2) = 1/8$, and the probability of AUA is $(1/2)(1/2)(1/2) = 1/8$. The total probability is thus 1/4.

c. The codons for leucine are UUA, UUG, CUU, CUC, CUA, and CUG, of which only UUA can exist. It has a probability of $(1/2)(1/2)(1/2) = 1/8$.

d. The codons for tyrosine are UAU and UAC, of which only UAU can exist. It has a probability of $(1/2)(1/2)(1/2) = 1/8$.

21. You have synthesized three different messenger RNAs with bases incorporated in random sequence in the following ratios: **(a)** 1 U:5 C's, **(b)** 1 A:1 C:4 U's, **(c)** 1 A:1 C:1 G:1 U. In a protein-synthesizing system in vitro, indicate the identities and proportions of amino acids that will be incorporated into proteins when each of these mRNAs is tested. (Refer to Figure 9-5.)

Answer:
a. 1 U : 5 C — The probability of a U is 1/6, and the probability of a C is 5/6.

Codon	Amino acid	Probability	Sum
UUU	Phe	$(1/6)(1/6)(1/6) = 0.005$	Phe = 0.028
UUC	Phe	$(1/6)(1/6)(5/6) = 0.023$	
CCC	Pro	$(5/6)(5/6)(5/6) = 0.578$	Pro 0.694
CCU	Pro	$(5/6)(5/6)(1/6) = 0.116$	
UCC	Ser	$(1/6)(5/6)(5/6) = 0.116$	Ser = 0.139
UCU	Ser	$(1/6)(5/6)(1/6) = 0.023$	
CUC	Leu	$(5/6)(1/6)(5/6) = 0.116$	Leu = 0.139
CUU	Leu	$(5/6)(1/6)(1/6) = 0.023$	

1 Phe : 25 Pro : 5 Ser : 5 Leu

b. Using the same method as above, the final answer is 4 stop : 80 Phe : 40 Leu : 24 Ile : 24 Ser : 20 Tyr : 6 Pro : 6 Thr : 5 Asn : 5 His : 1 Lys : 1 Gln.

c. All amino acids are found in the proportions seen in the code table.

22. In the fungus *Neurospora,* some mutants were obtained that lacked activity for a certain enzyme. The mutations were found, by mapping, to be in either of two unlinked genes. Provide a possible explanation in reference to quaternary protein structure.

Answer: Quaternary structure is due to the interactions of subunits of a protein. In this example, the enzyme activity being studied may be from a protein consisting of two different subunits. Both subunits are required for activity. The polypeptides of the subunits are encoded by separate and unlinked genes.

23. A mutant is found that lacks all detectable function for one specific enzyme. If you had a labeled antibody that detects this protein in a Western blot (see Chapter 1), would you expect there to be any protein detectable by the antibody in the mutant? Explain.

Answer: There are a number of mutational changes that can lead to the absence of enzymatic function in the product of a gene. Some of these changes would result in the complete absence of protein product and therefore also the absence of a detectable band on a Western blot. Mutations such as deletions of the gene, for example, would result in the lack of detectable protein. Other mutations that destroy function (missense, for example) may not alter the production of the protein and would be detected on a Western blot. Still other mutations (nonsense, frameshift) could alter the size of the protein yet would still lead to detectable protein.

24. In a Western blot (see Chapter 1), the enzyme tryptophan synthetase usually shows two bands of different mobility on the gel. Some mutants with no

enzyme activity showed exactly the same bands as the wild type. Other mutants with no activity showed just the slow band; still others, just the fast band.

a. Explain the different types of mutants at the level of protein structure.
b. Why do you think there were no mutants that showed no bands?

Answer:
a. Tryptophan synthetase is a heterotetramer of two copies each of two different polypeptides, each encoded by a separate gene. Mutations that prevent the synthesis of one subunit would lead to the loss of one of the bands on a Western blot. Mutants that still make both subunits (those with exactly the same bands as wild type) might have mutations that prevent the subunits from interacting or disrupt the active site of the enzyme.

b. Because the two subunits are encoded by separate genes, the absence of both bands simultaneously would require two independent and rare mutagenic events.

25. In the Crick–Brenner experiments described in this chapter, three "insertions" or three "deletions" restored the normal reading frame and the deduction was that the code was read in groups of three. Is this deduction really proved by the experiments? Could a codon have been composed of six bases, for example?

Answer: Yes. It was not known at the time what number of bases the "plus" and "minus" mutations actually were. If each mutation was two bases, then a codon would have been six bases. Since the mutations were actually adding or subtracting single bases, the codon is indeed three bases.

26. A mutant has no activity for the enzyme isocitrate lyase. Does this result prove that the mutation is in the gene encoding isocitrate lyase?

Answer: No. The enzyme may require post-translational modification to be active. Mutations in the enzymes required for these modifications would not map to the isocitrate lyase gene.

27. A certain nonsense suppressor corrects a nongrowing mutant to a state that is near, but not exactly, wild type (it has abnormal growth). Suggest a possible reason why the reversion is not a full correction.

Answer: A nonsense suppressor is a mutation in a tRNA such that its anticodon can base-pair with a stop codon. In this way, a mutant stop codon (nonsense mutation) can be read through and the polypeptide can be fully synthesized. However, the mutant tRNA may be for an amino acid that was not encoded in that position in the original gene. For example, the codon UCG (serine) is instead UAG in the nonsense mutant. The suppressor mutation could be in tRNA for tryptophan such that its anticodon now recognizes UAG instead of UGG. During translation in the double mutant, the machinery puts tryptophan into the location of the mutant stop codon. This allows translation to continue but does place tryptophan into a position that was serine in the wild-type gene. This may create a protein that is not as active and a cell that is "not exactly wild type." Another explanation is that translation of the mutant gene is not as efficient and that premature termination still occurs some of the time. This would lead to less product and, again, a state that is "not exactly wild type."

28. In bacterial genes, as soon as any partial mRNA transcript is produced by the RNA polymerase system, the ribosome jumps on it and starts translating. Draw a diagram of this process, identifying 5′ and 3′ ends of mRNA, the COOH and NH_2 ends of the protein, the RNA polymerase, and at least one ribosome. (Why couldn't this system work in eukaryotes?)

Answer:

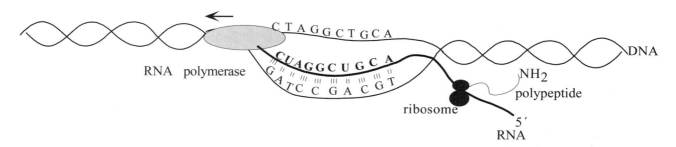

In eukaryotes, transcription occurs within the nucleus while translation occurs in the cytoplasm. Thus, the two processes cannot occur together.

29. In a haploid, a nonsense suppressor *su1* acts on mutation 1 but not on mutation 2 or 3 of gene *P*. An unlinked nonsense suppressor *su2* works on *P* mutation 2 but not on 1 or 3. Explain this pattern of suppression in regard to the nature of the mutations and the suppressors.

Answer: Assuming that the three mutations of gene *P* are all nonsense mutations, there are three different possible stop codons that might be the cause (amber, ochre, or opal). A suppressor mutation would be specific to one type of nonsense codon. For example, amber suppressors would suppress amber mutants but not opal or ochre.

30. In vitro translation systems have been developed in which specific RNA molecules can be added to a test tube containing a bacterial cell extract that includes all the components needed for translation (ribosomes, tRNAs, amino acids). If a radioactively labeled amino acid is included, any protein translated

from that RNA can be detected and displayed on a gel. If a eukaryotic mRNA is added to the test tube, would radioactive protein be produced? Explain.

Answer: Initiation of translation in prokaryotes requires specific base-pairing between the 3′ end of the 16s rRNA and a Shine–Dalgarno sequence found in the 5′ untranslated region of the mRNA. A Shine–Dalgarno sequence would not be expected (unless by chance) in a eukaryotic mRNA and therefore initiation of translation would not occur.

31. In a eukaryotic translation system (containing a cell extract from a eukaryotic cell) comparable with that in Problem 30, would a protein be produced by a bacterial RNA? If not, why not?

Answer: Initiaton of translation in eukaryotes requires initiation factors (eIF4a, b, and G) that associate with the 5′ cap of the mRNA. Because prokaryotic mRNAs are not capped, translation would not initiate.

32. Would a chimeric translation system containing the large ribosomal subunit from *E. coli* and the small ribosomal subunit from yeast (a unicellular eukaryote) be able to function in protein synthesis? Explain why or why not.

Answer: Not likely. Although the steps of translation and the components of ribosomes are similar in both eukaryotes and prokaryotes, the ribosomes are not identical. The sizes of both subunits are larger in eukaryotes and the many specific and intricate interactions that must take place between the small and large subunits would not be possible in a chimeric system.

33. Mutations that change a single amino acid in the active site of an enzyme can result in the synthesis of wild-type amounts of an inactive enzyme. Can you think of other regions in a protein where a single amino acid change might have the same result?

Answer: Single amino acid changes can result in changes in protein folding, protein targeting, or post-translational modifications. Any of these changes could give the results indicated.

34. What evidence supports the view that ribosomal RNAs are a more important component of the ribosome than the ribosomal proteins?

Answer: The first indication of rRNAs importance was the discovery of ribozymes. Recently, structural studies have shown that both the decoding center in the 30S subunit and the peptidyl transferase center in the 50S subunit are composed entirely of rRNA and that the important contacts in these centers are all tRNA/rRNA contacts.

35. Explain why antibiotics, such as erythromycin and Zithromax, that bind the large ribosomal subunit do not harm us.

Answer: Antibiotics need to selectively target bacterial structures and functions that are essential for life but unique or sufficiently different from the equivalent structure and functions of their animal hosts. The large bacterial ribosomal subunit fits these criteria as its function is obviously essential yet its structure is sufficiently different from the large eukaryotic ribosomal subunit. While the steps of protein synthesis are similar overall, eukaryotic ribosomes have larger and more numerous components. These differences make it possible to develop drugs that specifically bind bacterial ribosomes but have little or no affinity for eukaryotic ribosomes.

36. Why do multicellular eukaryotes need to have hundreds of kinase-encoding genes?

Answer: Recent studies indicate that most proteins function by interacting with other proteins. (The complete set of such interactions is called the interactome.) Most of these essential protein-protein interactions are regulated by phosphorylation/dephosphorylation modifications. Kinases are the enzymes that catalyze phosphorylations. Therefore, the complexity of the interactome necessary for the complexity of multicellularity requires the very large number of kinase-encoding genes.

37. Our immune system makes many different proteins that protect us from viral and bacterial infection. Biotechnology companies must produce large quantities of these immune proteins for human testing and eventual sale to the public. To this end, their scientists engineer bacterial or human cell cultures to express these immune proteins. Explain why proteins isolated from bacterial cultures are often inactive, whereas the same proteins isolated from human cell cultures are active (functional).

Answer: Bacterial and human cell cultures are both capable of producing the same polypeptide from the same mRNA, but that does not mean that the resulting protein will be active. Many proteins require posttranslational processing to become functional, and the enzymes and control necessary for such processing is not universal. The proteins produced and isolated from a human cell culture system will contain the necessary posttranslational modifications necessary for human protein function.

CHALLENGING PROBLEMS

38. A single nucleotide addition and a single nucleotide deletion approximately 15 sites apart in the DNA cause a protein change in sequence from

Lys–Ser–Pro–Ser–Leu–Asn–Ala–Ala–Lys

to

Lys–Val–His–His–Leu–Met–Ala–Ala–Lys

a. What are the old and new mRNA nucleotide sequences? (Use the codon dictionary in Figure 9-5.)

b. Which nucleotide has been added and which has been deleted?

(Problem 38 is from W. D. Stansfield, *Theory and Problems of Genetics.* McGraw-Hill, 1969.)

Answer:

a. and b. The goal of this type of problem is to align the two sequences. You are told that there is a single nucleotide addition and single nucleotide deletion, so look for single base differences that effect this alignment. These should be located where the protein sequence changes (i.e., between Lys-Ser and Asn-Ala). Remember also that the genetic code is redundant. (N = any base)

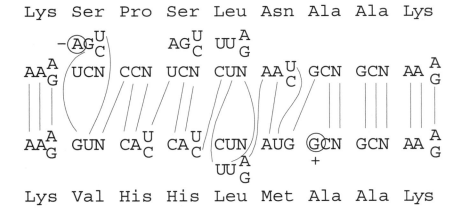

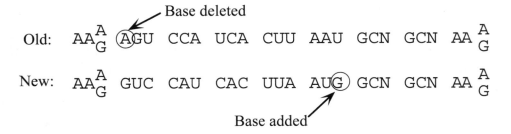

39. You are studying an *E. coli* gene that specifies a protein. A part of its sequence is

–Ala–Pro–Trp–Ser–Glu–Lys–Cys–His–

You recover a series of mutants for this gene that show no enzymatic activity. By isolating the mutant enzyme products, you find the following sequences:

Mutant 1:
    –Ala–Pro–Trp–Arg–Glu–Lys–Cys–His–
Mutant 2:
    –Ala–Pro–
Mutant 3:
    –Ala–Pro–Gly–Val–Lys–Asn–Cys–His–
Mutant 4:
    –Ala–Pro–Trp–Phe–Phe–Thr–Cys–His–

What is the molecular basis for each mutation? What is the DNA sequence that specifies this part of the protein?

Answer:
Mutant 1: A simple substitution of Arg for Ser exists, suggesting a nucleotide change. Two codons for Arg are AGA and AGG, and one codon for Ser is AGU. The U of the Ser codon could have been replaced by either an A or a G.

Mutant 2: The Trp codon (UGG) changed to a stop codon (UGA or UAG).

Mutant 3: Two frameshift mutations occurred:

5′–GCN CCN (–U)GGA GUG AAA AA(+U or C) UGU(or C) CAU(or C)–3′.

Mutant 4: An inversion occurred after Trp and before Cys. The DNA original sequence (with both strands shown for the area of inversion) was

3′–CGN GGN ACC TCA CTT TTT ACA(or G) GTA(or G) –5′
5′–   –AGT GAA AAA–   –3′

Therefore, the complementary RNA sequence was

5′–GCN CCN UGG AGU GAA AAA UGU/C CAU/C–3′

The DNA inverted sequence became

3′–CGN GGN ACC AAA AAG TGA ACA/G GTA/G–5′
          ^   ^

Therefore, the complementary RNA sequence was

5′–GCN CCN UGG UUU UUC ACU UGU/C CAU/C–3′
          ^   ^

40. Suppressors of frameshift mutations are now known. Propose a mechanism for their action.

Answer: If the anticodon on a tRNA molecule also was altered by mutation to be four bases long, with the fourth base on the 5′ side of the anticodon, it would suppress the insertion. Alterations in the ribosome can also induce frameshifting.

41. Consider the gene that specifies the structure of hemoglobin. Arrange the following events in the most likely sequence in which they would take place.

  a. Anemia is observed.
  b. The shape of the oxygen-binding site is altered.
  c. An incorrect codon is transcribed into hemoglobin mRNA.
  d. The ovum (female gamete) receives a high radiation dose.
  e. An incorrect codon is generated in the DNA of the hemoglobin gene.
  f. A mother (an X-ray technician) accidentally steps in front of an operating X-ray generator.
  g. A child dies.
  h. The oxygen-transport capacity of the body is severely impaired.
  i. The tRNA anticodon that lines up is one of a type that brings an unsuitable amino acid.
  j. Nucleotide-pair substitution occurs in the DNA of the gene for hemoglobin.

Answer: f, d, j, e, c, i, b, h, a, g

42. An induced cell mutant is isolated from a hamster tissue culture because of its resistance to α-amanitin (a poison derived from a fungus). Electrophoresis shows that the mutant has an altered RNA polymerase; *just one* electrophoretic band is in a position different from that of the wild-type polymerase. The cells are presumed to be diploid. What do the results of this experiment tell you about ways in which to detect recessive mutants in such cells?

Answer: Cells in long-established culture lines usually are not fully diploid. For reasons that are currently unknown, adaptation to culture frequently results in both karyotypic and gene dosage changes. This can result in hemizygosity for some genes, which allows for the expression of previously hidden recessive alleles.

43. A double-stranded DNA molecule with the sequence shown here produces, in vivo, a polypeptide that is five amino acids long.

TACATGATCATTTCACGGAATTTCTAGCATGTA
ATGTACTAGTAAAGTGCCTTAAAGATCGTACAT

a. Which strand of DNA is transcribed and in which direction?

b. Label the 5′ and the 3′ ends of each strand.

c. If an inversion occurs between the second and the third triplets from the left and right ends, respectively, and the same strand of DNA is transcribed, how long will the resultant polypeptide be?

d. Assume that the original molecule is intact and that the bottom strand is transcribed from left to right. Give the base sequence, and label the 5′ and 3′ ends of the anticodon that inserts the *fourth* amino acid into the nascent polypeptide. What is this amino acid?

Answer:

a. and b. The sequence of double-stranded DNA is as follows:

5′—TAC ATG ATC ATT TCA CGG AAT TTC TAG CAT GTA—3′
3′—ATG TAC TAG TAA AGT GCC TTA AAG ATC GTA CAT—5′

First look for stop codons. Next, look for the initiating codon, AUG (3′—TAC—5′ in DNA). Only the upper strand contains the necessary codons.

DNA	3′ TAC	GAT	CTT	TAA	GGC	ACT	5′
RNA	5′ AUG	CUA	GAA	AUU	CCG	UGA	3′
protein	Met	Leu	Glu	Ile	Pro	stop	

The DNA strand is read from right to left as written in your text and is written above in reverse order from your text.

c. Remember that polarity must be taken into account. The inversion is

DNA 5′ TAC	ATG	CTA	GAA	ATT	CCG	TGA	AAT	GAT	CAT	GTA 3′
RNA 3′		–GAU	CUU	UAA	GGC	ACU	UUA	CUA	GUA–	5′
amino acids	HOOC	7	6	5	4	3	2	1	–NH3	

d.

DNA	3′ATG	TAC	TAG	TAA	AGT	GCC	TTA	AAG	ATC	GTA	CAT 5′
mRNA	5′UAC	AUG	AUC	AUU	UCA	CGG	AAU	UUC	UAG 3′		
	1	2	3	4	5	6	7	stop			

Codon 4 is 5′–UCA–3′, which codes for Ser. Anticodon 4 would be 3′–AGU–5′ (or 3′–AGI–5′ given wobble).

44. One of the techniques used to decipher the genetic code was to synthesize polypeptides in vitro, with the use of synthetic mRNA with various repeating base sequences—for example, (AGA) $_n$, which can be written out as AGAAGAAGAAGAAGA. . . . Sometimes the synthesized polypeptide contained just one amino acid (a homopolymer), and sometimes it contained more than one (a heteropolymer), depending on the repeating sequence used. Furthermore, sometimes different polypeptides were made from the same synthetic mRNA, suggesting that the initiation of protein synthesis in the system in vitro does not always start on the end nucleotide of the messenger. For example, from $(AGA)_n$, three polypeptides may have been made: aa_1 homopolymer (abbreviated aa_1–aa_1), aa_2 homopolymer (aa_2–aa_2), and aa_3 homopolymer (aa_3–aa_3). These polypeptides probably correspond to the following readings derived by starting at different places in the sequence:

AGA AGA AGA AGA . . .
GAA GAA GAA GAA . . .
AAG AAG AAG AAG . . .

The following table shows the actual results obtained from the experiment done by Khorana.

Synthetic mRNA	Polypeptide(s) synthesized
$(UC)_n$	(Ser–Leu)
$(UG)_n$	(Val–Cys)
$(AC)_n$	(Thr–His)
$(AG)_n$	(Arg–Glu)
$(UUC)_n$	(Ser–Ser) and (Leu–Leu) and (Phe–Phe)
$(UUG)_n$	(Leu–Leu) and (Val–Val) and (Cys–Cys)
$(AAG)_n$	(Arg–Arg) and (Lys–Lys) and (Glu–Glu)
$(CAA)_n$	(Thr–Thr) and (Asn–Asn) and (Gln–Gln)
$(UAC)_n$	(Thr–Thr) and (Leu–Leu) and (Tyr–Tyr)
$(AUC)_n$	(Ile–Ile) and (Ser–Ser) and (His–His)
$(GUA)_n$	(Ser–Ser) and (Val–Val)
$(GAU)_n$	(Asp–Asp) and (Met–Met)
$(UAUC)_n$	(Tyr–Leu–Ser–Ile)
$(UUAC)_n$	(Leu–Leu–Thr–Tyr)
$(GAUA)_n$	None
$(GUAA)_n$	None

Note: The order in which the polypeptides or amino acids are listed in the table is not significant except for $(UAUC)_n$ and $(UUAC)_n$.

a. Why do $(GUA)_n$ and $(GAU)_n$ each encode only two homopolypeptides?

b. Why do $(GAUA)_n$ and $(GUAA)_n$ fail to stimulate synthesis?

c. Assign an amino acid to each triplet in the following list. Bear in mind that there are often several codons for a single amino acid and that the first two letters in a codon are usually the important ones (but that the third letter is occasionally significant). Also, remember that some very different-looking codons sometimes encode the same amino acid. Try to carry out this task without consulting Figure 9-5.

AUG	GAU	UUG	AAC
GUG	UUC	UUA	CAA
GUU	CUC	AUC	AGA
GUA	CUU	UAU	GAG
UGU	CUA	UAC	GAA
CAC	UCU	ACU	UAG
ACA	AGU	AAG	UGA

To solve this problem requires both logic and trial and error. Don't be disheartened: Khorana received a Nobel Prize for doing it. Good luck!

(Problem 44 is from J. Kuspira and G. W. Walker, *Genetics: Questions and Problems*. McGraw-Hill, 1973.)

Answer:

a. $(GAU)_n$ codes for Asp_n $(GAU)_n$, Met_n $(AUG)_n$, and $stop_n$ $(UGA)_n$. $(GUA)_n$ codes for Val_n $(GUA)_n$, Ser_n $(AGU)_n$, and $stop_n$ $(UAG)_n$.
One reading frame in each contains a stop codon.

b. Each of the three reading frames contains a stop codon.

c. The way to approach this problem is to focus initially on one amino acid at a time. For instance, line 4 indicates that the codon for Arg might be AGA or GAG. Line 7 indicates it might be AAG, AGA, or GAA. Therefore, Arg is at least AGA. That also means that Glu is GAG (line 4). Lys and Glu can be AAG or GAA (line 7). Because no other combinations except the ones already mentioned result in either Lys or Glu, no further decision can be made with respect to them. However, taking wobble into consideration, Glu may also be GAA, which leaves Lys as AAG.

Next, focus on lines 1 and 5. Ser and Leu can be UCU and CUC. Ser, Leu, and Phe can be UUC, UCU, and CUU. Phe is not UCU, which is seen in both lines. From line 14, CUU is Leu. Therefore, UUC is Phe, and UCU is Ser.

The footnote states that lines 13 and 14 are in the correct order. In line 13, if UCU is Ser (see above), then Ile is AUC, Tyr is UAU, and Leu is CUA.

Continued application of this approach will allow the assignment of an amino acid to each codon.

10

Gene Isolation and Manipulation

WORKING WITH THE FIGURES

1. Figure 10-1 shows that specific DNA fragments can be synthesized in vitro prior to cloning. What are two ways to synthesize DNA inserts for recombinant DNA in vitro?

    Answer: DNA can be amplified from genomic sequences *in vitro* using the polymerase chain reaction (PCR) or by copying the mRNA sequences into cDNA using reverse transcriptase.

2. In Figure 10-4, why is cDNA made only from mRNA and not also from tRNAs and ribosomal RNAs?

    Answer: cDNA is made from mRNA and not from tRNAs or rRNAs because polyT primers are used to prime the first DNA strand synthesis. Only the polyadenylated mRNAs will anneal to the primers.

3. Redraw Figure 10-6 with the goal of adding one *Eco*RI end and one *Xho*I end. Below is the *Xho*I recognition sequence.

    > Recognition sequence:
    > . . . C T C G A G . . .
    > . . . G A G C T C . . .
    >
    > After cut:
    > . . . C T C G A G . . .
    > . . . G A G C T C . . .

Answer:

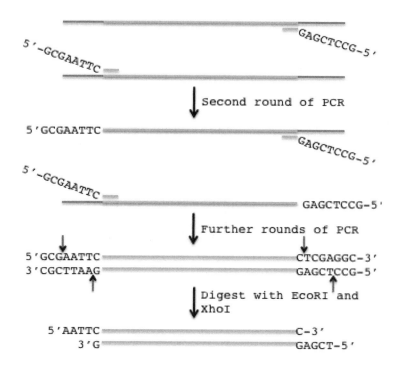

4. Redraw Figure 10-7 so that the cDNA can insert into an *Xho*I site of a vector rather than into an *Eco*RI site as shown.

Answer:

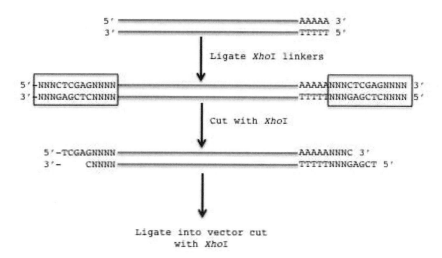

5. In Figure 10-11, determine approximately how many BAC clones are needed to provide 1 × coverage of

a. the yeast genome (12 Mbp).
b. the *E. coli* genome (4.1 Mbp).
c. the fruit-fly genome (130 Mbp).

Answer:
a. The yeast genome is approximately 12 Mb, if the average insert size is 100-200 kb, then 60-120 BAC clones will give 1X coverage.
b. For *E. coli*, only 21 to 41 clones will be required.
c. For fruitfly, 650-1300 clones will be required.

6. In Figure 10-15, why does DNA migrate to the anode (+ pole)?

Answer: DNA moves toward the positive pole in agarose gel electrophoresis because it is negatively charged.

7. In Figure 10-18(a), why are DNA fragments of different length and all ending in an A residue synthesized?

Answer: As DNA is synthesized in the sequencing reaction, the enzyme will randomly insert either dATP or ddATP across from T residues. If ddATP is selected, then the chain will terminate, as there is no 3′OH available for addition of the next nucleotide. As a result, the reaction will include fragments that have terminated at every position where an A is required.

8. As you will see in Chapter 15, most of the genomes of higher eukaryotes (plants and animals) are filled with DNA sequences that are present in hundreds, even thousands, of copies throughout the chromosomes. In the chromosome-walking procedure shown in Figure 10-20, how would the experimenter know whether the fragment he or she is using to "walk" to the next BAC or phage is repetitive? Can repetitive DNA be used in a chromosome walk?

Answer: The researcher would find that the fragment he or she was using as a probe to identify adjacent sequences would hybridize to many non-overlapping clones. Such repetitive sequences could not be used in a chromosome walk, since the repeated sequence would be present in many different genomic locations and it would be impossible to determine which was the correct one.

9. Redraw Figure 10-24 to include the positions of the single and double crossovers.

Answer:

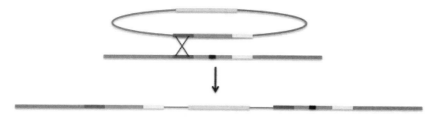

The position of the single crossover is as follows:

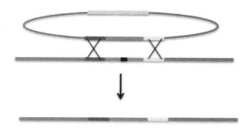

The positions of the double crossovers are as follows:

10. In Figure 10-26, why do only plant cells that have T-DNA inserts in their chromosomes grow on the agar plates? Do all of the cells of a transgenic plant grown from one clump of cells contain T-DNA? Justify your answer.

Answer: The T-DNA carries the kanamycin resistance gene; therefore, only cells that have acquired T-DNA inserts will grow on the agar plates containing the drug that selects for Kan^R.

11. In Figure 10-28, what is the difference between extra-chromosomal DNA and integrated arrays of DNA? Are the latter ectopic? What is distinctive about the syncitial region that makes it a good place to inject DNA?

Answer: Extrachromosomal arrays are maintained independently of the C. elegans chromosomes, while the integrated arrays become incorporated into the genome. The integrated arrays are ectopic, as they do not integrate into the homologous sequences in their normal chromosomal locus. The syncitial regio is a good place to inject DNA because there are a large number of nuclei in shared cytoplasm, any of which can take up the injected DNA. In addition, these cells will become egg or sperm, so the introduced genes will be passed on to individuals in the next generation.

BASIC PROBLEMS

12. From this chapter, make a list of all the examples of

(a) the hybridization of single-stranded DNAs and
(b) proteins that bind to DNA and then act on it.

Answer:

a. The following are examples of hybridization of single-stranded DNAs discussed in this chapter: use of primers (DNA sequencing, PCR), hybridization of sticky ends in cloning, probe hybridization (Southern blotting, Northern blotting, diagnosis of mutations, identification of clones, and *in situ* hybridization).

b. The following are examples of proteins that bind to and act on DNA discussed in this chapter: restriction enzymes, DNA polymerase (*Taq* polymerase), reverse transcriptase, and DNA ligase.

13. Compare and contrast the use of the word *recombinant* as used in the phrases
(a) "recombinant DNA" and
(b) "recombinant frequency."

Answer:

a. Recombinant DNA is the term used to describe techniques used to generate hybrid molecules consisting of DNA from two different sources. In this use, recombinant refers to the artificial joining of DNA from different sources.

b. Recombinant frequency is used to describe the statistical likelihood that two loci in the same organism will assort together or separately during meiosis. Here, recombinant refers to the combinations of loci that are different from the parent.

14. Why is ligase needed to make recombinant DNA? What would be the immediate consequence in the cloning process if someone forgot to add it?

Answer: Ligase is an essential enzyme within all cells that seals breaks in the phosphate-sugar backbone of DNA. During DNA replication it joins Okazaki fragments to create a continuous strand, and in cloning, it is used to join the various DNA fragments with the vector. If it was not added, the vector and cloned DNA would simply fall apart.

15. In the PCR process, if we assume that each cycle takes 5 minutes, how manyfold amplification would be accomplished in 1 hour?

Answer: Each cycle takes 5 minutes and doubles the DNA. In 1 hour there would be 12 cycles, so the DNA would be amplified $2^{12} = 4096$-fold amplification.

16. The position of the gene for the protein actin in the haploid fungus *Neurospora* is known from the complete genome sequence. If you had a slow-growing mutant that you suspected of being an actin mutant and you wanted to verify that it was one, would you
(a) clone the mutant by using convenient restriction sites flanking the actin gene and then sequence it or
(b) amplify the mutant sequence by using PCR and then sequence it?

Answer: The great advantage of PCR is that fewer procedures are necessary compared with cloning. However, it requires that the sequence be known, and that the primers are each present once in the genome and are sufficiently close (less than 2 kb). If these conditions are met, the primers determine the specificity of the DNA segment amplified and thus would be most efficient.

17. You obtain the DNA sequence of a mutant of a 2-kb gene in which you are interested and it shows base differences at three positions, all in different codons. One is a silent change, but the other two are missense changes (they encode new amino acids). How would you demonstrate that these changes are real mutations and not sequencing errors? (Assume that sequencing is about 99.9 percent accurate.)

Answer: Resequencing the relevant gene should be done, as this will tell you if the original sequence was correct. You can then use the mutant sequence in a gene replacement experiment (depending on the organism) to see if the mutant phenotype is actually the result of the sequence variation you detected.

18. In a T-DNA transformation of a plant with a transgene from a fungus (not found in plants), the presumptive transgenic plant does not express the expected phenotype of the transgene. How would you demonstrate that the transgene is in fact present? How would you demonstrate that the transgene was expressed?

Answer: You could isolate DNA from the suspected transgenic plant and probe for the presence of the transgene by Southern hybridization.

19. How would you produce a mouse that is homozygous for a rat growth-hormone transgene?

Answer: Inject the rat growth hormone gene (RGH) into fertilized eggs of mice. These eggs are then implanted into a surrogate mother and any resulting offspring mated to test their offspring to see if any are transgenic for RGH. Because RGH will be inherited in a dominant fashion, any large and transgenic offspring will be heterozygous at this point. The transgenic siblings will need to be mated to each other in order to generate mice that are homozygous for the RGH transgene.

20. Why was cDNA and not genomic DNA used in the commercial cloning of the human insulin gene?

Answer: The commercial cloning of insulin was into bacteria. Bacteria are not capable of processing introns. Genomic DNA would include the introns, while cDNA is a copy of processed (and thus intron-free) mRNA.

21. After *Drosophila* DNA has been treated with a restriction enzyme, the fragments are inserted into plasmids and selected as clones in *E. coli*. With the use of this "shotgun" technique, every DNA sequence of *Drosophila* in a library can be recovered.

   a. How would you identify a clone that contains DNA encoding the protein actin, whose amino acid sequence is known?
   b. How would you identify a clone encoding a specific tRNA?

Answer:
   a. Because the actin protein sequence is known, a probe could be synthesized by "guessing" the DNA sequence based on the amino acid sequence. (This works best if there is a region of amino acids that can be coded with minimal redundancy.) Alternatively, the gene for actin cloned in another species can be used as a probe to find the homologous gene in *Drosophila*. If an expression vector was used, it might also be possible to detect a clone coding for actin by screening with actin antibodies.

   b. Hybridization using the specific tRNA as a probe could identify a clone coding for itself.

22. In any particular transformed eukaryotic cell (say, of *Saccharomyces cerevisiae*), how could you tell if the transforming DNA (carried on a circular bacterial vector)

   a. replaced the resident gene of the recipient by double crossing over or single crossing over?
   b. was inserted ectopically?

Answer:
   a. The transformed phenotype would map to the same locus. If gene replacement occurred by a double crossing-over event, the transformed cells would not contain vector DNA. If a single crossing-over took place, the entire vector would now be part of the linear *Saccharomyces* chromosome.

b. The transformed phenotype would map to a different locus than that of the auxotroph if the transforming gene was inserted ectopically (i.e., at another location).

23. In an electrophoretic gel across which is applied a powerful electrical alternating pulsed field, the DNA of the haploid fungus *Neurospora crassa* ($n = 7$) moves slowly but eventually forms seven bands, which represent DNA fractions that are of different sizes and hence have moved at different speeds. These bands are presumed to be the seven chromosomes. How would you show which band corresponds to which chromosome?

    Answer: Size, translocations between known chromosomes, and hybridization to probes of known location can all be useful in identifying which band on a PFGE gel corresponds to a particular chromosome.

24. The protein encoded by the cystic-fibrosis gene is 1480 amino acids long, yet the gene spans 250 kb. How is this difference possible?

    Answer: Conservatively, the amount of DNA necessary to encode this protein of 445 amino acids is $445 \times 3 = 1335$ base pairs. When compared with the actual amount of DNA used, 60 kb, the gene appears to be roughly 45 times larger than necessary. This "extra" DNA mostly represents the introns that must be correctly spliced out of the primary transcript during RNA processing for correct translation. (There are also comparatively very small amounts of both 5′ and 3′ untranslated regions of the final mRNA that are necessary for correct translation encoded by this 60-kb of DNA.)

25. In yeast, you have sequenced a piece of wild-type DNA and it clearly contains a gene, but you do not know what gene it is. Therefore, to investigate further, you would like to find out its mutant phenotype. How would you use the cloned wild-type gene to do so? Show your experimental steps clearly.

    Answer: The typical procedure is to "knock out" the gene in question and then see if there is any observable phenotype. One methodology to do this is described in the companion text. A recombinant vector carrying a selectable gene within the gene of interest is used to transform yeast cells. Grown under appropriate conditions, yeast that have incorporated the marker gene will be selected. Many of these will have the gene of interest disrupted by the selectable gene. The phenotype of these cells would then be assessed to determine gene function.

CHALLENGING PROBLEMS

26. Prototrophy is often the phenotype selected to detect transformants. Prototrophic cells are used for donor DNA extraction; then this DNA is cloned and the clones are added to an auxotrophic recipient culture. Successful transformants are identified by plating the recipient culture on minimal medium and looking for colonies. What experimental design would you use to make sure that a colony that you hope is a transformant is not, in fact,

    a. a prototrophic cell that has entered the recipient culture as a contaminant?
    b. a revertant (mutation back to prototrophy by a second mutation in the originally mutated gene) of the auxotrophic mutation?

    Answer:
    a. To ensure that a colony is not, in fact, a prototrophic contaminant, the prototrophic line should be sensitive to a drug to which the recipient is resistant. A simple additional marker would also achieve the same end.

    b. Use a nonrevertible auxotroph as the recipient (such as one containing a deletion.)

27. A cloned fragment of DNA was sequenced by using the dideoxy chain-termination method. A part of the autoradiogram of the sequencing gel is represented here.

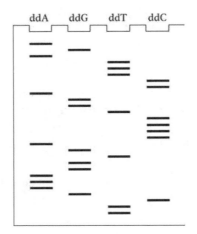

    a. Deduce the nucleotide sequence of the DNA nucleotide chain synthesized from the primer. Label the 5′ and 3′ ends.
    b. Deduce the nucleotide sequence of the DNA nucleotide chain used as the template strand. Label the 5′ and 3′ ends.
    c. Write out the nucleotide sequence of the DNA double helix (label the 5′ and 3′ ends).
    d. How many of the six reading frames are "open" as far as you can tell?

Answer:

a. The gel can be read from the bottom to the top in a 5′-to-3′ direction. The sequence is

5′ T T C G A A A G G T G A C C C C T G G A C C T T T A G A 3′

b. By complementarity, the template was

3′ A A G C T T T C C A C T G G G G A C C T G G A A A T C T 5′

c. The double helix is

5′ T T C G A A A G G T G A C C C C T G G A C C T T T A G A 3′
3′ A A G C T T T C C A C T G G G G A C C T G G A A A T C T 5′

d. Open reading frames have no stop codons. There are three frames for each strand, for a total of six possible reading frames. For the strand read from the gel, the transcript would be

5′ U C **UAA** A G G U C C A G G G G U C A C C U U U C G A A 3′

And for the template strand

5′ U U C G A A A G G **UGA** C C C C U G G A C C U U **UAG** A 3′

Stop codons are in bold and underlined. The two stop codons in the mRNA read from the template strand are both in the same reading frame. There are a total of four open reading frames of the six possible.

28. The cDNA clone for the human gene encoding tyrosinase was radioactively labeled and used in a Southern analysis of *Eco*RI-digested genomic DNA of wild-type mice. Three mouse fragments were found to be radioactive (were bound by the probe). When albino mice were used in this Southern analysis, no genomic fragments bound to the probe. Explain these results in relation to the nature of the wild-type and mutant mouse alleles.

Answer: The region of DNA that encodes tyrosinase in "normal" mouse genomic DNA contains two *Eco*RI sites. Thus, after *Eco*RI digestion, three different-sized fragments hybridize to the cDNA clone. When genomic DNA from certain albino mice is subjected to similar analysis, there are no DNA fragments that contain complementary sequences to the same cDNA. This indicates that these mice lack the ability to produce tyrosinase because the DNA that encodes the enzyme must be deleted.

29. Transgenic tobacco plants were obtained in which the vector Ti plasmid was designed to insert the gene of interest plus an adjacent kanamycin-resistance gene. The inheritance of chromosomal insertion was followed by testing progeny for kanamycin resistance. Two plants typified the results obtained generally. When plant 1 was backcrossed with wild-type tobacco, 50 percent of the progeny were kanamycin resistant and 50 percent were sensitive. When plant 2 was backcrossed with the wild type, 75 percent of the progeny were kanamycin resistant and 25 percent were sensitive. What must have been the difference between the two transgenic plants? What would you predict about the situation regarding the gene of interest?

Answer: Plant 1 shows the typical inheritance for a dominant gene that is heterozygous. Assuming kanamycin resistance is dominant to kanamycin sensitivity, the cross can be outlined as follows:

$$kan^R/kan^S \quad \times \quad kan^S/kan^S$$
$$\downarrow$$
$$1/2 \; kan^R/kan^S$$
$$1/2 \; kan^S/kan^S$$

This would suggest that the gene of interest would be inserted once into the genome.

Plant 2 shows a 3:1 ratio in the progeny of the backcross. This suggests that there have been two unlinked insertions of the kan^R gene and presumably the gene of interest as well.

$$kan^{R1}/kan^{S1} \; ; \; kan^{R2}/kan^{S2} \quad \times \quad kan^{S1}/kan^{S1} \; ; \; kan^{S2}/kan^{S2}$$
$$\downarrow$$
$$1/4 \; kan^{R1}/kan^{S1} \; ; \; kan^{R2}/kan^{S2}$$
$$1/4 \; kan^{R1}/kan^{S1} \; ; \; kan^{S2}/kan^{S2}$$
$$1/4 \; kan^{S1}/kan^{S1} \; ; \; kan^{R2}/kan^{S2}$$
$$1/4 \; kan^{S1}/kan^{S1} \; ; \; kan^{S2}/kan^{S2}$$

30. A cystic-fibrosis mutation in a certain pedigree is due to a single nucleotide-pair change. This change destroys an *Eco*RI restriction site normally found in this position. How would you use this information in counseling members of this family about their likelihood of being carriers? State the precise experiments needed. Assume that you find that a woman in this family is a carrier, and it transpires that she is married to an unrelated man who also is a heterozygote for cystic fibrosis, but, in his case, it is a different mutation in the same gene. How would you counsel this couple about the risks of a child's having cystic fibrosis?

Answer: Assuming that the DNA from this region is cloned, it could be used as a probe to detect this RFLP on Southern blots. DNA from individuals within this pedigree would be isolated (typically from blood samples containing white blood cells) and restricted with *Eco*RI, and Southern blots would be performed. Individuals with this mutant CF allele would have one band that would be larger (owing to the missing *Eco*RI site) when compared with wild type. Individuals that inherited this larger *Eco*RI fragment would, at minimum, be carriers for cystic fibrosis. In the specific case discussed in this problem, a woman that is heterozygous for this specific allele marries a man that is heterozygous for a different mutated CF allele. Just knowing that both are heterozygous, it is possible to predict that there is a 25 percent chance of their child's having CF. However, because the mother's allele is detectable on a Southern blot, it would be possible to test whether the fetus inherited this allele. DNA from the fetus (through either CVS or amniocentesis) could be isolated and tested for this specific *Eco*RI fragment. If the fetus did not inherit this allele, there would be a 0 percent chance of its having CF. On the other hand, if the fetus inherited this allele, there would be a 50 percent chance the child will have CF.

31. Bacterial glucuronidase converts a colorless substance called X-Gluc into a bright blue indigo pigment. The gene for glucuronidase also works in plants if given a plant promoter region. How would you use this gene as a reporter gene to find the tissues in which a plant gene that you have just cloned is normally active? (Assume that X-Gluc is easily taken up by the plant tissues.)

Answer: The promoter and control regions of the plant gene of interest must be cloned and joined in the correct orientation with the glucuronidase gene. This places the reporter gene under the same transcriptional control as the gene of interest. The companion text discusses the methodology used to create transgenic plants. Transform plant cells with the reporter gene construct, and as discussed in the text, grow into transgenic plants. The glucuronidase gene will now be expressed in the same developmental pattern as the gene of interest and its expression can easily be monitored by bathing the plant in an X-Gluc solution and assaying for the blue reaction product.

32. The plant *Arabidopsis thaliana* was transformed by using the Ti plasmid into which a kanamycin-resistance gene had been inserted in the T-DNA region. Two kanamycin-resistant colonies (A and B) were selected, and plants were regenerated from them. The plants were allowed to self-pollinate, and the results were as follows:

Plant A selfed → 3/4 progeny resistant to kanamycin
                     1/4 progeny sensitive to kanamycin
Plant B selfed → 5/16 progeny resistant to kanamycin
                     1/16 progeny sensitive to kanamycin

a. Draw the relevant plant chromosomes in both plants.
b. Explain the two different ratios.

Answer:

a. and b. During Ti plasmid transformation, the kanamycin gene will insert randomly into the plant chromosomes. Colony A, when selfed, has 3/4 kanamycin-resistant progeny, and colony B, when selfed, has 15/16 kanamycin-resistant progeny. This suggests that there was a single insertion into one chromosome in colony A and two independent insertions on separate chromosomes in colony B. This can be schematically represented by showing a single insertion within one of the pair of chromosome "A" for colony A.

Chromosomes "A"

and two independent insertions into one of each of the pairs of chromosomes "B" and "C" for colony B.

Chromosomes "B"

Chromosomes "C"

Genetically, this can be represented as

Colony A kan^{RA}/kan^{SA}
Colony B kan^{RB}/kan^{SB} ; kan^{RC}/kan^{SC}

When these are selfed $kan^{RA}/kan^{SA} \times kan^{RA}/kan^{SA}$
$$\downarrow$$
$1/4\ kan^{RA}/kan^{RA}$
$1/2\ kan^{RA}/kan^{SA}$
$1/4\ kan^{SA}/kan^{SA}$

kan^{RB}/kan^{SB} ; $kan^{RC}/kan^{SC} \times kan^{RB}/kan^{SB}$; kan^{RC}/kan^{SC}
$$\downarrow$$
$9/16\ kan^{RB}/-$; $kan^{RC}/-$
$3/16\ kan^{RB}/-$; kan^{SC}/kan^{SC}
$3/16\ kan^{SB}/kan^{SB}$; $kan^{RC}/-$
$1/16\ kan^{SB}/kan^{SB}$; kan^{SC}/kan^{SC}

11

Regulation of Gene Expression in Bacteria and Their Viruses

WORKING WITH THE FIGURES

1. Compare the structure of IPTG shown in Figure 11-7 with the structure of galactose shown in Figure 11-5. Why is IPTG bound by the *lac* repressor but not broken down by β-galactosidase?

   Answer: The sulfur atom in IPTG prevents hydrolysis by the beta-galactosidase enzyme.

2. Looking at Figure 11-9, why were partial diploids essential for establishing the trans-acting nature of the *lac* repressor? Could one distinguish cis-acting from transacting genes in haploids?

   Answer: Partial diploids were essential for distinguishing *cis*-acting from *trans*-acting mutants, because by definition one must introduce a second copy of the locus *in trans* to test this property of the mutants.

3. Why do promoter mutations cluster at positions –10 and –35 as shown in Figure 11-11?

   Answer: Promoter mutations cluster around the –10 and –35 positions because these are the DNA sequences recognized and bound by the sigma subunit of the polymerase. Alterations to these sequences will affect the ability of the RNA polymerase holoenzyme to recognize the promoter.

4. Looking at Figure 11-16, how large is the overlap between the operator and the *lac* transcription unit?

   Answer: The operator sequence overlaps the *lac* operon transcription unit by 24 base pairs.

5. Examining Figure 11-21, what effect do you predict *trpA* mutations will have on tryptophan levels?

Answer: Mutations that inactivate the *trpA* gene will block the synthesis of tryptophan in the cells, resulting in trp⁻ auxotrophs.

6. Examining Figure 11-21, what effect do you predict *trpA* mutations have on *trp* mRNA expression?

Answer: Because the *trpA* gene is the last to be transcribed, mutations in the *trpA* would not be predicted to affect transcription of the operon.

BASIC PROBLEMS

7. Explain why I^- alleles in the *lac* system are normally recessive to I^+ alleles and why I^+ alleles are recessive to IS alleles.

Answer: The *I* gene determines the synthesis of a repressor molecule, which blocks expression of the *lac* operon and which is inactivated by the inducer. The presence of the repressor I^+ will be dominant to the absence of a repressor I^-. I^s mutants are unresponsive to an inducer. For this reason, the gene product cannot be stopped from interacting with the operator and blocking the *lac* operon. Therefore, I^s is dominant to I^+.

8. What do we mean when we say that O^C mutations in the *lac* system are cis-acting?

Answer: O^c mutants are changes in the DNA sequence of the operator that impair the binding of the *lac* repressor. Therefore, the *lac* operon associated with the O^c operator cannot be turned off. Because an operator controls only the genes on the same DNA strand, it is cis (on the same strand) and dominant (cannot be turned off).

Unpacking the Problem

9. The genes shown in the following table are from the *lac* operon system of *E. coli*. The symbols *a, b,* and *c* represent the repressor (*I*) gene, the operator (*O*) region, and the structural gene (*Z*) for β-galactosidase, although not necessarily in that order. Furthermore, the order in which the symbols are written in the genotypes is not necessarily the actual sequence in the *lac* operon.

Activity (+) or inactivity (−) of *Z* gene

Genotype	Inducer absent	Inducer present
$a^- b^+ c^+$	+	+
$a^+ b^+ c^-$	+	+
$a^+ b^- c^-$	−	−
$a^+ b^- c^+/a^- b^+ c^-$	+	+
$a^+ b^+ c^+/a^- b^- c^-$	−	+
$a^+ b^+ c^-/a^- b^- c^+$	−	+
$a^- b^+ c^+/a^+ b^- c^-$	+	+

a. State which symbol (*a, b,* or *c*) represents each of the *lac* genes *I, O,* and *Z.*

b. In the table, a superscript minus sign on a gene symbol merely indicates a mutant, but you know that some mutant behaviors in this system are given special mutant designations. Use the conventional gene symbols for the *lac* operon to designate each genotype in the table.

(Problem 9 is from J. Kuspira and G. W. Walker, *Genetics: Questions and Problems.* Copyright 1973 by McGraw-Hill.)

Answer:

a. You are told that *a, b,* and *c* represent *lacI, lacO,* and *lacZ,* but you do not know which is which. Both a^- and c^- have constitutive phenotypes (lines 1 and 2) and therefore must represent mutations in either the operator (*lacO*) or the repressor (*lacI*). b^- (line 3) shows no ß-gal activity and by elimination must represent the *lacZ* gene.

Mutations in the operator will be cis-dominant and will cause constitutive expression of the *lacZ* gene only if it's on the same chromosome. Line 6 has c^- on the same chromosome as b^+ but the phenotype is still inducible (owing to c^+ in trans). Line 7 has a^- on the same chromosome as b^+ and is constitutive even though the other chromosome is a^+. Therefore *a* is *lacO, c* is *lacI,* and *b* is *lacZ.*

b. Another way of labeling mutants of the operator is to denote that they lead to a constitutive phenotype; *lacO⁻* (or a^-) can also be written as *lacOᶜ*. There are also mutations of the repressor that fail to bind inducer (allolactose) as opposed to fail to bind DNA. These two classes have quite different phenotypes and are distinguished by *lacIˢ* (fails to bind allolactose and leads to a dominant uninducible phenotype in the presence of a wild-type operator) and *lacI⁻* (fails to bind DNA and is recessive). It is possible that line 3, line 4, and line 7 have *lacIˢ* mutations (because dominance cannot be ascertained in a cell that is also *lacOᶜ*), but the other c^- alleles must be *lacI⁻*.

10. The map of the *lac* operon is

POZY

The promoter (P) region is the start site of transcription through the binding of the RNA polymerase molecule before actual mRNA production. Mutationally altered promoters (P^-) apparently cannot bind the RNA polymerase molecule. Certain predictions can be made about the effect of P^- mutations. Use your predictions and your knowledge of the lactose system to complete the following table. Insert a "+" where an enzyme is produced and a "–" where no enzyme is produced. The first one has been done as an example.

Genotype	β-Galactosidase		Permease	
	No lactose	Lactose	No lactose	Lactose
$I^+ P^+ O^+ Z^+ Y^+ / I^+ P^+ O^+ Z^+ Y^+$	–	+	–	+
a. $I^- P^+ O^C Z^+ Y^- / I^+ P^+ O^+ Z^- Y^+$				
b. $I^+ P^- O^C Z^- Y^+ / I^- P^+ O^C Z^+ Y^-$				
c. $I^S P^+ O^+ Z^+ Y^- / I^+ P^+ O^+ Z^- Y^+$				
d. $I^S P^+ O^+ Z^+ Y^- / I^- P^+ O^+ Z^+ Y^+$				
e. $I^- P^+ O^C Z^+ Y^- / I^- P^+ O^+ Z^- Y^+$				
f. $I^- P^- O^+ Z^+ Y^+ / I^- P^+ O^C Z^+ Y^-$				
g. $I^+ P^+ O^+ Z^- Y^+ / I^- P^+ O^+ Z^+ Y^-$				

Answer:

	ß-Galactosidase		**Permease**	
Part	No lactose	Lactose	No lactose	Lactose
a	+	+	–	+
b	+	+	–	–
c	–	–	–	–
d	–	–	–	–
e	+	+	+	+
f	+	+	–	–
g	–	+	–	+

a. The O^C mutation leads to the constitutive synthesis of ß-galactosidase because it is cis to a $lacZ^+$ gene, but the permease is inducible because the $lacY^+$ gene is cis to a wild-type operator.

b. The $lacP^-$ mutation prevents transcription so only the genes cis to $lacP^+$ will be transcribed. These genes are also cis to O^C so the $lacZ^+$ gene is transcribed constitutively.

c. The $lacI^s$ is a trans-dominant mutation and prevents transcription from either operon.

d. Same as part **c**.

e. There is no functional repressor made (and one operator is mutant as well).

f. Same as part **b**.

g. Both operators are wild type and the one functional copy of *lacI* will direct the synthesis of enough repressor to control both operons.

11. Explain the fundamental differences between negative control and positive control.

Answer: A gene is turned off or inactivated by the "modulator" (usually called a *repressor*) in negative control, and the repressor must be removed for transcription to occur. A gene is turned on by the "modulator" (usually called an *activator*) in positive control, and the activator must be added or converted to an active form for transcription to occur.

12. Mutants that are *lacY⁻* retain the capacity to synthesize β-galactosidase. However, even though the *lacI* gene is still intact, β-galactosidase can no longer be induced by adding lactose to the medium. Explain.

Answer: The *lacY* gene produces a permease that transports lactose into the cell. A cell containing a *lacY⁻* mutation cannot transport lactose into the cell, so ß-galactosidase will not be induced. (In wild-type cells, even when the repressor is present, there is still a small amount of transcription. This allows a small amount of baseline permease (and β-gal) expression. It is this low level of permease that allows trace amounts of lactose to enter the cell and initiate induction of the *lac* operon.)

13. What are the analogies between the mechanisms controlling the *lac* operon and those controlling phage λ genetic switches?

Answer: Both the *lac* operon and phase λ genetic switches can be interpreted as a simple switch between two states. For λ, there is competition between Cro and cI proteins binding at an operator to control the choice between the lysogenic and lytic life cycles. For the *lac* operon, a repressor binds at the operator to prevent transcription of the structural genes in the absence of lactose. Both also are capable of interpreting their environments and directing an "appropriate" response. The *lac* operon responds to the presence or absence of lactose and glucose. For λ, if resources are abundant, or the bacterial cell sustains DNA damage, the lytic cycle prevails, but if resources are not abundant, λ will enter the lysogenic cycle.

14. Compare the arrangement of cis-acting sites in the control regions of the *lac* operon and phage λ.

Answer: In the *lac* operon, the promoter is located between the CAP binding site and the repressor binding site (the operator). The *lac* repressor binds at the operator and blocks transcription, and CAP binds at its binding site and activates transcription. Complete induction of the *lac* structural genes requires the binding of CAP and the absence of binding of repressor. The λ genetic switch consists of an operator (with three binding sites for λ repressor (cI) and Cro) that overlaps two promoters, P_R and P_{RM}. P_R promotes transcription of lytic genes and P_{RM} promotes transcription of the cI gene. When repressor is bound, transcription from P_R is blocked, but when Cro is bound, transcription from P_{RM} is blocked.

CHALLENGING PROBLEMS

15. An interesting mutation in *lacI* results in repressors with 110-fold increased binding to both operator and nonoperator DNA. These repressors display a "reverse" induction curve, allowing β-galactosidase synthesis in the absence of an inducer (IPTG) but partly repressing β-galactosidase expression in the presence of IPTG. How can you explain this? (Note that, when IPTG binds a repressor, it does not completely destroy operator affinity, but rather it reduces affinity 1100-fold. Additionally, as cells divide and new operators are generated by the synthesis of daughter strands, the repressor must find the new operators by searching along the DNA, rapidly binding to nonoperator sequences and dissociating from them.)

    Answer: Normally, the repressor searches for the operator by rapidly binding and dissociating from nonoperator sequences. Even for sequences that mimic the true operator, the dissociation time is only a few seconds or less. Therefore, it is easy for the repressor to find new operators as new strands of DNA are synthesized. However, when the affinity of the repressor for DNA and operator is increased, it takes too long for the repressor to dissociate from sequences on the chromosome that mimic the true operator, and as the cell divides and new operators are synthesized, the repressor never quite finds all of them in time, leading to a partial synthesis of ß-galactosidase. This explains why, in the absence of IPTG, there is some elevated ß-galactosidase synthesis. When IPTG binds to the repressors with increased affinity, it lowers the affinity back to that of the normal repressor (without IPTG bound). Then, the repressor can rapidly dissociate from sequences in the chromosome that mimic the operator and find the true operator. Thus, ß-galactosidase is repressed in the presence of IPTG in strains with repressors that have greatly increased affinity for operator. In summary, because of a kinetic phenomenon, we see a reverse induction curve.

16. In *Neurospora*, all mutants affecting the enzymes carbamyl phosphate synthetase and aspartate transcarbamylase map at the *pyr*-3 locus. If you induce *pyr*-3 mutations by ICR-170 (a chemical mutagen), you find that either both enzyme functions are lacking or only the transcarbamylase function is lacking;

in no case is the synthetase activity lacking when the transcarbamylase activity is present. (ICR-170 is assumed to induce frameshifts.) Interpret these results in regard to a possible operon.

Answer: If there is an operon governing both genes, then a frameshift mutation could cause the stop codon separating the two genes to be read as a sense codon. Therefore, the second gene product will be incorrect for almost all amino acids. However, there are no known polycistronic messages in eukaryotes. The alternative, and better, explanation is that both enzymatic functions are performed by the same gene product. Here, a frameshift mutation beyond the first function, carbamyl phosphate synthetase, will result in the second half of the protein molecule being nonfunctional.

17. Certain *lacI* mutations eliminate operator binding by the *lac* repressor but do not affect the aggregation of subunits to make a tetramer, the active form of the repressor. These mutations are partly dominant over wild type. Can you explain the partly dominant I^- phenotype of the I^-/I^+ heterodiploids?

Answer: Because very small amounts of the repressor are made, the system as a whole is quite responsive to changes in repressor concentration. In the heterodiploids, repressor heterotetramers may form by association of polypeptides encoded by both I^- and I^+. Only homotetramers of I^+ work properly, and their concentration will be reduced by the various heterotetramer combinations of I^- and I^+. This will result in some expression of the *lac* genes in the absence of lactose.

18. You are examining the regulation of the lactose operon in the bacterium *Escherichia coli*. You isolate seven new independent mutant strains that lack the products of all three structural genes. You suspect that some of these mutations are *lacI*^{S} mutations and that other mutations are alterations that prevent the binding of RNA polymerase to the promoter region. Using whatever haploid and partial diploid genotypes that you think are necessary, describe a set of genotypes that will permit you to distinguish between the *lacI* and *lacP* classes of uninducible mutations.

Answer: The key to this question is to remember that *lacI* mutations will be trans-acting as they produce a protein product, and that *lacP* mutations (like *lacO* mutations) will be cis-acting as they are affecting a binding site for RNA polymerase. For the purposes of this problem, designate the uninducible *lacI* mutations as i^u mutations, and the uninducible *lacP* mutations as p^u mutations. There are quite a number of satisfactory genotypes which can serve to distinguish between the i^u and pu mutations. Here are a few examples:

| Genotype | Enzyme activity | |
	ß-galactosidase	Permease
1 $i^u p^+ o^+ z^+ y^+$	absent	absent
2 $i^+ p^u o^+ z^+ y^+$	absent	absent
3 $i^u p^+ o^+ z^+ y^- / i^+ p^+ o^+ z^- y^+$	absent	absent
4 $i^+ p^u o^+ z^+ y^- / i^+ p^+ o^+ z^- y^+$	absent	inducible
5 $i^u p^+ o^c z^+ y^- / i^+ p^+ o^+ z^- y^+$	constitutive	absent
6 $i^+ p^u o^c z^+ y^- / i^+ p^+ o^+ z^- y^+$	absent	inducible

Genotypes 1 and 2 are simply symbolic restatements of the phenotypes of the uninducible mutations. Genotypes 3 and 4 are straightforward tests to distinguish the cis-acting p^u mutations from the trans-acting i^u lesions. The results of genotype 3 reflect the expectation that i^u mutations would be trans-acting and dominant to i^+. This is expected because the i^u-encoded repressor protein molecules would be incapable of being inactived by binding to inducer; the presence or absence of normal repressor protein is irrelevant. The results of genotype 4, on the other hand, reflect the expectation that p^u mutations would only be cis-acting. Hence, any genes in cis to the *pu* allele would be inactive, while any genes in cis to the normal p^+ allele would potentially be transcribed normally (if all other regulatory functions were normal). In a similar fashion, genotypes 5 and 6 distinguish the cis versus trans action of i^u and p^u mutations. In genotype 5, i^u remains trans-dominant to i^+, but this dominance is overcome by the cis-acting o^c mutation (compare genotypes 3 and 5). In genotype 6, the presence of o^c is irrelevant, as it is in a *lac* operon that contains the p^u mutation preventing RNA polymerase binding (compare genotypes 4 and 6).

19. You are studying the properties of a new kind of regulatory mutation of the lactose operon. This mutation, called *S*, leads to the complete repression of the *lacZ*, *lacY*, and *lacA* genes, regardless of whether inducer (lactose) is present. The results of studies of this mutation in partial diploids demonstrate that this mutation is completely dominant over wild type. When you treat bacteria of the *S* mutant strain with a mutagen and select for mutant bacteria that can express the enzymes encoded by *lacZ*, *lacY*, and *lacA* genes in the presence of lactose, some of the mutations map to the *lac* operator region and others to the *lac* repressor gene. On the basis of your knowledge of the lactose operon, provide a molecular genetic explanation for all these properties of the *S* mutation. Include an explanation of the constitutive nature of the "reverse mutations."

Answer: The *S* mutation is an alteration in *lacI* such that the repressor protein binds to the operator, regardless of whether inducer is present or not. In other words, it is a mutation that inactivates the allosteric site which binds to inducer, while not affecting the ability of the repressor to bind to the operator site. The dominance of the *S* mutation is due to the binding of the mutant repressor, even under circumstances when normal repressor does not bind to DNA (that is, in the presence of inducer). The constitutive reverse mutations that map to *lacI* are mutational events which inactivate the ability of this repressor to bind to the operator. The constitutive reverse mutations that map to the operator alter the operator DNA sequence such that it will not permit binding to any repressor molecules (wild-type or mutant repressor).

20. The *trp* operon in *E. coli* encodes enzymes essential for the biosynthesis of tryptophan. The general mechanism for controlling the *trp* operon is similar to that observed with the *lac* operon: when the repressor binds to the operator, transcription is prevented; when the repressor does not bind the operator, transcription proceeds. The regulation of the *trp* operon differs from the regulation of the *lac* operon in the following way: the enzymes encoded by the *trp* opoeron are not synthesized when tryptophan is present but rather when it is absent. In the *trp* operon, the repressor has two binding sites: one for DNA and the other for the effector molecule, tryptophan. The *trp* repressor must first bind to a molecule of tryptophan before it can bind effectively to the *trp* operator.

   a. Draw a map of the tryptophan operon, indicating the promoter (P), the operator (O), and the first structural gene of the tryptothan operon (*trpA*). In your drawing, indicate where on the DNA the repressor protein binds when it is bound to tryptophan.

   b. The *trpR* gene encodes the repressor; *trpO* is the operator; *trpA* encodes the enzyme tryptophan synthetase. A *trpR*^{2} repressor cannot bind tryptophan, a *trpO*^{2} operator cannot be bound by the repressor, and the enzyme encoded by a *trpA*^{2} mutant gene is completely inactive. Do you expect to find active tryptophan synthetase in each of the following mutant strains when the cells are grown in the presence of tryptophan? In its absence?

   i. $R^+ O^+ A^+$ (wild type)
   ii. $R^- O^+ A^+ / R^+ O^+ A^-$
   iii. $R^+ O^- A^+ / R^+ O^+ A^-$

   Answer:
   a.

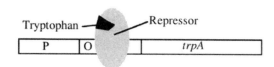

   b. i. *trpA* is not synthesized in the presence of tryptophan; it is synthesized in the absence of tryptophan.

   ii. *trpA* is not synthesized in the presence of tryptophan; it is synthesizedin the absence of tryptophan. The second operon contains a *trpA*$^-$ mutation and is unable to make any active tryptophan synthetase enzyme. However, this operon contains a wild-type *trpR* gene, which encodes a functional repressor molecule.

Because this gene product is a diffusible molecule (trans-acting), it will act on the other DNA molecule containing the mutant *trpR⁻* gene and bind to the operator on that molecule. Thus, *trpA* gene expression will be repressed when tryptophan binds the repressor, causing the repressor to bind the operator. In the absence of tryptophan, the repressor will not bind to either operator, and transcription will proceed.

iii. *trpA* is synthesized both in the presence and in the absence of tryptophan. Because the second operon contains a mutant *trpA⁻* gene, no functional tryptophan synthetase enzyme will be made from this DNA molecule. The first operon contains a wild-type *trpA* gene and will be responsible for the intracellular supply of tryptophan synthetase and ultimately, tryptophan. However, this operon contains a mutant (*trpO⁻*) operator region that cannot be bound by the repressor molecules. Because the operator region is cis-acting and does not encode a diffusible gene product, the wild-type *trpO* gene on the other DNA molecule cannot substitute for the mutant operator. Therefore, the *trpA* gene will always be transcribed (constitutive) regardless of the levels of tryptophan in the cell.

21. The activity of the enzyme β-galactosidase produced by wild-type cells grown in media supplemented with different carbon sources is measured. In relative units, the following levels of activity are found:

Glucose	Lactose	Lactose + glucose
0	100	1

Predict the relative levels of β-galactosidase activity in cells grown under similar conditions when the cells are *lacI⁻*, *lacIˢ*, *lacO⁻*, and *crp⁻*.

Answer:

	Glucose	Lactose	Lactose + glucose
wild-type	0	100	1
lacI⁻	1	100	1
lacIˢ	0	0	0
lacO⁻	1	100	1
crp⁻	0	1	1

lacI⁻—leads to absence of negative control by repressor binding but still under the positive control of CAP-cAMP binding

lacIˢ—because the repressor does not bind lactose but still binds DNA, transcription will be blocked under all conditions

lacO⁻—leads to absence of negative control by repressor binding but still under the positive control of CAP-cAMP binding

crp^- —leads to absence of positive control by CAP-cAMP binding but still under the negative control of repressor binding

22. A bacteriophage λ is found that is able to lysogenize its *E. coli* host at 30°C but not at 42°C. What genes may be mutant in this phage?

Answer: Mutations in *cI, cII, cIII* would all effect lysogeny. *cI* encodes the repressor, *cII* encodes an activator of P_{RE}, and *cIII* encodes a protein that protects *cII* from degradation. Mutations in *N* (an antiterminator) would also effect lysogeny as its function is required for transcription of the *cII* and *cIII* genes, but it is also necessary for genes involved in lysis. Mutations in the gene encoding the integrase (*int*) would also effect the ability of a mutant phage to lysogenize.

23. What would happen to the ability of bacteriophage λ to lyse a host cell if it acquired a mutation in the O_R binding site for the Cro protein? Why?

Answer: A mutation that prevents Cro binding to O_R would prevent the lytic cycle. *cI* binding would still allow lysogeny. Cro binding is necessary to block the maintenance of *cI* transcription and allow transcription of genes for the lytic cycle.

24. Contrast the effects of mutations in genes encoding sporulation-specific σ factors with mutations in the −35 and −10 regions of the promoters of genes in their regulons.

a. Would functional mutations in the σ-factor genes or in the individual promoters have the greater effect on sporulation?
b. On the basis of the sequences shown in Figure 11-30b, would you expect all point mutations in −35 or −10 regions to affect gene expression?

Answer:
a. Sporulation-specific σ factors control many genes. For example, σ^E binds to at least 121 promoters, within 34 operons and 87 individual genes, to regulate more than 250 genes. A loss of this σ factor would prevent the proper transcription of all of these more than 250 genes. The loss of any one of the 121 promoters controlled by this σ factor would only affect the operon or individual gene of that promoter.

b. If you look carefully at the −35 and −10 sequences shown in Figure 11-30b, you will observe that there are differences in these sequences among the genes controlled by the same σ factor. For example, the −35 region of the promoter for the *ybaN* gene is TTATATT, but the same region for the *ydcA* gene is TACTATT, yet both still function correctly. When comparing all promoters regulated by a particular σ factor, a "consensus" sequence can be determined, but there will be variations observed as well. Given this, you would not expect all point mutations in this region to effect gene expression. Some variation is possible.

12

Regulation of Gene Expression in Eukaryotes

WORKING WITH THE FIGURES

1. In Figure 12-4, certain mutations decrease the relative transcription rate of the β-globin gene. Where are these mutations located, and how do they exert their effects on transcription?

    Answer: The mutations that decrease transcription all fall within the promotor-proximal and promotor elements, which are located upstream of the coding region and serve as binding sites for RNA PolII and general transcription factors. Mutations in these regions hinder protein-DNA interactions and thereby reduce transcription. The figure shows that this reduction can be substantial when even a single nucleotide is changed.

2. Based on the information in Figure 12-6, how does Gal4 regulate four different *GAL* genes at the same time? Contrast this mechanism with how the Lac repressor controls the expression of three genes.

    Answer: Gal4 has a DNA binding domain that targets a specific UAS sequence. This sequence is present in the enhancer region of four different GAL genes. Gal4 is able to activate transcription from the four separate GAL promoters through its capacity to bind with all four UASs. In prokaryotes, the Lac repressor controls expression of three genes by blocking or unblocking access of RNA Polymerase to a single promotor region that is common to all three genes. These two regulatory systems are similar in that genes involved in metabolic pathways for a specific sugar are coordinately induced by the presence of the sugar.

3. In any experiment, controls are essential in order to determine the specific effect of changing some parameter. In Figure 12-7, which constructs are the "controls" that serve to establish the principle that activation domains are modular and interchangeable?

Answer: Controls lack certain experimental treatments, to establish a baseline for resolution of treatment effects. In this figure, one control (we could call this a positive control) is illustrated in panel *a*, which demonstrates normal expression of the lacZ reporter when the Gal4 DNA binding domain and activation domain are both present. The other set of controls (negative controls) consists of DNA binding domains from LexA and Gal4 that lack activation domains. They show that in the absence of an activation domain, the reporter gene lacZ is not expressed. Together, these controls allow a robust interpretation of panel d, where a LexA-Gal4 hybdrid is used, which is that the match between the activation domain and the promotor region activates transcription independently of the DNA binding domain.

4. Contrast the role of the MCM1 protein in different yeast cell types shown in Figure 12-10. How are the **a**-specific genes controlled differently in different cell types?

Answer: In **a** cells, MCM1 protein acts alone to activate **a**-specific genes. **α**-specific genes are not activated because the necessary helper proteins are not present in **a** cells. In **α** cells, **α**1 protein binds with MCM1 to activate **α**-specific genes. Another protein, **α**2 binds with MCM1 to prevent activation of **a**-specific genes. In **a/α** hybrids, neither the **a**-specific nor the **α**-specific genes are activated because MCM1 is prevented from activating **a**-specific genes by the presence of **α**2 protein, and **α**1, which is necessary for activation of **α**-specific genes, is repressed by an **a**1/**α**2 complex. In **a** cells, **a**-specific genes are activated by MCM1 alone. In other cells, the activation of **a**-specific genes is prevented by **α**2/MCM1.

5. In Figure 12-11b, in what chromosomal region are you likely to find the most H1 histone protein?

Answer: Figure 12-11c illustrates variation in the density of condensation along a chromosome. H1 histone is evenly distributed along an uncondensed chromatin fiber (shown in 12-11a), and the density increases with the increasing condensation of the chromosome. The most H1 histone would be found where the chromosome is most condensed, in the heterochromatin (depicted in red in the figure).

6. What is the conceptual connection between Figures 12-13 and 12-19?

Answer: Figure 12-13 depicts a general model for how chromatin remodeling can uncover previously hidden regions along the chromosome. Figure 12-19 shows the details of how SWI-SNF participates in chromatin remodeling to expose the TATA box of β-interferon, allowing binding of the TATA binding protein and initiation of transcription. Both illustrate the same basic process, but 12-13 is a general model and 12-19 shows the details for a specific gene.

7.   In Figure 12-19, where is the TATA box located before the enhanceosome forms at the top of the figure?

Answer: Before the enhanceosome forms, the TATA box is located within the region of the chromosome bound to the core histone proteins. It is inaccessible to the TATA binding protein so transcription cannot be initiated. Formation of the enhanceosome uncovers the TATA box, allowing transcription to be initiated.

8.   Let's say that you have incredible skill and can isolate the white and red patches of tissue from the *Drosophila* eyes shown in Figure 12-24 in order to isolate mRNA from each tissue preparation. Using your knowledge of DNA techniques from Chapter 10, design an experiment that would allow you to determine whether RNA is transcribed from the white gene in the red tissue or the white tissue or both. If you need it, you have access to radioactive white-gene DNA.

Answer: If sufficient RNA could be isolated from the different tissue patches, a Northern blot could be done, in which RNA from each tissue patch is fractionated by gel electrophoresis, then probed with radioactively labeled white-gene DNA. Presence of a radioactive band would indicate mRNA from the white gene, and would be expected in the red tissue. Lack of a radioactive band would indicate no white-gene mRNA and would be expected in the white tissue. Control lanes on the gel would consist of RNA from white gene wildtypes, which would have all red eyes, and white gene mutants, which would have all white eyes. The more likely case is that very little RNA would be present. In this event, cDNA could be synthesized from the RNA using reverse transcriptase. The cDNA could then be amplified with PCR using primers unique to the white gene. Amplification of white-gene DNA would indicate the presence of a white gene transcript, and would be expected from red tissue. Lack of amplification would mean the white gene transcript was absent and would be expected in the white tissue. In this experiment, radioactive white-gene DNA would not be needed because detection of the white-gene mRNA would be based on whether or not the DNA amplified.

9.   In Figure 12-26, provide a biochemical mechanism for why HP-1 can bind to the DNA only on the left side of the barrier insulator. Similarly, why can HMTase bind only to the DNA on the left of the barrier insulator?

Answer: The barrier insulator recruits HAT, which hyperacetylates nearby histones. This inhibits binding of HMTase, and prevents methylation of histones. Without methylation, HP-1 cannot bind and drive the formation of heterochromatin. In this way, heterochromatin-promoting proteins are prevented from binding near or to the right of the barrier insulator.

10. In reference to Figure 12-28, draw the outcome if there is a mutation in gene B.

Answer: If a mutant B gene is inherited from the father, the phenotypic outcome will be unaffected because that gene is already silenced by imprinting. If the mutant B gene is inherited from the mother, the mutant phenotype will be expressed because the paternal allele has been silenced and cannot mask the mutation. The mutation will appear to be dominant even though it is not.

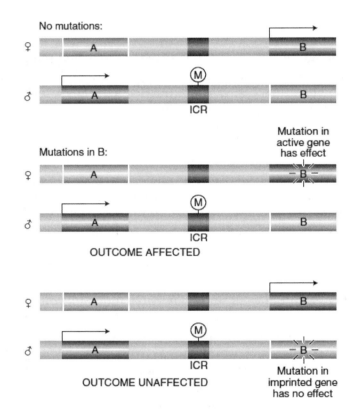

BASIC PROBLEMS

11. What analogies can you draw between transcriptional trans-acting factors that activate gene expression in eukaryotes and the corresponding factors in bacteria? Give an example.

Answer: Activation of gene expression by trans-acting factors occurs in both prokaryotes and eukaryotes. In both cases, the trans-acting factors interact with specific DNA sequences that control expression of cis genes.

In prokaryotes, proteins bind to specific DNA sequences, which in turn regulate one or more downstream genes.

In eukaryotes, highly conserved sequences such as CCAAT and various enhancers in conjunction with trans-acting binding proteins increase transcription controlled by the downstream TATA box promoter. Several proteins have been found that bind to the CCAAT sequence, upstream GC boxes, and the TATA sequence in *Drosophila*, yeast, and other organisms. Specifically, the Sp1 protein recognizes the upstream GC boxes of the SV40 promoter and many other genes; GCN4 and GAL4 proteins recognize upstream sequences in yeast; and many hormone receptors bind to specific sites on the DNA (e.g., estrogen complexed to its receptor binding to a sequence upstream of the ovalbumin gene in chicken oviduct cells). Additionally, the structure of some of these trans-acting DNA-binding proteins is quite similar to the structure of binding proteins seen in prokaryotes. Further, protein-protein interactions are important in both prokaryotes and eukaryotes. For the above reasons, eukaryotic regulation is now thought to be very close to the model for regulation of the bacterial *ara* operon.

12. Contrast the states of genes in bacteria and eukaryotes with respect to gene activation.

   Answer: In general, the ground state of a bacterial gene is "on." Thus, transcription initiation is prevented or reduced if the binding of RNA polymerase is blocked. In contrast, the ground state in eukaryotes is "off." Thus, the transcriptional machinery (including RNA polymerase II and associated general transcription factors) cannot bind to the promoter in the absence of other regulatory proteins.

13. Predict and explain the effect on *GAL1* transcription, in the presence of galactose alone, of the following mutations:

   a. Deletion of one Gal4-binding site in the *GAL1* UAS element.
   b. Deletion of all four Gal4-binding sites in the *GAL1* UAS element.
   c. Deletion of the Mig1-binding site upstream of *GAL1*.
   d. Deletion of the Gal4 activation domain.
   e. Deletion of the *GAL80* gene.
   f. Deletion of the *GAL1* promoter.
   g. Deletion of the *GAL3* gene.

   Answer:
   a. Generally, transcription factors bind to multiple sites in a cooperative manner and their effect on transcription is synergistic. The *GAL1* UAS element has four binding sites for the Gal4 protein. The UAS elements near the *GAL2* and *GAL7* genes each have two Gal4 binding sites. Loss of one Gal4 binding site from the *GAL1* UAS element may have a small

effect on *GAL1* transcription, but you would still expect to see Gal4-specific transcriptional activation due to the three remaining binding sites.

b. The deletion of all four Gal4-binding sites from the UAS would prevent transcriptional activation of the *GAL1* gene. Experiments have shown that Gal4-DNA binding at the UAS is necessary for its function.

c. The binding of Mig1 at its binding site is necessary for the repression of *GAL1* in the presence of glucose. In this case, the absence of the Mig1 binding site will have no effect on *GAL1* transcription in the presence of galactose and the absence of glucose. It would effect *GAL1* transcription if both sugars were present as transcription would still be activated.

d. If the Gal4 protein is missing its activation domain, it would still bind at the UAS but it would not promote transcription. The activation domain helps recruit the transcriptional machinery to the promoter.

e. The product of the *GAL80* gene inhibits Gal4 function by binding to a region within the Gal4 activation domain. Therefore, the deletion of *GAL80* will have no effect on the transcription of *GAL1* in the presence of galactose.

f. The deletion of the *GAL1* promoter would prevent transcription of the gene. The promoter is required for the binding of the TATA-binding protein and then by extension, the binding of RNA polymerase II.

g. The product of the *GAL3* gene is both a sensor and inducer. It binds galactose (and ATP) causing an allosteric change that allows Gal3 to bind Gal80, which in turn releases Gal4 from Gal80. When Gal80 is bound to Gal4, Gal4 cannot activate transcription so its release by Gal80 is necessary for gene activation. In the absence of Gal3 protein, Gal4 will remain bound to Gal80 and will not activate *GAL1* transcription.

14. How is the activation of the *GAL1* gene prevented in the presence of galactose and glucose?

Answer: In the presence of both galactose and glucose, the activation of the *GAL1* gene is prevented by the Mig1 protein. Mig1 is a sequence-specific DNA-binding protein that binds to a site between the UAS element and the promoter of the *GAL1* gene. Mig1 recruits a protein complex called Tup1 that contains a histone deacetylase and represses gene transcription.

15. What are the roles of histone deacetylation and histone acetylation in gene regulation, respectively?

Answer: The acetylation of histones is one of several types of covalent modifications that are part of what is called a histone code. Histones associated with nucleosomes of active genes are hyperacetylated, whereas they are hypoacetylated in nucleosomes associated with inactive genes. Given this, histone deacetylation plays a key role in gene repression while histone acetylation plays a key role in gene activation.

16. An α strain of yeast that cannot switch mating type is isolated. What mutations might it carry that would explain this phenotype?

Answer: Among the mutations that would prevent an α strain of yeast from switching mating type would be mutations in the *HO* and *HMRa* genes. The *HO* gene encodes an endonuclease that cuts the DNA to initiate switching, and the *HMRa* locus contains the "cassette" of unexpressed genetic information for the MATa mating type.

17. What genes are regulated by the α1 and α2 proteins in an a cell?

Answer: In a haploid α cell, the α1 protein, in a complex with MCM1 protein, binds to specific DNA sites and activates the transcription of α-specific genes. The α2 protein, in a complex with MCM1 protein, binds to specific DNA sites and represses a-specific genes.

18. What are Sir proteins? How do mutations in *SIR* genes affect the expression of mating-type cassettes?

Answer: In yeast, the Sir proteins (**s**ilent **i**nformation **r**egulators) are necessary to keep the genetic information in the *HMR* and *HML* cassettes silent. Their functions facilitate the condensation of chromatin and help lock up *HMR* and *HML* in chromatin domains that are inaccessible to transcriptional activators. Mutations in *SIR* genes are sterile since both *a* and α information is expressed in the absence of this specific gene silencing.

19. What is meant by the term *epigenetic inheritance*? What are two examples of such inheritance?

Answer: The term epigenetic inheritance is used to describe heritable alterations in which the DNA sequence itself is not changed. It can be defined operationally as the inheritance of chromatin states from one cell generation to next. Genomic imprinting, X-chromosome inactivation, and position-effect variegation are several such examples.

20. What is an enhanceosome? Why could a mutation in any one of the enhanceosome proteins severely reduce the transcription rate?

Answer: An enhancersome is a large protein complex that acts synergistically to activate transcription. Because this is a case where the whole is greater than the sum of its parts, loss of any one of the required proteins may severely impact the required synergy.

21. Why are mutations in imprinted genes usually dominant?

Answer: Imprinted genes are functionally hemizygous. Maternally imprinted genes are inactive when inherited from the mother, and paternally imprinted genes are inactive when inherited from the father. A mutation in one of these genes is dominant when an offspring inherits a mutant allele from one parent and a "normal" but inactivated allele from the other parent.

22. What has to happen for the expression of two different genes on two different chromosomes to be regulated by the same miRNA?

Answer: Complementarity between the seed region of the miRNA and the 3′UTR of the mRNA is the basis for regulation by miRNA. For one miRNA to coordinately regulate two genes, the seed region would have to be complementary to both 3′UTRs. This is probably a common occurrence because many miRNAs have seed regions complementary to the 3′UTR of several genes. It would not matter whether the genes in question were on the same chromosome or not because miRNA is a trans-acting agent targeting the mRNA during translation.

23. What mechanisms are thought to be responsible for the inheritance of epigenetic information?

Answer: The inheritance of chromatin structure is thought to be responsible for the inheritance of epigenetic information. This is due to the inheritance of the histone code and may also include inheritance of DNA methylation patterns.

24. What is the fundamental difference in how bacterial and eukaryotic genes are regulated?

Answer: Many DNA-protein interactions are shared by prokaryotes and eukaryotes, but the mechanisms by which proteins bound to DNA at great distances from the start of transcription affect that transcription is unique to eukaryotes. Also, mechanisms of gene regulation based on chromatin structure are distinctly eukaryotic.

25. Why is it said that transcriptional regulation in eukaryotes is characterized by combinatorial interactions?

Answer: Transcription factors can be broadly divided into two classes: recruitment and activation of the transcriptional apparatus, and chromatin remodeling or modifying enzymes. Importantly, many of the transcriptional complexes that form share many subunits and appear to be assembled in a modular fashion. The many different combinatorial interactions that are possible can result in distinct patterns of gene expression.

26. The following diagram represents the structure of a gene in *Drosophila melanogaster*; blue segments are exons, and yellow segments are introns.

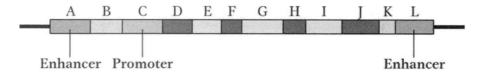

a. Which segments of the gene will be represented in the initial RNA transcript?
b. Which segments of the gene will be removed by RNA splicing?
c. Which segments would most likely bind proteins that interact with RNA polymerase?

Answer:
a. D through J — the primary transcript will include all exons and introns.
b. E, G, I — all introns will be removed.
c. A, C, L — the promoter and enhancer regions will bind various transcription factors that may interact with RNA polymerase.

CHALLENGING PROBLEMS

27. The transcription of a gene called *YFG* (your favorite gene) is activated when three transcription factors (TFA, TFB, TFC) interact to recruit the co-activator CRX. TFA, TFB, TFC, and CRX and their respective binding sites constitute an enhanceosome located 10 kb from the transcription start site. Draw a diagram showing how you think the enhanceosome functions to recruit RNA polymerase to the promoter of *YFG*.

Answer: There are numerous possibilities. As one example:

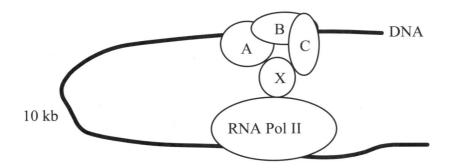

28. A single mutation in one of the transcription factors in Problem 27 results in a drastic reduction in *YFG* transcription. Diagram what this mutant interaction might look like.

Answer: The following represents a cell that is mutant for TFC. In this case, CRX does not bind and the activation of transcription does not occur.

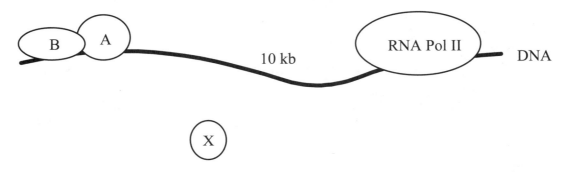

29. Diagram the effect of a mutation in the binding site for one of the transcription factors in Problem 27.

Answer: The same as 18 but now the binding site for TFC is mutant.

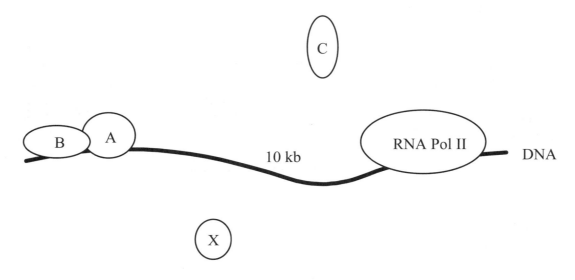

30. How does an epigenetically silenced gene differ from a mutant gene (a null allele of the same gene)?

Answer: A gene not expressed due to alteration of its DNA sequence will never be expressed and the inactive form will be inherited generation to generation. An epigenetically inactivated gene may still be regulated. Chromatin structure can change in the course of the cell cycle, for example, when transcription factors modify the histone code. Also, unlike mutational inactivation, epigenetic inactivation may change from generation to generation.

31. What are epigenetic marks? Which are associated with heterochromatin? How are epigenetic marks thought to be interpreted into chromatin structure?

Answer: Epigenetic marks include both DNA methylation and histone modifications. Heritable epigenetic marks on histones and DNA have profound effects on chromatin structure and gene expression. Heterochromatin is associated with nucleosomes containing methylated histone H3. Proteins, such as HP-1 (heterochromatin protein-1) interact with specific epigenetic marks and are required in some way to produce or maintain heterochromatin. Other proteins act in concert with specific DNA sequences to insulate regions of euchromatin from regions of heterochromatin. Since epigenetic marks are inherited, chromatin structure is also inherited. Marked histones from nucleosomes of the parental strands are mixed with new, unmarked histones so that both daughter strands become associated with nucleosomes that contain both marked and unmarked histones. The code carried by the old histones most likely guides the modification of the new histones, ensuring that the chromatin state is remembered and passed on in future divisions.

32. You receive four strains of yeast in the mail and the accompanying instructions state that each strain contains a single copy of transgene *A*. You grow the four strains and determine that only three strains express the protein product of transgene *A*. Further analysis reveals that transgene *A* is located at a different position in the yeast genome in each of the four strains. Provide a hypothesis to explain this result.

Answer: Chromatin structure has a profound effect on gene expression. Transgenes inserted into regions of euchromatin would more likely be capable of expression than those inserted into regions of heterochromatin.

33. In *Neurospora*, all mutants affecting the enzymes carbamyl phosphate synthetase and aspartate transcarbamylase map at the *pyr*-3 locus. If you induce *pyr*-3 mutations by ICR-170 (a chemical mutagen), you find that either both enzyme functions are lacking or only the transcarbamylase function is lacking;

in no case is the synthetase activity lacking when the transcarbamylase activity is present. (ICR-170 is assumed to induce frameshifts.) Interpret these results in regard to a possible operon.

Answer: If there is an operon governing both genes, then a frameshift mutation could cause the stop codon separating the two genes to be read as a sense codon. Therefore, the second gene product will be incorrect for almost all amino acids. However, there are no known polycistronic messages in eukaryotes. The alternative, and better, explanation is that both enzymatic functions are performed by the same gene product. Here, a frameshift mutation beyond the first function, carbamyl phosphate synthetase, will result in the second half of the protein molecule being nonfunctional.

34. You wish to find the cis-acting regulatory DNA elements responsible for the transcriptional responses of two genes, *c-fos* and *globin*. Transcription of the *c-fos* gene is activated in response to fibroblast growth factor (FGF), but it is inhibited by cortisol (Cort). On the other hand, transcription of the globin gene is not affected by either FGF or cortisol, but it is stimulated by the hormone erythropoietin (EP). To find the cis-acting regulatory DNA elements responsible for these transcriptional responses, you use the following clones of the *c-fos* and *globin* genes, as well as two "hybrid" combinations (fusion genes), as shown in the diagram below. The letter A represents the intact *c-fos* gene, D represents the intact *globin* gene, and B and C represent the *c-fos–globin* gene fusions. The *c-fos* and *globin* exons (E) and introns (I) are numbered. For example, E3(f) is the third exon of the *c-fos* gene and I2(g) is the second intron of the *globin* gene. (These labels are provided to help you make your answer clear.) The transcription start sites (black arrows) and polyadenylation sites (red arrows) are indicated.

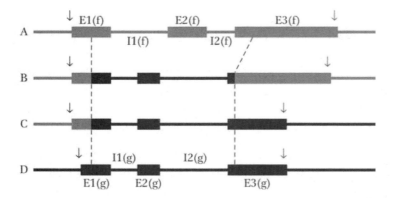

You introduce all four of these clones simultaneously into tissue-culture cells and then stimulate individual aliquots of these cells with one of the three factors. Gel analysis of the RNA isolated from the cells gives the following results:

The levels of transcripts produced from the introduced genes in response to various treatments are shown; the intensity of these bands is proportional to the amount of transcript made from a particular clone. (The failure of a band to appear indicates that the level of transcript is undetectable.)

a. Where is the DNA element that permits activation by FGF?
b. Where is the DNA element that permits repression by Cort?
c. Where is the DNA element that permits induction by EP? Explain your answer.

Answer:
a. Clone A is activated by FGF, but B, C, and D are not. This indicates that the DNA binding site for this activation is located somewhere between the 3′ end of exon 1 [E1(f)] and the 5′ end of exon 3 [E3(f)]. This is the region of DNA found only in clone A.

b. Cortisol represses transcription of clone A and B, but not C. (You do not expect any effect on D, the intact globin gene.) Comparing these clones indicates that the DNA site involved in this repression must be located in the 3′ region of E3(f) or the 3′ flanking sequences of this gene.

c. Activation by EP is seen in clones C and D, but not B. (Again, you do not expect A, the intact *c-fos* gene to respond to EP.) This indicates that the DNA site involved in this activation must be localized to the 3′ side of E3(g) or the 3′ flanking regions of the globin gene.

13

The Genetic Control of Development

WORKING WITH THE FIGURES

1. In Figure 13-2, the transplantation of certain regions of embryonic tissue induces the development of structures in new places. What are these special regions called, and what are the substances they are proposed to produce?

   Answer: Special regions that induce development of specific tissues are called organizers because they can direct the development of surrounding tissues. The cells in the organizers are proposed to produce morphogens, which induce specific concentration-dependent developmental affects in surrounding tissue.

2. In Figure 13-5, two different methods are illustrated for visualizing gene expression in developing animals. Which method would allow one to detect where within a cell a protein is localized?

   Answer: Immunolocalization of protein expression allows fluorescent visualization of the protein in a cell and visual detection of where the protein is localized. *In situ* hybridization allows visualization of mRNA, but it does not provide information about the location of the protein because mRNA and the protein will not necessarily be in the same location.

3. Figure 13-7 illustrates the expression of the Ultrabithorax (Ubx) Hox protein in developing flight appendages. What is the relationship between where the protein is expressed and the phenotype resulting from the loss of its expression (shown in Figure 13-1)?

   Answer: Wild-type Ubx expresses a Hox protein in the developing hind wings that represses normal wing formation and leads to the development of halteres. A Ubx mutant lacks expression of the Hox protein in the hind wings, leading to their subsequent development as wings.

4. In Figure 13-11, what is the evidence that vertebrate *Hox* genes govern the identity of serially repeated structures?

Answer: The normal number of lumbar vertebrae in mice is six. When the function of one *Hox* gene is lost, one lumbar vertabra is added, to make seven, and one sacral vertebra is lost. When the function of two *Hox* genes is lost, the number of lumbar vertebrae is increased to eight and two sacral vertebrae are lost. This indicates that *Hox* proteins govern the identity of the vertebral structures such that lack of *Hox* leads to formation of a lumbar vertebra in that segment.

5.   As shown in Figure 13-14, what is the fundamental distinction between a pair-rule gene and a segment-polarity gene?

Answer: Developing segments exist as pairs, and pair-rule mutants affect only one region of each segment pair, for example only the anterior or only the posterior member of the pair. In that way, pair-rule genes control double-segment periodicity. Segment-polarity genes have effects on all segments, in both members of the pair.

6.   In Table 13-1, what is the most common function of proteins that contribute to pattern formation? Why is this the case?

Answer: Most pattern formation proteins act as transcription factors. This allows them to coordinately regulate expression of many genes by binding specific sequences common to the enhancers of the different genes.

7.   In Figure 13-20, which gap protein regulates the posterior boundary of *eve* stripe 2? Describe how it does so in molecular terms.

Answer: The Giant protein, with a concentration curve shown in orange, regulates the posterior boundary of *eve* stripe 2. As the concentration of Giant increases, the expression of *eve* decreases. This produces the posterior boundary of stripe 2. From a molecular perspective, Giant is a transcription factor that binds with the stripe 2 enhancer of *eve* to repress transcription of *eve* posterior to stripe 2.

8.   As shown in Figure 13-22, how many different transcription factors govern where the *Distal-less* (*Dll*) gene will be expressed?

Answer: Six transcription factors, two Hox and four others, govern where *Dll* will be expressed. Mutations in both Hox binding sites cause full derepression of the *Dll* in A1-A7. Mutations in binding sites for other transcription factors, or combinations of transcription factors, cause different patterns of *Dll* derepression.

9.   As shown in Figure 13-26, the *Sonic hedgehog* gene is expressed in many places in a developing chicken. Is the identical Sonic hedgehog protein expressed in each tissue? If so, how do the tissues develop into different structures? If not, how are different Sonic hedgehog proteins produced?

Answer: *Sonic hedgehog* encodes a signaling protein and acts to induce other genes by activating transcription. Its developmental effect depends on the genes it induces, which differ from tissue to tissue. This leads to *Shh* activity promoting the development of feathers from a feather bud, or limbs from a limb bud. It would be possible for a gene involved in developmental regulation, such as *Shh*, to act differently in different tissues by producing different proteins. Alternate splicing of development gene mRNA could produce different isoforms in different tissues. The different isoforms could have tissue-specific effects. Sex determination in *Drosophila* is regulated through alternate splicing of genes involved in developmental pathways.

BASIC PROBLEMS

10.   *Gooseberry, runt, knirps,* and *Antennapedia.* To a *Drosophila* geneticist, what are they? How do they differ?

Answer: These are names of genes that are required for normal *Drosophila* development. The mutant phenotypes associated with these genes provide clues to their different roles: *knirps* is a gap gene; *runt* is a pair-rule gene; *gooseberry* is a segment polarity gene; and *antennapedia* is a segment identity (homeotic) gene.

11.   Describe the expression pattern of the *Drosophila* gene *eve* in the early embryo.

Answer: The primary pair-rule gene *eve* (*even-skipped*) would be expressed in seven stripes along the A-P axis of the late blastoderm.

12.   Contrast the function of homeotic genes with that of pair-rule genes.

Answer: Pair-rule genes encode transcription factors expressed in repeating patterns of seven stripes. They are necessary to divide the embryo into the correct number of segments. Homeotic genes are also transcription factors, but they are essential for segment identity and do not affect segment number.

13.   When an embryo is a homozygous mutant for the gap gene *Kr*, the fourth and fifth stripes of the pair-rule gene *ftz* (counting from the anterior end) do not form normally. When the gap gene *kni* is mutant, the fifth and sixth *ftz* stripes

do not form normally. Explain these results in regard to how segment number is established in the embryo.

Answer: Proper *ftz* expression requires *Kr* in the fourth and fifth segments, and *kni* in the fifth and sixth segments. Each of the seven stripes of pair-rule gene expression is controlled independently through distinct cis-acting regulatory elements and unique combinations of trans-acting transcription factors.

14. Some of the mammalian *Hox* genes have been shown to be more similar to one of the insect *Hox* genes than to the others. Describe an experimental approach that would enable you to demonstrate this finding in a functional context.

Answer: A number of experiments could be devised. A comparison of amino acid sequence between mammalian gene products and insect gene products would indicate which genes are most similar to each other. Using cloned cDNA sequences from mammalian genes for hybridization to insect DNA would also indicate which genes are most similar to each other. The comparison of the gene and protein expression along the anterior-posterior axis in the two species may provide information relative to their deeply conserved functions. Finally, transgenic experiments could be performed to test functional equivalency. For example, when the mouse *HoxB6* gene is inserted in *Drosophila*, it can substitute for *Antennapedia*.

15. The three homeodomain proteins ABD-B, ABD-A, and UBX are encoded by genes within the *Bithorax* complex of *Drosophila*. In wild-type embryos, the *Abd-B* gene is expressed in the posterior abdominal segments, *Abd-A* in the middle abdominal segments, and *Ubx* in the anterior abdominal and posterior thoracic segments. When the *Abd-B* gene is deleted, *Abd-A* is expressed in both the middle and the posterior abdominal segments. When *Abd-A* is deleted, *Ubx* is expressed in the posterior thorax and in the anterior and middle abdominal segments. When *Ubx* is deleted, the patterns of *Abd-A* and *Abd-B* expression are unchanged from wild type. When both *Abd-A* and *Abd-B* are deleted, *Ubx* is expressed in all segments from the posterior thorax to the posterior end of the embryo. Explain these observations, taking into consideration the fact that the gap genes control the initial expression patterns of the homeotic genes.

Answer: If you diagram these results, you will see that deletion of a gene that functions posteriorly allows the next-most anterior segments to extend in a posterior direction. Deletion of an anterior gene does not allow extension of the next-most posterior segment in an anterior direction. The gap genes activate *Ubx* in both thoracic and abdominal segments, whereas the *abd-A* and *Abd-B* genes are activated only in the middle and posterior abdominal segments. The functioning of the *abd-A* and *Abd-B* genes in those segments somehow prevents *Ubx* expression. However, if the *abd-A* and *Abd-B* genes are deleted, *Ubx* can be expressed in these regions.

16. How can you tell if a gene is required zygotically and if it has a maternal effect?

Answer: For zygotically required recessive mutations, crossing $+/m \times +/m$ parents will produce offspring in the expected Mendelian ratio of 3 wild type : 1 mutant. Only homozygous mutant offspring will be mutant. For genes that exhibit a maternal effect, it is the genotype of the mother that determines the phenotype of the offspring. So, for the cross outlined above ($+/m \times +/m$), none of the offspring will be phenotypically mutant. However, a cross *of m/m* females to $+/+$ males will produce all mutant offspring, while a cross of $+/+$ females to *m/m* males will produce only wild-type offspring.

17. In considering the formation of the A–P and D–V axes in *Drosophila*, we noted that, for mutations such as *bcd*, homozygous mutant mothers uniformly produce mutant offspring with segmentation defects. This outcome is always true regardless of whether the offspring themselves are bcd^+/bcd or *bcd/bcd*. Some other maternal-effect lethal mutations are different, in that the mutant phenotype can be "rescued" by introducing a wild-type allele of the gene from the father. In other words, for such rescuable maternal-effect lethals, mut^+/mut animals are normal, whereas *mut/mut* animals have the mutant defect. Explain the difference between rescuable and nonrescuable maternal-effect lethal mutations.

Answer: The nonrescuable maternal-effect lethal mutations may produce a product that is required very early in development, before the developing fly is producing any proteins, while the rescuable maternal-effect lethal mutations may act later in development when embryo protein production can compensate for the maternal mutation.

18. Suppose you isolate a mutation affecting A–P patterning of the *Drosophila* embryo in which every other segment of the developing mutant larva is missing.

   a. Would you consider this mutation to be a mutation in a gap gene, a pair-rule gene, a segment-polarity gene, or a segment-identity gene?

   b. You have cloned a piece of DNA that contains four genes. How could you use the spatial-expression pattern of their mRNA in a wild-type embryo to identify which represents a candidate gene for the mutation described?

   c. Assume that you have identified the candidate gene. If you now examine the spatial-expression pattern of its mRNA in an embryo that is homozygous mutant for the gap gene *Krüpel* , would you expect to see a normal expression pattern? Explain.

Answer:
   a. A pair-rule gene.

b. Look for expression of the mRNA from the candidate gene in a repeating pattern of seven stripes along the A–P axis of the developing embryo.

c. No. An embryo mutant for the gap gene *Krüpel* would be missing many anterior segments. This effect would be epistatic to the expression of a pair-rule gene.

19. How does the Bicoid protein gradient form?

Answer: Biocoid mRNA is tethered to the anterior pole of the egg. Upon translation, BCD protein diffuses in the common syncytial cytoplasm creating the observed gradient. As a transcription factor, the gradient of BCD protein provides positional information along the anteroposterior axis.

20. In an embryo from a homozygous *Bicoid* mutant female, which class(es) of gene expression is (are) abnormal?

a. Gap genes
b. Pair-rule genes
c. Segment-polarity genes
d. *Hox* genes

Answer: All of the classes listed would show abnormal gene expression. A homozygous *bcd* mutant female would produce embryos lacking functional BCD protein. Functional BCD protein is necessary for correct regulation and expression of gap genes. Proper expression of gap genes is necessary for proper expression of pair-rule genes and homeotic genes, and the proper expression of pair-rule genes is necessary for the proper expression of segment-polarity genes.

CHALLENGING PROBLEMS

21. **a.** The *eyeless* gene is required for eye formation in *Drosophila*. It encodes a homeodomain. What would you predict about the biochemical function of the Eyeless protein?

b. Where would you predict that the *eyeless* gene is expressed in development? How would you test your prediction?

c. The *Small eye* and *Aniridia* genes of mice and humans, respectively, encode proteins with very strong sequence similarity to the fly Eyeless protein, and they are named for their effects on eye development. Devise one test to examine whether the mouse and human genes are functionally equivalent to the fly *eyeless* gene.

Answer:

a. The homeodomain is a conserved protein domain containing 60 amino acids found in a significant number of transcription factors. Any protein that contains a functional homeodomain is almost certainly a sequence-specific DNA-binding transcription factor.

b. The *eyeless* gene (named for its mutant phenotype) regulates eye development in *Drosophila*. You would expect that it is expressed only in those cells that will give rise to the eyes. To test this prediction, visualization of the location of *eyeless* mRNA expression by *in situ* hybridization and the eyeless protein by immunological methods should be performed. Through genetic manipulation, it is possible to express the *eyeless* gene in tissues where it is not ordinarily expressed. For example, when *eyeless* is turned on in cells destined to form legs, eyes form on the legs!

c. Transgenic experiments have shown that the mouse *Small eye* gene and the *Drosophila eyeless* gene are so similar that the mouse gene can substitute for *eyeless* when introduced into *Drosophila*. As above, when the mouse *Small eye* gene is expressed in *Drosophila*, even in cells destined to form legs, eyes form on the legs! (Of course, the "eyes" are not mouse eyes as these genes act as a master switch to turn on the entire cascade of genes needed to build the eye, and in this case, the *Drosophila* set to build a *Drosophila* eye.)

22. Gene X is expressed in the developing brain, heart, and lungs of mice. Mutations that selectively affect gene X function in these three tissues map to three different regions (A, B, and C, respectively) 5′ of the X coding region.

a. Explain the nature of these mutations.

b. Draw a map of the X locus consistent with the preceding information.

c. How would you test the function of the A, B, and C regions?

Answer:

a. The mutations are in distinct cis-acting regulatory elements of gene X. The A element controls expression of gene X in the brain, the B element controls heart expression, and the C element controls lung expression.

b.

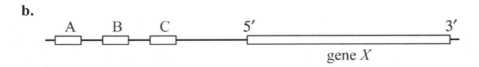

c. Transgenic experiments using the three cis-acting regulatory elements hooked up to a reporter gene (for example, GFP, green fluorescent protein from jellyfish) will allow visualization of localized expression. In this case, you would expect a transgenic fly with the element A-GFP construct to show GFP expression in the brain. Other constructs should confirm the other expression patterns.

23. Why are regulatory mutations at the *Sonic hedgehog* gene dominant and viable? Why do coding mutations cause more widespread defects?

Answer: *Sonic hedgehog* (*Shh*) is expressed in several tissues of developing vertebrates, for example, the ZPA (zone of polarizing activity) of developing limb buds, the floor plate of the neural tube, and developing feather buds in chickens. These serve to illustrate the different roles played by toolkit genes, like *Shh*, at different places and times in development. Since these genes have multiple roles in different tissues and cell types, mutations in coding regions will likely have widespread defects. Mutations in regulatory regions will be dominant and viable because the wild-type allele of *Shh* will still supply the necessary functions while the mixexpressed mutant allele will cause the unexpected phenotype.

24. A mutation occurs in the *Drosophila doublesex* gene that prevents Tra from binding to the *dsx* RNA transcript. What would be the consequences of this mutation for Dsx protein expression in males? In females?

Answer: In *Drosophila*, the *dsx* transcript is alternatively spliced in the two sexes. The product of the *tra* gene is required for the female-specific RNA splicing of *dsx*. In the presence of Tra (and a related protein Tra2), the pre-mRNA of *dsx* is spliced to create the *dsx*^{F} transcript. In its absence, the *dsx*^{M} transcript is the default product of splicing. If a mutation prevents Tra from binding the *dsx* transcript, the default male-specific splicing will occur. In males, this will have no effect, but XX females homozygous for this mutation will be transformed into the male phenotype.

25. You isolate a *glp-1* mutation of *C. elegans* and discover that the DNA region encoding the spatial control region (SCR) has been deleted. What will the GLP-1 protein expression pattern be in a four-cell embryo in mutant heterozygotes? In mutant homozygotes?

Answer: GLP-1 protein is localized to the two anterior cells of the four-cell *C. elegans* embryo by repressing its translation in the two posterior cells. The repression of GLP-1 translation requires the 3′ UTR spatial control region (SCR). Deletion of the SCR will allow *glp-1* expression in both anterior and posterior cells. In both heterozygous and homozygous mutants, you would expect GLP-1 protein expression in all cells.

14

Genomes and Genomics

WORKING WITH THE FIGURES

1. Based on Figure 14-2, why must the DNA fragments sequenced overlap in order to obtain a genome sequence?

   Answer: Sequence overlap is required to align the sequenced segments relative to each other.

2. Filling gaps in draft genome sequences is a major challenge. Based on Figure 14-6, can paired-end reads from a library of 2-kb fragments fill a 10-kb gap?

   Answer: No, because typically these gaps consist of repetitive sequences. The sequencing reads in these regions cannot be aligned with certainty. Alternatively, gaps arise because the sequences in the region of the gap cannot be easily cloned in *E. coli* and would therefore not be represented in the libraries.

3. In Figure 14-9, how are the positions of codons determined?

   Answer: Several methods for identifying protein-coding sequences are discussed in the companion text. Computer analysis of the nucleotide sequence to look for open reading frames can identify at least some of the exons. These searches can be refined to include the detection of codon bias, binding site predictions, and conservation of predicted amino acid sequences. However, comparison of cDNA sequences to the genomic sequence will directly reveal the sequences comprising the coding portion of the gene.

4. In Figure 14-10, expressed sequence tags (ESTs) are aligned with genomic sequence. How are ESTs helpful in genome annotation?

Answer: The ESTs delineate the 5′ and 3′ ends of a transcription unit, indicating that the exons must lie between these boundaries.

5.   In Figure 14-10, cDNA sequences are aligned with genomic sequence. How are cDNA sequences helpful in genome annotation? Are cDNAs more important for bacterial or eukaryotic genome annotations?

Answer: The cDNA defines the exons of a particular gene. The sequences in between, therefore, represent the introns. Consequently, cDNAs are more important for eukaryotic genome annotation because bacterial genes lack introns.

6.   Figure 14-16 shows syntenic regions of mouse chromosome 11 and human chromosome 17. What do these syntenic regions reveal about the genome of the last common ancestor of mice and humans?

Answer: Because these genes in these large conserved blocks are found in the same order in mice and in humans, they were found in the same order in the last common ancestor of these two species.

7.   Based on Figure 14-17 and the features of ultraconserved elements, what would you predict you'd observe if you injected a reporter-gene construct of the rat ortholog of the *ISL1* ultraconserved element into fertilized mouse oocytes and examined reporter gene expression in the developing embryo?

Answer: Because these regulatory elements are so highly conserved that the expression pattern conferred by the human element mimics the expression of the native mouse *ISL1* locus, you would expect the pattern conferred by the rat element to look like the image seen in Figure 14-17a.

8.   The genomes of two *E. coli* strains are compared in Figure 14-18. Would you expect any third strain to contain more of the blue, tan, or red regions shown in Figure 14-18? Explain.

Answer: The blue regions represent the genetic sequences common to both *E. coli* strains, K-12 and O157:H7. These common sequences contain those that define these strains as *E. coli*, and therefore, would be more likely to be shared with a third *E. coli* strain.

9.   Figure 14-21 depicts the Gal4-based two-hybrid system. Why don't the "bait" proteins fused to the Gal4 DNA-binding protein activate reporter-gene expression?

Answer: The "bait" proteins do not activate transcription because they do not contain activation domains. If a "bait" activated transcription on its own, the assay would not work.

BASIC PROBLEMS

10.   The word *contig* is derived from the word *contiguous*. Explain the derivation.

Answer: *Contig* describes a set of adjacent DNA sequences or clones assembled using overlapping sequences or restriction fragments. Because the pieces are assembled into a continuous whole, it makes sense that contig is derived from the word contiguous.

11.   Explain the approach that you would apply to sequencing the genome of a newly discovered bacterial species.

Answer: Because bacteria have relatively small genomes (roughly three megabase pairs) and essentially no repeating sequences, the whole genome shotgun approach would be used.

12.   Terminal-sequencing reads of clone inserts are a routine part of genome sequencing. How is the central part of the clone insert ever obtained?

Answer: Terminal sequence reads of cloned inserts, such as those generated during whole genome shotgun sequencing, are assembled into a scaffold by matching homologous sequences shared by reads from overlapping clones. In essence, the central sequence of any single clone will be generated from the terminal sequences of overlapping clones.

13.   What is the difference between a contig and a scaffold?

Answer: A scaffold is also called a supercontig. Contigs are sequences of overlapping reads assembled into units, and a scaffold is a collection of joined-together contigs.

14.   Two particular contigs are suspected to be adjacent, possibly separated by repetitive DNA. In an attempt to link them, end sequences are used as primers to try to bridge the gap. Is this approach reasonable? In what situation will it not work?

Answer: Yes. If the gap is short, PCR fragments can be generated from primers based on the ends of your contigs, and these fragments can be directly sequenced without a cloning step. The process does not work if the gap is too long.

15. A segment of cloned DNA containing a protein-encoding gene is radioactively labeled and used in an in situ hybridization to chromosomes. Radioactivity was observed over five regions on different chromosomes. How is this result possible?

Answer: The clone may contain DNA that hybridizes to a small family of repetitive DNA that is adjacent to the gene being studied or within one of its introns. Alternatively, the clone may share enough homology with members of a small gene family to hybridize to all or there may be four pseudogenes that are descendants of the unique gene being studied.

16. In an in situ hybridization experiment, a certain clone bound to only the X chromosome in a boy with no disease symptoms. However, in a boy with Duchenne muscular dystrophy (X-linked recessive disease), it bound to the X chromosome and to an autosome. Explain. Could this clone be useful in isolating the gene for Duchenne muscular dystrophy?

Answer: Yes. The clone hybridizes to and spans a translocation breakpoint that involves the X chromosome and an autosome. Because the DMD gene normally maps to the X chromosome, it is of interest that a translocation of the X is found in a patient with DMD. If the translocation is the cause of DMD mutation, the clone identifies a least a portion and/or location of the gene.

17. In a genomic analysis looking for a specific disease gene, one candidate gene was found to have a single-base-pair substitution resulting in a nonsynonymous amino acid change. What would you have to check before concluding that you had identified the disease-causing gene?

Answer: You still need to verify that the gene with the amino acid change is the correct candidate for the phenotype under study. For example, you might transform mutants with a wild-type copy of the candidate gene and see if the transgene reverts the mutant phenotype (functional complementation). Alternatively, you can attempt to compare the phenotype of the mutant with the known function of the candidate gene. A third alternative solution is to determine if the presence of the amino acid change (or homozygosity for the amino acid change if the phenotype is recessive) correlates with the phenotype, and whether the change is absent in the general population that do not have the phenotype (or transmit the phenotype—if recessive).

18. Is a bacterial operator a binding site?

Answer: Yes. The operator is the location at which repressor functionally binds through interactions between the DNA sequence and the repressor protein.

19. A certain cDNA of size 2 kb hybridized to eight genomic fragments of total size 30 kb and contained two short ESTs. The ESTs were also found in two of the genomic fragments each of size 2 kb. Sketch a possible explanation for these results.

Answer: There are numerous possible drawings that can be made. The goal is to indicate that at least one exon be present in all eight genomic fragments and that the ESTs define the 5′ and 3′ ends of the transcript.

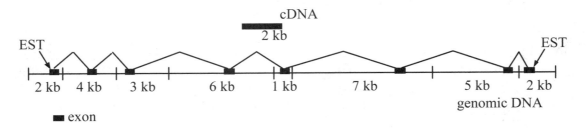

20. A sequenced fragment of DNA in *Drosophila* was used in a BLAST search. The best (closest) match was to a kinase gene from *Neurospora*. Does this match mean that the *Drosophila* sequence contains a kinase gene?

Answer: A level of 35 percent or more amino acid identity at comparable positions in two polypeptides is indicative of a common three-dimensional structure. It also suggests that the two polypeptides are likely to have at least some aspect of their function in common. However, it does not prove that your particular sequence does in fact encode a kinase.

21. In a two-hybrid test, a certain gene *A* gave positive results with two clones, M and N. When M was used, it gave positives with three clones, A, S, and Q. Clone N gave only one positive (with A). Develop a tentative interpretation of these results.

Answer: The yeast two-hybrid test detects possible physical interactions between two proteins. These results indicate that gene *A* codes for a protein that interacts with proteins encoded by clones M and N, and further, that clone M encodes a protein that also interacts with proteins encoded by clones S and Q. For example:

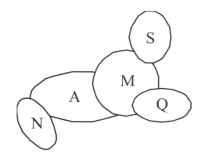

22. You have the following sequence reads from a genomic clone of the *Drosophila melanogaster* genome:

    Read 1: TGGCCGTGATGGGCAGTTCCGGTG
    Read 2: TTCCGGTGCCGGAAAGA
    Read 3: CTATCCGGGCGAACTTTTGGCCG
    Read 4: CGTGATGGGCAGTTCCGGTG
    Read 5: TTGGCCGTGATGGGCAGTT
    Read 6: CGAACTTTTGGCCGTGATGGGCAGTTCC

    Use these six sequence reads to create a sequence contig of this part of the *D. melanogaster* genome.

    Answer: This is just a matter of aligning the sequences to determine their overlap.

Read 1:   TGGCCGTGATGGGCAGTTCCGGTG
Read 2:   TTCCGGTGCCGGAAAGA
Read 3: CTATCCGGGCGAACTTTTGGCCG
Read 4:   CGTGATGGGCAGTTCCGGTG
Read 5:   TTGGCCGTGATGGGCAGTT
Read 6:   CGAACTTTTGGCCGTGATGGGCAGTTCC

    And this creates the contig:

    CTATCCGGGCGAACTTTTGGCCGTGATGGGCAGTTCCGGTGCCGGAAAGA

23. Sometimes, cDNAs turn out to be "monsters"; that is, fusions of DNA copies of two different mRNAs accidentally inserted adjacent to each other in the same clone. You suspect that a cDNA clone from the nematode *Caenorhabditis elegans* is such a monster because the sequence of the cDNA insert predicts a protein with two structural domains not normally observed in the same protein. How would you use the availability of the entire genomic sequence to assess if this cDNA clone is a monster or not?

    Answer: You can determine whether the cDNA clone was a monster or not by alignment of the cDNA sequence against the genomic sequence. (There are computer programs available to do this.) Is it derived from two different sites? Does the cDNA map within one [gene-sized] region in the genome or to two different regions? Of course, introns may complicate the issue.

24. In browsing through the human genome sequence, you identify a gene that has an apparently long coding region, but there is a two-base-pair deletion that disrupts the reading frame.

    a. How would you determine whether the deletion was correct or an error in the sequencing?

    b. You find that the exact same deletion exists in the chimpanzee homolog of the gene but that the gorilla gene reading frame is intact. Given the phylogeny of great apes below, what can you conclude about when in ape evolution the mutation occurred?

    Answer:
    a.   To determine whether the 2-bp deletion was a sequencing error, you would simply resequence that region of the genome, or a cDNA copy of the gene.
    b.   You would conclude that the 2-bp deletion occurred in the common ancestor of both chimpanzee and human after the divergence from the ancestor of gorillas.

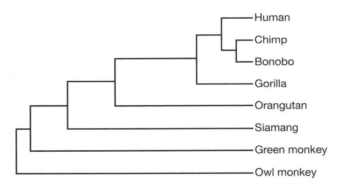

25. In browsing through the chimpanzee genome, you find that it has three homologs of a particular gene, whereas humans have only two.

    a. What are two alternative explanations for this observation?

    b. How could you distinguish between these two possibilities?

    Answer:
    a. It may be that the ancestor of both chimpanzee and humans had three copies of the gene and one was lost after the human ancestors diverged from

chimpanzee, or conversely that the common ancestor had two and a duplication has occurred in the chimpanzee line.

b. One would determine how many copies are present in the other great apes. This would allow one to determine when in the evolution of apes the change occurred.

26. The platypus is one of the few venomous mammals. The male platypus has a spur on the hind foot through which it can deliver a mixture of venom proteins. Looking at the phylogeny in Figure 14-15, how would you go about determining whether these venom proteins are unique to the platypus?

Answer: One would perform an analysis similar to the one described for vitellogenin in the companion text. One would search the genome sequences of other mammals and at least one evolutionary outgroup for orthologs of the venom protein genes.

27. You have sequenced the genome of the bacterium *Salmonella typhimurium*, and you are using BLAST analysis to identify similarities within the *S. typhimurium* genome to known proteins. You find a protein that is 100 percent identical in the bacterium *Escherichia coli*. When you compare nucleotide sequences of the *S. typhimurium* and *E. coli* genes, you find that their nucleotide sequences are only 87 percent identical.

a. Explain this observation.
b. What do these observations tell you about the merits of nucleotide- versus protein-similarity searches in identifying related genes?

Answer:
a. Because the triplet code is redundant, changes in the DNA nucleotide sequence, (especially at those nucleotides coding for the third position of a codon) can occur without change to its encoded protein.

b. Protein sequences are expected to evolve and diverge more slowly than the genes that encode them. Due to the flexibility in the third positions of most codons, the DNA sequence can accumulate changes without affecting protein structure. Protein sequences will evolve and diverge more slowly because amino acid changes may change the structure and function of the protein. Since the amino acid sequence is functionally constrained, natural selection will eliminate many deleterious amino acid changes. This will reduce the rate of change in the amino acid sequence and lead to greater sequence conservation of the amino acid sequence compared to the nucleotide sequence.

28. To inactivate a gene by RNAi, what information do you need? Do you need the map position of the target gene?

Answer: Gene position is not a necessary requirement for using RNAi. What is needed is the sequence of the gene that you intend to inactivate, as you need to start with double-stranded RNA that is made with sequences that are homologous to part of the gene.

29. What is the purpose of generating a phenocopy?

Answer: Phenocopying is the mimicking of a mutant phenotype by inactivating the gene product rather than the gene itself. It can be used regardless of how well-developed the genetic technology has been elaborated for a particular organism and can provide meaningful results when gene knockout or site-directed mutagenesis is not possible. Three methods of producing phenocopies are: the prevention of gene-specific translation by the introduction of anti-sense RNA (RNA complementary to the gene-specific mRNA); the introduction of double-stranded RNA with sequences homologous to part of a specific mRNA resulting in major reduction in the level of the mRNA (called double-stranded RNA interference); and inhibition of a specific protein through high-affinity binding of compound(s) identified through chemical genetics.

30. What is the difference between forward and reverse genetics?

Answer: Forward genetics identifies heritable differences by their phenotypes and map locations and precedes the molecular analysis of the gene products. Reverse genetics starts with an identified protein or RNA and works toward mutating the gene that encodes it (and in the process, discovered the phenotype when the gene is mutated).

CHALLENGING PROBLEMS

31. You have the following sequence reads from a genomic clone of the *Homo sapiens* genome:

Read 1: ATGCGATCTGTGAGCCGAGTCTTTA
Read 2: AACAAAAATGTTGTTATTTTTATTTCAGATG
Read 3: TTCAGATGCGATCTGTGAGCCGAG
Read 4: TGTCTGCCATTCTTAAAAACAAAAATGT
Read 5: TGTTATTTTTATTTCAGATGCGA
Read 6: AACAAAAATGTTGTTATT

a. Use these six sequence reads to create a sequence contig of this part of the *H. sapiens* genome.

b. Translate the sequence contig in all possible reading frames.

c. Go to the BLAST page of the National Center for Biotechnology Information, or NCBI (http://www.ncbi.nlm.nih.gov/BLAST/, Appendix B) and see if you can identify the gene of which this sequence is a part by using each of the reading frames as a query for protein–protein comparison (BLASTp).

Answer:
a. This is just a matter of aligning the sequences to determine their overlap.

```
Read 1:                                                     ATGCGATCTGTGAGCCGAGTCTTTA
Read 2:                      AACAAAAATGTTGTTATTTTTATTTCAGATG
Read 3:                                        TTCAGATGCGATCTGTGAGCCGAG
Read 4:  TGTCTGCCATTCTTAAAAACAAAAATGT
Read 5:                             TGTTATTTTTATTTCAGATGCGA
Read 6:                      AACAAAAATGTTGTTATT
```

And this creates the contig

```
5´-TGTCTGCCATTCTTAAAAACAAAAATGTTGTTATTTTTATTTCAGATGCGATCTGTGAGCCGAGTCTTTA-3´
```

Since you do not know if the DNA sequence represents the template strand for the mRNA or is the complementary strand, there are two possible transcripts.

```
5´-UGUCUGCCAUUCUUAAAAACAAAAAUGUUGUUAUUUUUAUUUCAGAUGCGAUCUGUGAGCCGAGUCUUUA-3´
```

and

```
3´-ACAGACGGUAAGAAUUUUUGUUUUUACAACAAUAAAAAUAAAGUCUACGCUAGACACUCGGCUCAGAAAU-5´
```

b. Translation of the first possible transcript starting at the first letter reads

CLPFLKTKMLLFLFQMRSVSRVF

Translation starting at the second letter reads

VCHS stop

Translation starting at the third reads

SAILKNKNVVIFISDAICRPSL

For the second possible transcript, starting at the first letter reads

stop

For the second possible transcript, starting at the second letter reads

KDSAHRSHLK stop

For the second possible transcript, starting at the third letter reads

```
KTRLTDRI stop
```

c. Using the nucleotide sequence of the contig and performing BLASTn or the possible translation products and performing tBLASTn, you will discover that this sequence and the translation product above listed first match perfectly with a region of exon 19 of the human CFTR gene.

32. Some sizable regions of different chromosomes of the human genome are more than 99 percent nucleotide identical with one another. These regions were overlooked in the production of the draft genome sequence of the human genome because of their high level of similarity. Of the techniques discussed in this chapter, which would allow genome researchers to identify the existence of such duplicate regions?

Answer: The correct assembly of large and nearly identical regions is problematic with either method of genomic sequencing. However, the whole genome shotgun method is less effective at finding these regions than the clone-based strategy. This method also has the added advantage of easy access to the suspect clone(s) for further analysis. (Comparative genomic hybridization has been used effectively and is superior to sequencing for this determination. See Chapter 17, Figure 17-36 in the companion text.)

33. Some exons in the human genome are quite small (less than 75 bp long). Identification of such "microexons" is difficult because these distances are too short to reliably use ORF identification or codon bias to determine if small genomic sequences are truly part of an mRNA and a polypeptide. What techniques of "gene finding" can be used to try to assess if a given region of 75 bp constitutes an exon?

Answer: Assessing whether a short sequence constitutes an exon is difficult. The best way to determine if a suspected micro-exon is actually used is to look for a cDNA or an EST that includes it. Alternatively, identification of consensus donor and acceptor splice site sequences can be tried and also the use of comparative genomics, that is, the conservation of the predicted amino acid encoded by the micro-exon in the same or other genomes.

34. You are studying proteins having roles in translation in the mouse. By BLAST analysis of the predicted proteins of the mouse genome, you identify a set of mouse genes that encode proteins with sequences similar to those of known eukaryotic translation-initiation factors. You are interested in determining the phenotypes associated with loss-of-function mutations of these genes.

a. Would you use forward- or reverse-genetics approaches to identify these mutations?

b. Briefly outline two different approaches that you might use to look for loss-of-function phenotypes in one of these genes.

Answer:

a. Forward genetics identifies heritable differences by their phenotypes and map locations and precedes the molecular analysis of the gene products. Reverse genetics starts with an identified protein or RNA and works toward mutating the gene that encodes it (and in the process, discovers the phenotype when the gene is mutated). Because you have known proteins and want to determine the phenotypes associated with loss-of-function mutations in the genes that encoded them, a reverse genetic approach is the answer.

b. The two general approaches would be either directed mutations in the gene of interest or the generation of phenocopies of the mutant phenotype by inactivating the gene product rather than the gene itself.

35. The entire genome of the yeast *Saccharomyces cerevisiae* has been sequenced. This sequencing has led to the identification of all the open reading frames (ORFs, gene-size sequences with appropriate translational initiation and termination signals) in the genome. Some of these ORFs are previously known genes with established functions; however, the remainder are unassigned reading frames (URFs). To deduce the possible functions of the URFs, they are being systematically, one at a time, converted into null alleles by in vitro knockout techniques. The results are as follows:

15 percent are lethal when knocked out.
25 percent show some mutant phenotype (altered morphology, altered nutrition, and so forth).
60 percent show no detectable mutant phenotype at all and resemble wild type.

Explain the possible molecular-genetic basis of these three mutant categories, inventing examples where possible.

Answer: Fifteen percent are essential gene functions (such as enzymes required for DNA replication or protein synthesis).

Twenty-five percent are auxotrophs (enzymes required for the synthesis of amino acids or the metabolism of sugars, genes required for normal growth rate and normal colony morphology or other traits evident in the laboratory, etc.).

Sixty percent are redundant or pathways not tested (genes for histones, tubulin, ribosomal RNAs, etc., are present in multiple copies; the yeast may require many genes under only unique or special situations or in other ways that are not necessary for life in the "lab"—where the yeast will be exposed to competition from other microbes and a wide array of conditions not found in the laboratory).

36. Different strains of *E. coli* are responsible for enterohemorrhagic and urinary tract infections. Based on the differences between the benign K-12 strain and the enterohemorrhagic O157:H7 strain, would you predict that there are obvious genomic differences:

    a. Between K-12 and uropathogenic strains?
    b. Between O157:H7 and uropathogenic strains?

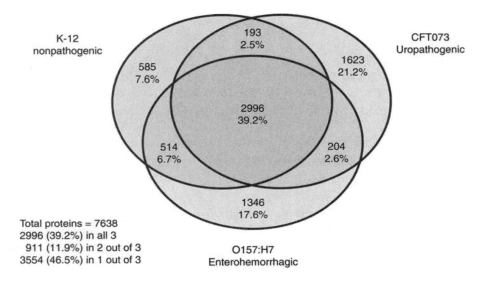

K-12
nonpathogenic

193
2.5%

CFT073
Uropathogenic

585
7.6%

1623
21.2%

2996
39.2%

514
6.7%

204
2.6%

1346
17.6%

Total proteins = 7638
2996 (39.2%) in all 3
  911 (11.9%) in 2 out of 3
3554 (46.5%) in 1 out of 3

O157:H7
Enterohemorrhagic

    c. What might explain the observed pair-by-pair differences in genome content?
    d. How might the function of strain-specific genes be tested?

Answer:
    a. The two *E. coli* strains K-12 and O157:H7 have 3574 protein-coding genes in common and among these orthologous genes, share 98.4 percent nucleotide identity. Even so, the complete genomes and proteomes of these species differ enormously. The genome of O157:H7 contains 1387 genes not found in the K-12 genome, and there are 528 genes in the K-12 genome not found in the O157:H7 genome. Most of the O157:H7-specific sequences are horizontally transferred foreign DNAs and would be expected to contain candidate genes that encode virulence factors, cell-invasion proteins, adherence proteins, secretion proteins, and possible metabolic genes required for nutrient transport, antibiotic resistance, and other activities that may confer the ability to survive in different hosts. It would certainly be expected that comparisons between K-12 and uropathogenic strains of *E. coli* would show similar enormous differences in their genomes and proteomes.

b. In comparing O157:H7 to uropathogenic strains of *E. coli*, you would expect that they would share the same backbone genes as K-12 and also possibly share some of the candidate genes that encode virulence factors, cell-invasion proteins, adherence proteins, secretion proteins, and other activities that help confer the ability to survive in different hosts. However, you would also expect significant differences between the two genomes as the urinary tract and colon represent separate niches that likely require unique gene products.

c. The various strains of *E. coli* have evolved through a complex process. The ancestral backbone genes that define *E. coli* have undergone slow accumulation of vertically acquired sequence changes. This can be seen in the 98.4 percent nucleotide identity in the orthologous genes shared by K-12 and O157:H7. However, the sequences not in common have been introduced via numerous, independent horizontal gene-transfer events at many discrete sites. These sequences are from the genomes of viruses and other bacterial species. Some differences between the strains may also be the result of deletion.

d. Functions of genes may be determined by various mechanisms. Comparison of the strain-specific genes to all other known genes using a BLAST search might provide insight to function. Reverse genetic analysis to disrupt gene function and observation of the resulting phenotype may allow you to assess the role of the normal gene product. Finally, transforming *E. coli* K-12 with pathogenic-specific genes and analysis of the resulting transgenic strains may also allow gene function to be assessed.

15

The Dynamic Genome

WORKING WITH THE FIGURES

1. In the chapter-opening photograph of kernels on an ear of corn, what is the genetic basis of the following (**Hint:** Refer to Figure 15-4 for some clues):

   a. the fully pigmented kernel?
   b. the unpigmented kernels? Note that they can arise in two different ways.

   Answer:
   a. The fully pigmented kernel results from the wild-type copy of the C gene expressed in all cells.

   b. The unpigmented kernels can result from a loss of the C gene by Ac-activated chromosome breakage at a *Ds* element, resulting in appearance of the recessive, colorless phenotype. Additionally, insertion of the *Ds* element into the C gene in the absence of an Ac element to mediate its excision will result in colorless kernels.

2. In Figure 15-3a, what would the kernel phenotype be if the strain was homozygous for all dominant markers on chromosome 9?

   Answer: The kernel would be pigmented, plump, and shiny.

3. For Figure 15-7, draw out a series of steps that could explain the origin of this large plasmid containing many transposable elements.

   Answer: One series of events leading to the generation of the R plasmid in Figure 15-7 is shown, although other answers are possible. For example, the insertion of Tn3 into Tn4 could have happened in a previous host genome or plasmid, and both elements could transfer together to this plasmid.

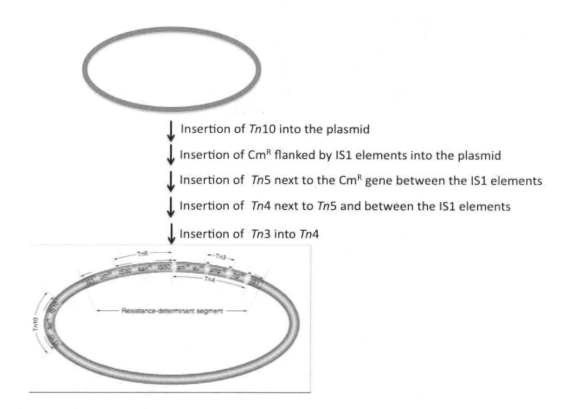

Insertion of *Tn*10 into the plasmid

Insertion of Cm^R flanked by IS1 elements into the plasmid

Insertion of *Tn*5 next to the Cm^R gene between the IS1 elements

Insertion of *Tn*4 next to *Tn*5 and between the IS1 elements

Insertion of *Tn*3 into *Tn*4

4. For Figure 15-8, draw a figure for the third mode of transposition, retrotransposition.

Answer:

Retrotransposition

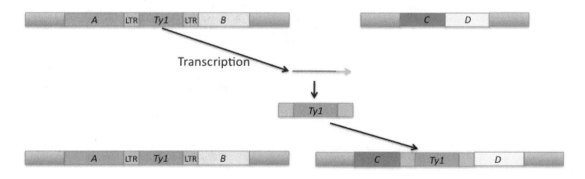

5. In Figure 15-10, show where the transposase would have to cut to generate a 6-bp target-site duplication. Also show the location of the cut to generate a 4-bp target-site duplication.

Answer:

The cuts indicated by the red arrows below would result in a 6-bp duplication.

The cuts indicated by the red arrows below would result in a 4-bp duplication.

6. If the transposable element in Figure 15-14 were a DNA transposon that had an intron in its transposase gene, would the intron be removed following transposition? Justify your answer.

Answer: No, the intron would not be removed during transposition. In order for the intron to be lost, the element must transpose via an RNA intermediate, the new DNA copy must be made from a spliced RNA template.

7. For Figure 15-22, draw the pre-mRNA that is transcribed from this gene and then draw its mRNA.

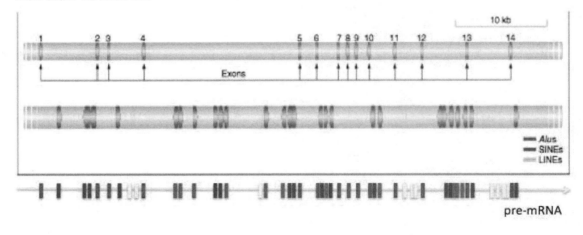

BASIC PROBLEMS

8.   Describe the generation of multiple-drug-resistant plasmids.

Answer: R plasmids are the main carriers of drug resistance. These plasmids are self-replicating and contain any number of genes for drug resistance, as well as the genes necessary for transfer by conjugation (called the RTF region). It is R plasmid's ability to transfer rapidly to other cells, even those of related species, that allows drug resistance to spread so rapidly. R plasmids acquire drug-resistance genes through transposition. Drug-resistance genes are found flanked by IR (inverted repeat) sequences and as a unit are known as transposons. Many transposons have been identified, and as a set they encode a wide range of drug resistances. Because transposons can "jump" between DNA molecules (e.g., from one plasmid to another or from a plasmid to the bacterial chromosome and vice versa), plasmids can continue to gain new drug-resistance genes as they mix and spread through different strains of cells. It is a classic example of evolution through natural selection. Those cells harboring R plasmids with multiple drug resistances survive to reproduce in the new environment of antibiotic use.

9.   Briefly describe the experiment that demonstrates that the transposition of the *Ty1* element in yeast takes place through an RNA intermediate.

Answer: Boeke, Fink, and their co-workers demonstrated that transposition of the *Ty* element in yeast involved an RNA intermediate. They constructed a plasmid using a *Ty* element that had a promoter that could be activated by galactose, and an intron inserted into its coding region. First, the frequency of transposition was greatly increased by the addition of galactose, indicating that an increase in transcription (and production of RNA) was correlated to rates of transposition. More importantly, after transposition they found that the newly transposed *Ty* DNA lacked the intron sequence. Because intron splicing occurs only during RNA processing, there must have been an RNA intermediate in the transposition event.

10.   Explain how the properties of *P* elements in *Drosophila* make gene-transfer experiments possible in this organism.

Answer: P elements are transposable elements found in *Drosophila*. Under certain conditions they are highly mobile and can be used to generate new mutations by random insertion and gene knockout. As such, they are a valuable tool to tag and then clone any number of genes. P elements can also be manipulated and used to insert almost any DNA (or gene) into the *Drosophila* genome. P element-mediated gene transfer requires inserting the DNA of interest between the inverted repeats necessary for P element transposition. This recombinant DNA, along with helper intact P element DNA (to supply the transposase), are then co-injected into very early embryos. The progeny of these embryos are then screened for those that contain the randomly inserted DNA of interest.

11. Although class 2 elements are abundant in the genomes of multicellular eukaryotes, class 1 elements usually make up the largest fraction of very large genomes such as those from humans (~2500 Mb), maize (~2500 Mb), and barley (~5000 Mb). Given what you know about class 1 and class 2 elements, what is it about their distinct mechanisms of transposition that would account for this consistent difference in abundance?

Answer: Most eukaryotic class 2 elements transpose by a "cut and paste" mechanism that involves their excision from one site and reinsertion into a new site elsewhere in the genome. Class 2 elements can increase their copy number in a few ways; one is by excising from a site that has been replicated into another site that has not yet been replicated. On the other hand, for class 1 elements, the RNA transcript is the transposition intermediate (it is copied into double-stranded DNA and reinserted into the genome). As such, a single class 1 element can give rise to many transcripts, and each can theoretically be copied into double stranded DNA and reinserted. However, because the reinsertion of so many DNA copies into the genome would almost certainly harm the host, mechanisms have evolved to prevent this from happening. For example, the transcription of Class 1 elements is usually repressed. In addition, in situations where transcription of Class 1 elements is not repressed, the host controls the conversion of the RNA transcript into double-stranded DNA.

12. As you saw in Figure 15-22, the genes of multicellular eukaryotes often contain many transposable elements. Why do most of these elements not affect the expression of the gene?

Answer: The transposable elements shown in Figure 15-22, like many in the human genome reside in non-coding sequences, such as introns.

13. What are safe havens? Are there any places in the much more compact bacterial genomes that might be a safe haven for insertion elements?

Answer: Some transposable elements have evolved strategies to insert into safe havens— regions of the genome where they will do minimal harm. Safe havens include duplicate genes (such as tRNA or rRNA genes) and other transposable elements. Safe havens in bacterial genomes might be very specific sequences between genes or the repeated rRNA genes.

14. Nobel prizes are usually awarded many years after the actual discovery. For example, James Watson, Francis Crick, and Maurice Wilkens were awarded the Nobel Prize in Medicine or Physiology in 1962, almost a decade after their discovery of the double-helical structure of DNA. However, Barbara McClintock was awarded the Nobel Prize in 1983, almost four decades after her discovery of transposable elements in maize. Why do you think it took this long?

Answer: Politics aside, Barbara McClintock was classically ahead of her time. She once wrote, "I stopped publishing detailed reports long ago when I realized, and acutely, the extent of disinterest and lack of confidence in the conclusions I was drawing from the studies." But beginning in the 1970's, transposition was discovered in bacteria and then in yeast. Transposable elements were subsequently found to be a significant component of most genomes, not just an oddity found only in maize. The importance of transposition in the transfer of drug resistance, cancer, and genetic engineering, just to name a few, was finally understood at the molecular level, and she was fully recognized for her lifetime of brilliant work.

CHALLENGING PROBLEMS

15. The insertion of transposable elements into genes can alter the normal pattern of expression. In the following situations, describe the possible consequences on gene expression.

a. A LINE inserts into an enhancer of a human gene.

b. A transposable element contains a binding site for a transcriptional repressor and inserts adjacent to a promoter.

c. An *Alu* element inserts into the 3′ splice (AG) site of an intron.

d. A *Ds* element that was inserted into the exon of a gene excises imperfectly and leaves 3 base pairs behind in the exon.

e. Another excision by that same *Ds* element leaves 2 base pairs behind in the exon.

f. A *Ds* element that was inserted into the middle of an intron excises imperfectly and leaves 5 base pairs behind in the intron.

Answer:

a. The consequences will be different for different genes and different insertions. In the simplest scenario, the insertion prevents the binding of transcriptional activators that are required for the ultimate binding of RNA polymerase to the promoter. In this case, the gene will not be expressed (no mRNA will be synthesized).

In more complex scenarios, the gene may be regulated by many enhancers (as is the case for most human genes). For example, one enhancer might be required for transcription in the liver, one required for transcription in muscle cells, etc. In this instance, insertion of the LINE into the liver enhancer may prevent transcription of the gene in the liver but not interfere with its transcription in muscle.

b. Again, the consequences will differ depending on the gene that has sustained the insertion. In the simplest scenario, the presence of the transposable element will provide a binding site for the transcriptional repressor to bind near the promoter and prevent the binding of RNA polymerase II.

c. The *Alu* element will be transcribed into RNA with the rest of the gene sequences and will prevent the splicing of the intron that it has inserted into. The insertion will almost certainly result in a null allele as the *Alu* sequence and the intron will now be translated. The intron, *Alu,* or both probably will contain stop codons.

d. The insertion and excision will result in a 3-bp indel in the exon and slightly alter the amino acid sequence of the protein, but it will not produce a frameshift mutation. The minor change in the amino acid sequence may or may not affect the function of the encoded protein.

e. This insertion and excision will produce a frameshift mutation and is more likely than (d) to impair protein function.

f. Chances are that this mutation will not affect gene expression as the intron will probably still be spliced correctly.

16. Before the integration of a transposon, its transposase makes a staggered cut in the host target DNA. If the staggered cut is at the sites of the arrows below, draw what the sequence of the host DNA will be after the transposon has been inserted. Represent the transposon as a rectangle.

AATTTGGCC → TAGTACTAATTGGTTGG
TTAAACCGGATCATGATT → AACCAACC

Answer: The staggered cut will lead to a nine base-pair target site duplication that flanks the inserted transposon.

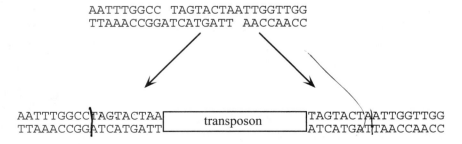

17. In *Drosophila*, M. Green found a *singed* allele (*sn*) with some unusual characteristics. Females homozygous for this X-linked allele have singed bristles, but they have numerous patches of sn^+ (wild-type) bristles on their heads, thoraxes, and abdomens. When these flies are mated with *sn* males, some females give only singed progeny, but others give both singed and wild-type progeny in variable proportions. Explain these results.

Answer: The sn^+ patches in an *sn* background and the occurrence of sn^+ progeny from an $sn \times sn$ mating indicate that sn^+ alleles are appearing at relatively high frequencies and that the *sn* alleles are unstable. High reversion rates suggest that the *sn* allele is the result of an insertion of a transposable element that is capable of excision in the germline as well as in the soma.

18. Consider two maize plants:

a. Genotype C/c^m ; Ac/Ac^+, where c^m is an unstable allele caused by *Ds* insertion

b. Genotype C/c^m, where c^m is an unstable allele caused by *Ac* insertion

What phenotypes would be produced and in what proportions when (1) each plant is crossed with a base-pair-substitution mutant *c/c* and (2) the plant in part *a* is crossed with the plant in part *b*? Assume that *Ac* and *c* are unlinked, that the chromosome-breakage frequency is negligible, and that mutant *c/C* is Ac^+.

Answer: In the *Ac-Ds* system, *Ac* can produce an unstable allele that is autonomous. *Ds* can revert only in the presence of *Ac* and is nonautonomous. In the following, $Ac+$ indicates the absence of the *Ac* regulator gene.

Cross 1:
P C/c^{Ds} ; $Ac/Ac^+ \times c/c$; Ac^+/Ac^+

F$_1$ 1/4 C/c ; Ac/Ac^+ (solid pigment)
1/4 C/c ; Ac^+/Ac^+ (solid pigment)
1/4 c^{Ds}/c ; Ac/Ac^+ (unstable colorless or spotted)
1/4 c^{Ds}/c ; Ac^+/Ac^+ (colorless)

Overall: 2 solid:1 spotted:1 colorless

Cross 2:
P $C/c^{Ac} \times c/c$

F$_1$ 1/2 C/c (solid pigment)
1/2 C/c^{Ac} (unstable colorless or spotted)
Overall: 1 solid:1 spotted

Cross 3:

P C/c^{Ds} ; Ac/Ac^+ × C/c^{Ac} ; Ac^+/Ac^+

F$_1$ 1/8 C/C ; Ac/Ac^+ (solid pigment)
1/8 C/c^{Ac} ; Ac/Ac^+ (solid pigment)
1/8 C/C ; Ac^+/Ac^+ (solid pigment)
1/8 C/c^{Ac} ; Ac^+/Ac^+ (solid pigment)
1/8 C/c^{Ds} ; Ac^+/Ac^+ (solid pigment)
1/8 C/c^{Ds} ; Ac^+/Ac (solid pigment)
1/8 c^{Ds}/c^{Ac} ; Ac^+/Ac^+ (unstable colorless or spotted)
1/8 c^{Ds}/c^{Ac} ; Ac^+/Ac (unstable colorless or spotted)

Overall: 3 solid:1 spotted

19. You meet your friend, a scientist, at the gym and she begins telling you about a mouse gene that she is studying in the lab. The product of this gene is an enzyme required to make the fur brown. The gene is called *FB* and the enzyme is called FB enzyme. When *FB* is mutant and cannot produce the FB enzyme, the fur is white. The scientist tells you that she has isolated the gene from two mice with brown fur and that, surprisingly, she found that the two genes differ by the presence of a 250-bp SINE (like the human *Alu* element) in the *FB* gene of one mouse but not in the gene of the other. She does not understand how this difference is possible, especially given that she determined that both mice make the FB enzyme. Can you help her formulate a hypothesis that explains why the mouse can still produce FB enzyme with a transposable element in its *FB* gene?

Answer: It would not be surprising to find a SINE element in an intron of a gene, rather than an exon. Processing of tbe pre-mRNA would remove the transposable element as part of the intron, and translation of the FB enzyme would not be effected.

20. The yeast genome has class 1 elements (*Ty1, Ty2,* and so forth) but no class 2 elements. Can you think of a possible reason why DNA elements have not been successful in the yeast genome?

Answer: There is no correct answer to this question. Yeast does have a very compact genome with almost 70 percent of its DNA representing exons, so it may not contain sufficient "safe havens" for DNA transposition. Transposons (and transposition events) that kill the host organism do not have a chance to spread. The *Ty* elements of yeast have evolved mechanisms that target their insertions so that they presumably do not harm their host. Perhaps it has only been chance that DNA transposons are not found in yeast but it is also possible that yeast has specific mechanisms that prevent their spread. One could look at related species of yeasts to determine if they have class 2 elements.

21. In addition to *Tc*1, the *C. elegans* genome contains other families of DNA transposons such as *Tc*2, *Tc*3, *Tc*4, and *Tc*5. Like *Tc*1, their transposition is repressed in the germ line but not in somatic cells. Predict the behavior of these elements in the mutant strains where *Tc*1 is no longer repressed due to mutations in the RNAi pathway. Justify your answer.

Answer: The RNAi pathway represses transposition of *Tc*1 in the germline by targeting the transposase mRNA for degradation. If transposition of *Tc*1 is no longer repressed in the germline, the element will transpose during the production of gametes in the worm. This will result in an increased probability of the insertion of the element into genes, causing mutations.

22. Based on the mechanism of gene silencing, what features of transposable elements does the RNAi pathway exploit to ensure that the host's own genes are not also silenced?

Answer: Transposable elements may insert near active genes, whose transcription "reads through" the element. Because the ends of the transposable elements consist of inverted repeat sequences, RNAs resulting from transcription that begins outside of the element will contain complementary sequences. These complementary sequences can base pair forming double-stranded RNAs that are recognized by the RNAi pathway.

23. What are the similarities and differences between retroviruses and retrotransposons? It has been hypothesized that retroviruses evolved from retrotransposons. Do you agree with this model? Justify your answer.

Answer: Retroviruses and retrotransposons insert themselves into the genome of the host as double-stranded DNA copies of an RNA intermediate. Both express an enzyme, reverse transcriptase, that can synthesize RNA from a DNA template. In addition, the structure of the transposons and the retroviruses are similar, containing long terminal repeats at their ends. Although the retrotransposons carry genes related to the viral *gag* and *pol* genes required for RNA processing and DNA synthesis, they do not carry the *env* gene required for the generation of the viral particle and cannot leave the cell. The conservation of the structure and replication of the elements suggests that they are evolutionarily related, however, without additional information, it is not possible to say which came first.

24. You have isolated a transposable element from the human genome and have determined its DNA sequence. How would you use this sequence to determine the copy number of the element in the human genome if you just had a computer with an Internet connection? (**Hint:** see Chapter 14.)

Answer: You would input your element's sequences as the query sequence in a BLAST search, limiting the search to human genome database. This would allow you to find not only exact matches, but partial matches as well.

25. Following up on the previous question, how would you determine whether other primates had a similar element in their genomes?

    Answer: You would perform a BLAST search as described above, however, you would expand your search to include the databases of all of the sequenced primate genomes.

26. Of all the genes in the human genome, the ones with the most characterized *Alu* insertions are those that cause hemophilia, including several insertions in the factor VIII and factor IX genes. Based on this fact, your colleague hypothesizes that the *Alu* element prefers to insert into these genes. Do you agree? What other reason can you provide that also explains these data?

    Answer: We do not have enough information to determine whether these genes are "hot spots" for insertion. While these are the most characterized insertions, it may be that insertions in this gene are observed simply because the insertions do not cause lethality and result in an obvious phenotype.

27. If all members of a transposable element family can be silenced by dsRNA synthesized from a single family member, how is it possible for one element family (like *Tc*1) to have 32 copies in the *C. elegans* genome while another family (*Tc*2) has fewer than 5 copies?

    Answer: The elements would be able to transpose until the host RNAi established repression. It is possible that the more abundant element transposed to more non-deleterious locations by chance before the repression was established. Alternatively, it may be that the expression of *Tc*1 was greater before repression was established resulting in a greater frequency of transposition.

16

Mutation, Repair, and Recombination

WORKING WITH THE FIGURES

1. In Figure 16-3a, what is the consequence of the new 5′ splice site on the open reading frame? In 16-3b, how big could the intron be to maintain the reading frame (let's say between 75 and 100 bp)?

   Answer: A mutation that generates a new 5′ splice site within an existing exon will result in the loss of information from the open reading frame, since some of the exon will be removed with the splicing of the intron. In addition, because the number of nucleotides deleted is not a multiple of 3 (64 nt) a frameshift mutation will result. In part *b* of the figure, the retained intron will maintain the reading frame as long as the length in nucleotides is divisible by three (ex. 99 bp) and it does not contain any stop codons in the same frame as exon 1.

2. Using Figure 16-4 as an example, compare the migration of RNA and protein for the wild-type gene and the mutation shown in Figure 16-3b. Assume that the retained intron maintains the reading frame.

   Answer: The RNA produced by the mutation in Figure 16-3b will be longer than the mature wild-type RNA and thus will not migrate as far in the gel. The same will be true of the protein (mutant protein will be larger than wild-type protein) if the reading frame is maintained and no stop codons are present in the sequence.

3. In the Ames test shown in Figure 16-17, what is the reason for adding the liver extract to each sample?

   Answer: Sometimes compounds only become mutagenic when processed by the enzymes in the vertebrate liver. Therefore, although exposure to these compounds in humans may cause cancer, they would not give a positive result in a bacterial mutation assay.

4. Based on the mode of action of aflatoxin (Figure 16-16), propose a scenario that explains its response in the Ames test (Figure 16-18).

Answer: Aflatoxin B_1 becomes covalently attached to guanine residues in the DNA. The addition of the aflatoxin adduct destabilizes the N-glycosidic bond, leaving an abasic site. The replicative polymerases cannot synthesize DNA across from an abasic site, requiring the use of a bypass polymerase. In the absence of complimentary base-pairing information, the polymerase will frequently incorporate the wrong nucleotide opposite the abasic site. This will result in daughter cells with base substitutions, as observed in the Ames test.

5. In Figure 16-22, point out the mutant protein(s) in patients with Cockayne syndrome. What protein(s) is/are mutant in patients with XP? How are these different mutations thought to account for the different disease symptoms?

Answer: The proteins CSA or CSB are mutant in Cockayne syndrome, while the XPB and XPD proteins are defective in patients with Xeroderma pigmentosum. The defective proteins in Cockayne syndrome cannot recognize RNA polymerase complexes stalled by DNA damage. This failure may trigger apoptosis. Defects observed in Cockayne syndrome may be the result of both apoptosis and a failure to repair lesions, while XP defects may result primarily from a failure to repair DNA.

6. The MutH protein nicks the newly synthesized strand (Figure 16-23). How does it "know" which strand this is?

Answer: The MutH protein nicks the unmethylated strand at hemimethylated GATC sequences. The transient delay in the methylation of the newly synthesized strand at these sequences therefore serves as a signal for repair.

7. What features of the bypass polymerase make it ideal for its role in translesion synthesis, shown in Figure 16-24?

Answer: The active sites of these polymerases can accommodate bulky adducts that the replicative polymerases cannot, and they lack the 3'-5' exonuclease "proofreading" activity. In addition, they are distributive polymerases, adding only a few nucleotides at a time before falling off of the substrate, limiting the amount of DNA synthesized by these error prone enzymes.

BASIC PROBLEMS

8. Consider the following wild-type and mutant sequences:

Wild-type   ....CTTGCAAGCGAATC....

Mutant   ....CTTGCTAGCGAATC....

The substitution shown seems to have created a stop codon. What further information do you need to be confident that it has done so?

Answer: You need to know the reading frame of the possible message.

9.   What type of mutation is depicted by the following sequences (shown as mRNA)?

Wild type   ....5′ AAUCCUUACGGA 3′....
Mutant   ....5′ AAUCCUACGGA 3′....

Answer: The mutant has a deletion of one base, and this will result in a frameshift (−1) mutation.

10.   Can a missense mutation of proline to histidine be made with a G·C → A·T transition-causing mutagen? What about a proline-to-serine missense mutation?

Answer: Proline can be coded for by CCN (N stands for any nucleotide) and histidine can be coded for by CAU or CAC. For a mutation to change a proline codon to a histidine codon requires a transverion (C to A) at the middle position. Therefore, a transition-causing mutagen cannot cause this change. Serine can be coded for by UCN. A change from C to U at the first position (a transition) would cause this missense mutation and would be possible with this mutagen.

11.   By base-pair substitution, what are all the synonymous changes that can be made starting with the codon CGG?

Answer: Assuming single base-pair substitutions, then CGG can be changed to CGU, CGA, CGC, or AGG and still would code for arginine.

12.   a. What are all the transversions that can be made starting with the codon CGG?
       b. Which of these transversions will be missense? Can you be sure?

Answer:
a. and b. By transversion, CGG (arginine) can be become AGG (arginine), GGG (glycine), CCG (proline), CUG (leucine), CGC (arginine), or CGU (arginine).

13. **a.** Acridine orange is an effective mutagen for producing null alleles by mutation. Why does it produce null alleles?

b. A certain acridine-like compound generates only single insertions. A mutation induced with this compound is treated with the same compound, and some revertants are produced. How is this outcome possible?

Answer:

a. Acridine orange causes frameshift mutations and frameshift mutations often result in null alleles.

b. A +1 frameshift mutation can be reverted by two further single insertions so that the reading frame is re-established.

14. Defend the statement "Cancer is a genetic disease."

Answer: The following is a list of observations that argue "cancer is a genetic disease:"

1. Certain cancers are inherited as highly penetrant simple Mendelian traits.

2. Most carcinogenic agents are also mutagenic.

3. Various oncogenes have been isolated from tumor viruses.

4. A number of genes that lead to the susceptibility to particular types of cancer have been mapped, isolated, and studied.

5. Dominant oncogenes have been isolated from tumor cells.

6. Certain cancers are highly correlated to specific chromosomal rearrangements. (See Chapter 17 of the companion text.)

15. Give an example of a DNA-repair defect that leads to cancer.

Answer: XP (xeroderma pigmentosum) patients lack nucleotide excision repair and are highly prone to developing pigmented skin cancers. Individuals with HNPCC (hereditary nonpolyposis colorectal cancer) are prone to colorectal cancer due to a loss of the mismatch repair system. Individuals homozygous for mutations in *BRCA1* or *BRCA2* (breast cancer predisposition genes 1 and 2) are prone to breast cancer due to the loss of repair of double-stranded breaks.

16. In mismatch repair in *E. coli*, only a mismatch in the newly synthesized strand is corrected. How is *E. coli* able to recognize the newly synthesized strand? Why does this ability make biological sense?

Answer: DNA in *E. coli* is methylated. To distinguish the old template strand from the newly synthesized strand, the mismatch repair mechanism takes advantage of a delay in the methylation of the new strand. This makes sense as replication errors produce mismatches only on the newly synthesized strand, so the mismatch repair system replaces the "wrong" base on that strand.

17. A mutational lesion results in a sequence containing a mismatched base pair:

5′ AGCT**G**CCTT 3′
3′ ACG**AT**GGAA 5′
        Codon

If mismatch repair occurs in either direction, which amino acids could be found at this site?

Answer: The mismatched "T" would be corrected to C and the resulting ACG, after transcription, would be 5′ UGC 3′ and code for cysteine. Or, if the other strand was corrected, ATG would be transcribed to 5′ UAC 3′ and code for tyrosine.

18. Under what circumstances could nonhomologous end joining be said to be error prone?

Answer: NHEJ (nonhomologous end-joining) is error prone as some sequence may be lost in the repair process. The consequences of imperfect repair may be far less harmful than leaving the lesion unrepaired. Presumably this repair pathway evolved because, unless repaired, the broken ends can degrade further, leading to loss of more genetic information. Also, these lesions can initiate potentially harmful chromosomal rearrangements that could lead to cell death.

19. Why are many chemicals that test positive by the Ames test also classified as carcinogens?

Answer: There is a very strong correlation between mutagens and carcinogens. As discussed in Problem 14, cancer is a genetic disease. Therefore, any chemical classified as a mutagen by the Ames test should also be considered a carcinogen.

20. The Spo11 protein is conserved in eukaryotes. Do you think it is also conserved in bacterial species? Justify your answer.

Answer: Meiotic recombination is initialized when the Spo11 protein makes double-strand cuts in one of the homologous chromosomes. It is highly conserved in eukaryotes, indicating that this mechanism to initiate recombination is also conserved. Bacterial species do not have reciprocal meiotic recombination, so you would not expect this function to be conserved.

21. Differentiate between the elements of the following pairs:

a. Transitions and transversions
b. Synonymous and neutral mutations
c. Missense and nonsense mutations
d. Frameshift and nonsense mutations

Answer:
a. A transition mutation is the substitution of a purine for a purine or the substitution of a pyrimidine for a pyrimidine. A transversion mutation is the substitution of a purine for a pyrimidine, or vice versa.

b. Both are base-pair substitutions. A synonymous mutation is one that does not alter the amino acid sequence of the protein product from the gene, because the new codon codes for the same amino acid as did the nonmutant codon. A neutral mutation results in a different amino acid that is functionally equivalent, and the mutation therefore has no known adaptive significance.

c. A missense mutation results in a different amino acid in the protein product of the gene. A nonsense mutation causes premature termination of translation, resulting in a shortened protein.

d. Frameshift mutations arise from addition or deletion of one or more bases in other than multiples of three, thus altering the reading frame for translation. Therefore, the amino acid sequence from the site of the mutation to the end of the protein product of the gene will be altered. Frameshift mutations can and often do result in premature stop codons in the new reading frame, leading to shortened protein products. A nonsense mutation causes premature termination of translation in the original reading frame, resulting in a shortened protein.

22. Describe two spontaneous lesions that can lead to mutations.

Answer: Depurination results in the loss of the adenine or guanine base from the DNA backbone. Because the resulting apurinic site cannot specify a complementary base, replication is blocked. Under certain conditions, replication proceeds with a near random insertion of a base opposite the apurinic site. In three-fourths of these insertions, a mutation will result.

Deamination of cytosine yields uracil. If left unrepaired, the uracil will be paired with adenine during replication, ultimately resulting in a transition mutation.

Deamination of 5-methylcytosine yields thymine and thus frequently leads to C to T transitions.

Oxidatively damaged bases, such as 8-OxodG (8-oxo-7-hydrodeoxyguanosine) can pair with adenine, resulting in a transversion.

Errors during DNA replication can lead to spontaneous indel mutations.

23. What are bypass polymerases? How do they differ from the replicative polymerases? How do their special features facilitate their role in DNA repair?

Answer: Translesion or bypass polymerases are able to replicate past damaged DNA that otherwise would stall replicative polymerases. They differ from replicative polymerases in that they can tolerate large adducts on the bases (as they have much larger active sites that can accommodate damaged bases), they are much more error-prone (as they lack the 3′ to 5′ proofreading function), and they can only add relatively few nucleotides before falling off. Their main function is to unblock the replication fork, not to synthesize long stretches of DNA that could contain many mismatches.

24. In adult cells that have stopped dividing, what types of repair systems are possible?

Answer: There are many repair systems that are available: direct reversal, excision repair, transcription-coupled repair, and non-homologous end-joining.

25. A certain compound that is an analog of the base cytosine can become incorporated into DNA. It normally hydrogen bonds just as cytosine does, but it quite often isomerizes to a form that hydrogen bonds as thymine does. Do you expect this compound to be mutagenic, and, if so, what types of changes might it induce at the DNA level?

Answer: Yes. It will cause CG-to-TA transitions.

26. Two pathways, homologous recombination and nonhomologous end joining (NHEJ), can repair double-strand breaks in DNA. If homologous recombination is an error-free pathway whereas NHEJ is not always error free, why is NHEJ used most of the time in eukaryotes?

Answer: Since cells of higher eukaryotes are usually not replicating their DNA, error-free repair is not possible because there are no undamaged strands or sister chromatids available as templates for new DNA synthesis.

27. Which repair pathway recognizes DNA damage during transcription? What happens if the damage is not repaired?

Answer: DNA damage that stalls transcription is repaired by TC-NER (transcription-coupled nucleotide excision repair). Humans lacking this pathway suffer from Cockayne syndrome. A consequence of this defect is that a cell is much more likely to activate its apoptosis (cell suicide) pathway. Affected individuals are very sensitive to sunlight and have short stature, the appearance of premature aging, and a variety of developmental disorders.

CHALLENGING PROBLEMS

28. **a.** Why is it impossible to induce nonsense mutations (represented at the mRNA level by the triplets UAG, UAA, and UGA) by treating wild-type strains with mutagens that cause only $A \cdot T \rightarrow G \cdot C$ transitions in DNA?

   b. Hydroxylamine (HA) causes only $G \cdot C \rightarrow A \cdot T$ transitions in DNA. Will HA produce nonsense mutations in wild-type strains?

   c. Will HA treatment revert nonsense mutations?

   Answer:
   a. Because 5′-UAA-3′ does not contain G or C, a transition to a GC pair in the DNA cannot result in 5′-UAA-3′. 5′-UGA-3′ and 5′-UAG-3′ have the DNA antisense-strand sequence of 3′-ACT-5′ and 3′-ATC-5′, respectively. A transition to either of these stop codons occurs from the nonmutant 3′-ATT-5′. However, a DNA sequence of 3′-ATT-5′ results in an RNA sequence of 5′-UAA-3′, itself a stop codon.

   b. Yes. An example is 5′-UGG-3′, which codes for *trp*, to 5′-UAG-3′.

   c. No. In the three stop codons the only base that can be acted upon is G (in UAG, for instance). Replacing the G with an A would result in 5′-UAA-3′, a stop codon.

29. Several auxotrophic point mutants in *Neurospora* are treated with various agents to see if reversion will take place. The following results were obtained (a plus sign indicates reversion; HA causes only G · C → A · T transitions).

Mutant	5-BU	HA	Proflavin	Spontaneous reversion
1	−	−	−	−
2	−	−	+	+
3	+	−	−	+
4	−	−	−	+
5	+	+	−	+

a. For each of the five mutants, describe the nature of the original mutation event (not the reversion) at the molecular level. Be as specific as possible.

b. For each of the five mutants, name a possible mutagen that could have caused the original mutation event. (Spontaneous mutation is not an acceptable answer.)

c. In the reversion experiment for mutant 5, a particularly interesting prototrophic derivative is obtained. When this type is crossed with a standard wild-type strain, the progeny consist of 90 percent prototrophs and 10 percent auxotrophs. Give a full explanation for these results, including a precise reason for the frequencies observed.

Answer:
a. and b. Mutant 1: most likely a deletion. It could be caused by radiation.

Mutant 2: because proflavin causes either additions or deletions of bases and because spontaneous mutation can result in additions or deletions, the most probable cause was a frameshift mutation by an intercalating agent.

Mutant 3: 5-BU causes transitions, which means that the original mutation was most likely a transition. Because HA causes GC-to-AT transitions and HA cannot revert it, the original must have been a GC-to-AT transition. It could have been caused by base analogs.

Mutant 4: the chemical agents cause transitions or frameshift mutations. Because there is spontaneous reversion only, the original mutation must have been a transversion. X-irradiation or oxidizing agents could have caused the original mutation.

Mutant 5: HA causes transitions from GC-to-AT, as does 5-BU. The original mutation was most likely an AT-to-GC transition, which could be caused by base analogs.

c. The suggestion is a second-site reversion linked to the original mutant by 20 map units and therefore most likely in a second gene. Note that auxotrophs equal half the recombinants.

Unpacking the Problem

30. You are using nitrosoguanidine to "revert" mutant *nic-2* (nicotinamide-requiring) alleles in *Neurospora*.

    You treat cells, plate them on a medium without nicotinamide, and look for prototrophic colonies. You obtain the following results for two mutant alleles. Explain these results at the molecular level, and indicate how you would test your hypotheses.

    a. With *nic-2* allele 1, you obtain no prototrophs at all.

    b. With *nic-2* allele 2, you obtain three prototrophic colonies A, B, and C, and you cross each separately with a wild-type strain. From the cross prototroph A × wild type, you obtain 100 progeny, all of which are prototrophic. From the cross prototroph B × wild type, you obtain 100 progeny, of which 78 are prototrophic and 22 are nicotinamide requiring. From the cross prototroph C × wild type, you obtain 1000 progeny, of which 996 are prototrophic and 4 are nicotinamide requiring.

    Answer:
    a. A lack of revertants suggests either a deletion or an inversion within the gene.

    b. To understand these data, recall that half the progeny should come from the wild-type parent.

        Prototroph A: because 100 percent of the progeny are prototrophic, a reversion at the original mutant site may have occurred.

        Prototroph B: half the progeny are parental prototrophs, and the remaining prototrophs, 28 percent, are the result of the new mutation. Notice that 28 percent is approximately equal to the 22 percent auxotrophs. The suggestion is that an unlinked suppressor mutation occurred, yielding independent assortment with the *nic* mutant.

        Prototroph C: there are 496 "revertant" prototrophs (the other 500 are parental prototrophs) and four auxotrophs. This suggests that a suppressor mutation occurred in a site very close to the original mutation and was infrequently separated from the original mutation by recombination [100% $(4 \times 2)/1000 = 0.8$ m.u.].

31. You are working with a newly discovered mutagen, and you wish to determine the base change that it introduces into DNA. Thus far, you have determined that the mutagen chemically alters a single base in such a way that its base-pairing properties are altered permanently. To determine the specificity of the alteration, you examine the amino acid changes that take place after mutagenesis. A sample of what you find is shown here:

Original:	Gln–His–Ile–Glu–Lys
Mutant:	Gln–His–Met–Glu–Lys
Original:	Ala–Val–Asn–Arg
Mutant:	Ala–Val–Ser–Arg
Original:	Arg–Ser–Leu
Mutant:	Arg–Ser–Leu–Trp–Lys–Thr–Phe

   What is the base-change specificity of the mutagen?

   Answer: Compare the original amino acid sequences to the mutant ones and list the changes.

   original: ile; mutant: met
   original: asn;   mutant: ser
   original: stop; mutant: trp

   Now compare the codons that must have been altered by this mutagen.

   original: ile AU<u>A</u>;   mutant: met AU<u>G</u>
   original: asn A<u>A</u>C or A<u>A</u>U;   mutant: ser A<u>G</u>C or A<u>G</u>U
   original: stop U<u>A</u>G or UG<u>A</u>; mutant: trp U<u>G</u>G

   All these mutations can be the result of T to C or A to G transitions in the DNA. The result would be an A to G change in the mRNA that explains all three codon changes. This mutagen, then, might work by altering the base-pairing specificity of T so that it now base pairs with G. Or, the mutagen could alter the pairing specificity of A so that it now pairs with C, which would have the same effect.

32. You now find an additional mutant from the experiment in Problem 31:

Original:	Ile–Leu–His–Gln
Mutant:	Ile–Pro–His–Gln

   Could the base-change specificity in your answer to Problem 31 account for this mutation? Why or why not?

Answer: Yes. Mutation of the double-stranded DNA sequence by either a T to C transition in the coding strand or an A to G transition in the template strand will result in the CUN to CCN in the mRNA.

Original: leu CUN; Mutant: pro CCN (where N = any base)

33. You are an expert in DNA-repair mechanisms. You receive a sample of a human cell line derived from a woman who has symptoms of xeroderma pigmentosum. You determine that she has a mutation in a gene that has not been previously associated with XP. How is this possible?

Answer: XP is a heterogeneous genetic disorder and is caused by mutations in any one of several genes involved in the process of NER (nucleotide excision repair). As you read in the text about the discovery of yet another protein involved in NHEJ through research on cell line 2BN, it is certainly possible that this new patient has a mutation in as yet an unknown gene that encodes a protein necessary for NER.

34. Ozone (O_3) is an important naturally occurring component in our atmosphere, where it forms a layer that absorbs UV radiation. A hole in the ozone layer was discovered in the 1970s over Antarctica and Australia. The hole appears seasonally and was found to be due to human activity. Specifically, ozone is destroyed by a class of chemicals (called CFCs for chlorofluorocarbons) that are found in refrigerants, air-conditioning systems, and aerosols.

As a scientist working on DNA-repair mechanisms, you discover that there has been a significant increase in skin cancer in the beach communities in Australia. A newspaper reporter friend offers to let you publish a short note (a paragraph) in which you are to describe the possible connection between the ozone hole and the increased skin cancers. On the basis of what you have learned about DNA repair in this chapter, write a paragraph that explains the mechanistic connection.

Answer: Sun worshippers beware! Science has established a clear link between UV exposure and skin cancer, and with the drastic changes occurring in the protective ozone layer, UV levels are rising. UV damages DNA causing our cells to struggle to repair this damage by a number of different mechanisms. While this damage may be so great that the cell dies (the peeling of dead cells after a sunburn), other times the damage is more insidious as it causes permanent changes in the instructions encoded in the DNA itself. These changes are sometimes caused by the cell attempting to repair the damage. In these cases, when the damage prevents the DNA from being "read," the cell may simply "guess" as to the meaning, changing the instructions. Eventually, the changed instructions may tell a cell to divide when it's not supposed to, or not die when it is supposed to. These are the very changes that make a cell cancerous.

17

Large-Scale Chromosomal Changes

WORKING WITH THE FIGURES

1. Based on Table 17-1, how would you categorize the following genomes? (Letters H through J stand for four different chromosomes.)

   HH II J KK
   HH II JJ KKK
   HHHH IIII JJJJ KKKK

   Answer: Monosomic ($2n-1$) 7 chromosomes
            Trisomic ($2n+1$)   9   -II-
            Tetraploid ($4n$)   16   -II-

2. Based on Figure 17-4, how many chromatids are in a trivalent?

   Answer: There are 6 chromatids in a trivalent.

3. Based on Figure 17-5, if colchicine is used on a plant in which $2n = 18$, how many chromosomes would be in the abnormal product?

   Answer: Colchicine prevents migration of chromatids, and the abnormal product of such treatment would keep all the chromatids ($2n = 18$) in one cell.

4. Basing your work on Figure 17-7, use colored pens to represent the chromosomes of the fertile amphidiploid.

   Answer: A fertile amphidiploids would be an organism produced from a hybrid with two different sets of chromosomes (n_1 and n_2), which would be infertile until some tissue undergoes chromosomal doubling ($2 n_1 + 2 n_2$) and such chromosomal set would technically become a diploid (each chromosome has its pair; therefore they could undergo meiosis and produce gametes). This could be a new species.
   Picture example: 3 pairs of chromosomes/ different lengths/ red for $n_1 = 3$

4 chromosomes/ different lengths/ green for $n_2 = 4$
Hybrid/ infertile: 7 chromosomes (n_1+n_2)
Amphidiploids/ fertile: 14 chromosomes (pairs of n_1 and n_2)

5.   If Emmer wheat (Figure 17-9) is crossed to another wild wheat CC (not shown), what would be the constitution of a sterile product of this cross? What amphidiploid could arise from the sterile product? Would the amphidiploid be fertile?

Answer: Emmer wheat was domesticated 10,000 years ago, as a tetraploid with two chromosome sets ($2n_1 + 2 n_2$ or $AA + BB$). If AA BB tetraploid genome is combined with another wild wheat (n_3 or C gamete), the product would be sterile. In such case, C chromosomes would not have homologous pairs in a hybrid parent wheat. Amphidiploid could occur, if chromosome doubling happens in a parental tissue (e.g., flower parts), and a fertile wheat species would be hexaploid (AA BB CC).

6.   In Figure 17-12, what would be the constitution of an individual formed from the union of a monosomic from a first-division nondisjunction in a female and a disomic from a second-division nondisjunction in a male, assuming the gametes were functional?

Answer: A gamete from a first-division nondisjunction would be an egg without the chromosome in question ($n - 1$); while a gamete/sperm from a second-division non-disjunction would be a ($n + 1$). If both gametes are functional, they would result in a euploid ($2n$) zygote, with two copies of a father's chromosome.

7.   In Figure 17-14, what would be the expected percentage of each type of segregation?

Answer: These are three equally possible combinations/ segregation of chromosomes in meiosis, in a trisomic individual. Therefore each type is expected at 33.3%

8.   In Figure 17-19, is there any difference between the inversion products formed from breakage and those formed from crossing over?

Answer: Inversion products look the same, having the sequence: 1-3-2-4, yet they could be genetically different. Breakage and rejoining (on the left) results in a rearrangement with possibly damaged gene sequence and yet the same length of the chromosome. Crossing over between repetitive DNA (blue) might generate an inversion with less damage to the gene sequence, since the crossing over happens in homologous regions of a repetitive sequence.

9. Referring to Figure 17-19, draw a diagram showing the process whereby an inversion formed from crossing over could generate a normal sequence.

    Answer: An inversion formed by a crossing over:_1 >   3 2   <   4_ Repetitive sequences are homologous (< and > oriented) and they could pair and form crossing over in the next generation of gametes, producing a normal segment again: _1 < 2 3   >   4 .

10. In Figure 17-21, would the recessive *fa* allele be expressed when paired with deletion 264-32? 258-11?

    Answer: Deletion 264-32 overlaps with the segment carrying *fa* allele (7) on polytene chromosome; this means that the *fa* allele would be pseudodominant (expressed) in such combination of chromosomes.
    Deletion 258-11 is covering segments before the *fa* allele and the gene would not be expressed. In this case a dominant allele (*fa+*) of the 258-11 chromosome will show in the phenotype.

11. Look at Figure 17-22 and state which bands are missing in the cri du chat deletion.

    Answer: Cri du chat syndrome in humans is caused by the deletion of the tip of the p5 (bands 15.3 and 15.2).

12. In Figure 17-25, which species is most closely related to the ancestral yeast strain? Why are genes 3 and 13 referred to as duplicate?

    Answer: *Kluyveromyces waitii* seems to be most similar to the common ancestor. When genome of *Saccharomyces* lineage doubled, genes such as 3 and 13 become duplicate in both species and in the same relative order. Some other genes were lost (2, 7, etc).

13. Referring to Figure 17-26, draw the product if breaks occurred within genes *A* and *B*.

    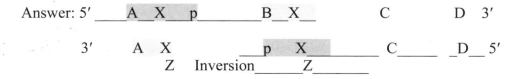

    Answer: 5'_____A X   p_____B__X__   C   D   3'

              3'   A X   __p X_____ C_____ _D_ 5'
                      Z   Inversion_____Z_____

After the joining of the breaks, genes *A* and *B* become disrupted: 5′ "A" "BA" "B" C D 3′ and the final product may have the fused parts of genes *A* and *B*: 5′ "A" B/A "B" C D   3′

14. In Figure 17-26, the bottom panel shows that genes *B* and *C* are oriented in a different direction (note the promoters). Do you think this difference in orientation would affect their functionality?

Answer: If promoter positions are changed, this could interfere with gene function because of the reading frame for RNA polymerase. An inversion of the gene order can alter normal expression of the genes by placing a gene in the new regulatory environment.

15. In Figure 17-28, what would be the consequence of a crossover between the centromere and locus *A*?

Answer: Inversions that include centromere are called pericentric. If the breakage happens between the centromere (o) and the locus *A*:
o--X--A B C D E
o --X-- A D C B E
Crossing over products might be both normal and inverted:
E B C D—o-- A

16. Based on Figure 17-30, are normal genomes ever formed from the two types of segregation? Are normal genomes ever formed from an adjacent-1 segregation?

Answer: With the adjacent 1 segregation final meiotic products are often inviable due to a deletion of a significant gene segment. In plants such gametes do not function, while in animals gametes might be fine (but not the zygotes).

17. Referring to Figure 17-32, draw an *inviable* product from the same meiosis.

Answer: This figure shows a translocation heterozygote:

____**a**_____**b**_____ Not linked
_____ **b+** **a+**_ Pseudo-linked

After meiosis some products would be inviable because they carry deletions of one or another gene segment, such as:

__**a**_____ or _____**b**___ , etc.

The only viable progeny are the ones with parental genotypes and pseudo-linkage is an indication of reciprocal translocation.

18. Based on Figure 17-35, write a sentence stating how translocation can lead to cancer. Can you think of another genetic cause of cancer?

Answer: This figure illustrates how relocation of an oncogene (via translocation) might cause cancer in somatic cells, such as Burkitt's lymphoma and chronic myologenous leukemia. In general, any chromosomal rearrangement that changes the regulatory environment of a gene could lead to cancer.
Other causes would be based on a mutation in a gene which might control cell division or a gene that suppresses growth of tumors. Finally, mutations in genes responsible for repair mechanisms at the molecular or cellular level might also be genetic causes of different types of cancer.

19. Looking at Figure 17-36, why do you think the signal ratio is so much higher in the bottom panel?

Answer: Comparative genomic hybridization technique compares mutant to wild type ratios and the ratio of values of each cDNA is calculated. For example, duplication would have a ratio higher than 1 (more gene product), while a deletion will have less then 1 (less product) in a chromosomal microarray. The last graph indicated tandem amplification of a gene (multiple copies) which lead to a very high ratio (close to 10).

20. Using Figure 17-37, calculate what percentage of conceptions are triploid. The same figure shows XO in the spontaneous-abortion category; however, we know that many XO individuals are viable. In which of the viable categories would XO be grouped?

Answer: Figure 17-37 shows the proportion of chromosomal mutations in human conceptions (zygotes). Triploid zygotes (3n) were not found in the live births, but only in the spontaneous abortions (12,750 per 1,000,000 conceptions). Therefore only around 1.275 % of conceptions were triploid (or with 69 chromosomes).
Turner syndrome (45 X0) is an aneuploidy (2n-1). It is found in both live births (some of the 422 females with sex chromosome aneuploidy) and in spontaneous abortions (13,500 per 1,000,000 conceptions). Therefore viable XO (Turner syndrome females) belong to the category of sex chromosome aneuploids, under live births with chromosome abnormalities. Other than Turner syndrome, sex chromosome aneuploid females might also be 47 XXX, so the percentage of living XO is not specified in this figure (out of 422 births).

BASIC PROBLEMS

21. In keeping with the style of Table 17-1, what would you call organisms that are MM N OO; MM NN OO; MMM NN PP?

Answer: MM N OO would be classified as $2n - 1$ (monosomic); MM NN OO would be classified as $2n$ (euploid); and MMM NN PP would be classified as $2n + 1$ (trisomic).

22. A large plant arose in a natural population. Qualitatively, it looked just the same as the others, except much larger. Is it more likely to be an allopolyploid or an autopolyploid? How would you test that it was a polyploid and not just growing in rich soil?

Answer: It would more likely be an autopolyploid. To make sure it was polyploid, you would need to microscopically examine stained chromosomes from mitotically dividing cells and count the chromosome number.

23. Is a trisomic an aneuploid or a polyploid?

Answer: Aneuploid. Trisomic refers to three copies of one chromosome. Triploid refers to three copies of all chromosomes.

24. In a tetraploid $B/B/b/b$, how many quadrivalent possible pairings are there? Draw them (see Figure 17-5).

Answer: There would be one possible quadrivalent with 50 percent of recombinant products (B/b).

25. Someone tells you that cauliflower is an amphidiploid. Do you agree? Explain.

Answer: No. Amphidiploid means "doubled diploid ($2n_1 + 2n_2$)." Because cauliflower has $n = 9$ chromosome, it could not have arose in this fashion. It has, however, contributed to other amphidiploid species, such as rutabaga.

26. Why is *Raphanobrassica* fertile, whereas its progenitor wasn't?

Answer: The progenitor had nine chromosomes from a cabbage parent and nine chromosomes from a radish parent. These chromosomes were different enough that pairs did not synapse and segregate normally at meiosis. By doubling the chromosomes in the progenitor ($2n = 36$), all chromosomes now had homologous partners and meiosis could proceed normally.

27. In the designation of wheat genomes, how many chromosomes are represented by the letter B?

Answer: In modern hexaploid wheat (*T. aestivum*) there are $2n$ or $6x$ chromosomes in its genome, with a total of 42. If each haploid set has the same number of chromosomes, B (or x) represents seven chromosomes.

28. How would you "re-create" hexaploid bread wheat from *Triticum tauschii* and Emmer?

Answer: Cross *T. tauschii* and Emmer to get ABD offspring. Treat the offspring with colchicine to double the chromosome number to AABBDD to get the hexaploid bread wheat.

29. How would you make a monoploid plantlet by starting with a diploid plant?

Answer: Cells destined to become pollen grains can be induced by cold treatment to grow into embryoids. These embryoids can then be grown on agar to form monoploid plantlets.

30. A disomic product of meiosis is obtained. What is its likely origin? What other genotypes would you expect among the products of that meiosis under your hypothesis?

Answer: The likely origin of a disomic ($n + 1$) gamete is nondisjunction during meiosis. Depending whether the nondisjunction took place during the first or second division, you would expect one nullosomic ($n - 1$), or two nullosomics and another disomic, respectively.

31. Can a trisomic *A/A/a* ever produce a gamete of genotype *a*?

Answer: Yes. You would expect that one-sixth of the gametes would be *a*. Also, two-sixths would be *A*, two-sixths would be *Aa*, and one-sixth would be *AA*.

32. Which, if any, of the following sex-chromosome aneuploids in humans are fertile: XXX, XXY, XYY, XO?

Answer: Both XYY (male) and XXX (female) would be fertile. XO (Turner syndrome) and XXY (Klinefelter syndrome) are known to be sterile.

33. Why are older expectant mothers routinely given amniocentesis or CVS?

Answer: Older mothers have an elevated risk of having a child with some chromosomal aberration, due to the age of their egg cells. Down syndrome and other aneuploidy due to the meiotic nondisjunction in mother's gametogeneseis are among the most common. However, age of the father also contributes to the increased risk of some chromosomal aberrations.

34. In an inversion, is a 5′ DNA end ever joined to another 5′ end? Explain.

Answer: No. The DNA backbone has strict 5′ to 3′ polarity, and 5′ ends can only be joined to 3′ ends.

35. If you observed a dicentric bridge at meiosis, what rearrangement would you predict had taken place?

Answer: A cross over within a paracentric inversion heterozygote results in a dicentric bridge (and an acentric fragment).

36. Why do acentric fragments get lost?

Answer: By definition, an acentric fragment has no centromere, so it cannot be aligned or moved during meiosis (or mitosis). Consequently, at the end of a cell division, it gets left in the cytoplasm where it is not replicated.

37. Diagram a translocation arising from repetitive DNA. Repeat for a deletion.

Answer:
Possible translocation:

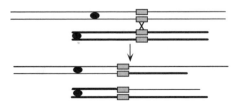

Possible deletion (and duplication):

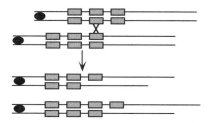

38. From a large stock of *Neurospora* rearrangements available from the fungal genetics stock center, what type would you choose to synthesize a strain that had a duplication of the right arm of chromosome 3 and a deletion for the tip of chromosome 4?

Answer: You could cross a strain with the appropriate 3; 4 reciprocal translocation to a wild-type strain to generate heterozygotes. One of the meiotic products of adjacent-1 segregation in these heterozygotes will have a duplication of the translocated portion of chromosome 3 and a deletion for the translocated portion of chromosome 4.

39. You observe a very large pairing loop at meiosis. Is it more likely to be from a heterozygous inversion or heterozygous deletion? Explain.

Answer: Very large deletions tend to be lethal. This is likely due to genomic imbalance or the unmasking of recessive lethal genes. Therefore, the observed very large pairing loop is more likely to be from a heterozygous inversion.

40. A new recessive mutant allele doesn't show pseudodominance with any of the deletions that span *Drosophila* chromosome 2. What might be the explanation?

Answer: Because the new mutant does not show pseudodominance with any of the deletions that span chromosome 2, it is likely that the mutation does not map to this chromosome, or it is located in a different region of the chromosome 2 in *Drosophila*.

41. Compare and contrast the origins of Turner syndrome, Williams syndrome, cri du chat syndrome, and Down syndrome. (Why are they called syndromes?)

Answer: Turner syndrome is a monosomy in X chromosomes (45 XO) due to a meiotic nondisjunction. Down syndrome (47 or trisomy 21) results from meiotic nondisjunction or from a Robertsonian translocation (with 46 chromosomes but a translocation between 21 and 14). Williams syndrome is the result of a deletion of the 7q11.23 region of chromosome 7. Cri du chat syndrome is the result of a deletion of a significant portion of the short arm of chromosome 5 (specifically bands 5p15.2 and 5p15.3). The term *syndrome* is used to describe a set of phenotypic changes (often complex and varied) that generally occur together with a specific human chromosomal aberration. All four of these syndromes often include mental retardation and unique body and facial features. In addition, fatality rates are low and such children in most cases reach adulthood.

42. List the diagnostic features (genetic or cytological) that are used to identify these chromosomal alterations:

    a. Deletions
    b. Duplications
    c. Inversions
    d. Reciprocal translocations

Answer:

    a. Cytologically, deletions lead to shorter chromosomes with missing bands (if banded) and an unpaired loop during meiotic pairing when heterozygous. Genetically, deletions are usually lethal when homozygous, do not revert, and when heterozygous, lower recombinational frequencies and can result in "pseudodominance" (the expression of recessive alleles on one homolog that are deleted on the other). Occasionally, heterozygous deletions express an abnormal (mutant) phenotype.

    b. Cytologically, duplications lead to longer chromosomes and, depending on the type, unique pairing structures during meiosis when heterozygous. These may be simple unpaired loops or more complicated twisted loop structures. Genetically, duplications can lead to asymmetric pairing and unequal crossing-over events during meiosis, and duplications of some regions can produce specific mutant phenotypes. As in deletions, we could detect duplications using their hybridization signals.

    c. Cytologically, inversions can be detected by banding, and when heterozygous, they show the typical twisted "inversion" loop during homologous pairing. Pericentric inversions can result in a change in the $p{:}q$ ratio (the position of the centromere). Genetically, no viable crossover products are seen from recombination within the inversion when heterozygous, and as a result, flanking genes show a decrease in RF.

    d. Cytologically, reciprocal translocations may be detected by banding, or they may drastically change the size of the involved chromosomes as well as the positions of their centromeres. Genetically, they establish new linkage relationships. When heterozygous, they show the typical cross structure during meiotic pairing and cause a diagnostic 50 percent reduction of viable gamete production, leading to semisterility.

43. The normal sequence of nine genes on a certain *Drosophila* chromosome is 123·456789, where the dot represents the centromere. Some fruit flies were found to have aberrant chromosomes with the following structures:

a. 123 · 476589
b. 123 · 46789
c. 1654 · 32789
d. 123 · 4566789

Name each type of chromosomal rearrangement, and draw diagrams to show how each would synapse with the normal chromosome.

Answer:
a. Paracentric inversion

b. Deletion

c. Pericentric inversion

d. Duplication

44. The two loci *P* and *Bz* are normally 36 m.u. apart on the same arm of a certain plant chromosome. A paracentric inversion spans about one-fourth of this region but does not include either of the loci. What approximate recombinant frequency between *P* and *Bz* would you predict in plants that are

a. heterozygous for the paracentric inversion?
b. homozygous for the paracentric inversion?

Answer:
a. The products of crossing-over within the inversion will be inviable when the inversion is heterozygous. This paracentric inversion spans 25 percent of the region between the two loci and therefore will reduce the observed

recombination between these genes by a similar percentage (i.e., 9 percent.) The observed RF will be 27 percent.

b. When the inversion is homozygous, the products of crossing-over within the inversion will be viable, so the observed RF will be 36 percent.

45. As stated in Solved Problem 2, certain mice called *waltzers* have a recessive mutation that causes them to execute bizarre steps. W. H. Gates crossed waltzers with homozygous normals and found, among several hundred normal progeny, a single waltzing female mouse. When mated with a waltzing male, she produced all waltzing offspring. When mated with a homozygous normal male, she produced all normal progeny. Some males and females of this normal progeny were intercrossed, and there were no waltzing offspring among their progeny. T. S. Painter examined the chromosomes of waltzing mice that were derived from some of Gates's crosses and that showed a breeding behavior similar to that of the original, unusual waltzing female. He found that these mice had 40 chromosomes, just as in normal mice or the usual waltzing mice. In the unusual waltzers, however, one member of a chromosome pair was abnormally short. Interpret these observations as completely as possible, both genetically and cytologically.

(Problem 45 is from A. M. Srb, R. D. Owen, and R. S. Edgar, *General Genetics,* 2nd ed. Copyright 1965, W. H. Freeman and Company.)

Answer: The following represents the crosses that are described in this problem:

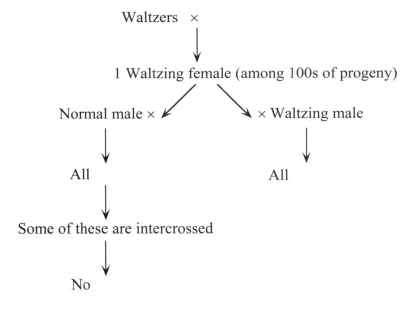

The single waltzing female that arose from a cross between waltzers and normals is expressing a recessive gene. It is possible that this represents a new "waltzer" mutation that was inherited from one of the "normal" mice, but given the cytological evidence (the presence of a shortened chromosome), it is more likely that this exceptional female inherited a deletion of the wild-type allele, which allowed expression of the mutant recessive phenotype.

When this exceptional female was mated to a waltzing male, all the progeny were waltzers; when mated to a normal male, all the progeny were normal. When some of these normal offspring were intercrossed, there were no progeny that were waltzers. If a "new" recessive waltzer allele had been inherited, all these "normal" progeny would have been w^+/w. Any intercross should have therefore produced 25 percent waltzers. On the other hand, if a deletion had occurred, half the progeny would be w^+/w and half would be $w+/w^{deletion}$. If $w^+/w^{deletion}$ are intercrossed, 25 percent of the progeny would not develop (the homozygous deletion would likely be lethal), and no waltzers would be observed. This is consistent with the data.

46. Six bands in a salivary-gland chromosome of *Drosophila* are shown in the following illustration, along with the extent of five deletions (Del 1 to Del 5):

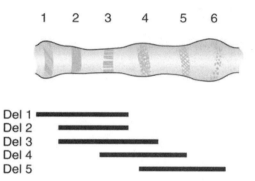

Recessive alleles *a, b, c, d, e,* and *f* are known to be in the region, but their order is unknown. When the deletions are combined with each allele, the following results are obtained:

	a	*b*	*c*	*d*	*e*	*f*
Del 1	−	−	−	+	+	+
Del 2	−	+	−	+	+	+
Del 3	−	+	−	+	−	+
Del 4	+	+	−	−	−	+
Del 5	+	+	+	−	−	−

In this table, a minus sign means that the deletion is missing the corresponding wild-type allele (the deletion uncovers the recessive), and a plus sign means that the corresponding wild-type allele is still present. Use these data to infer which salivary band contains each gene. (Problem 46 is from D. L. Hartl, D. Friefelder, and L. A. Snyder, *Basic Genetics*, Jones and Bartlett, 1988.)

Answer: This problem uses a known set of overlapping deletions to order a set of mutants. This is called deletion mapping and is based on the expression of the recessive mutant phenotype when heterozygous with a deletion of the corresponding allele on the other homolog. For example, mutants a, b, and c are all expressed when heterozygous with Del1. Thus it can be assumed that these genes are deleted in Del1. When these results are compared with the crosses with Del2 and it is discovered that these progeny are b^+, the location of gene b is mapped to the region deleted in Del1 that is not deleted in Del2. This logic can be applied in the following way:

Compare deletions 1 and 2: this places allele b more to the left than alleles a and c. The order is b (a, c), where the parentheses indicate that the order is unknown.

Compare deletions 2 and 3: this places allele e more to the right than (a, c). The order is b (a, c) e.

Compare deletions 3 and 4: allele a is more to the left than c and e, and d is more to the right than e. The order is b a c e d.

Compare deletions 4 and 5: allele f is more to the right than d. The order is b a c e d f.

Allele	Band
b	1
a	2
c	3
e	4
d	5
f	6

47. A fruit fly was found to be heterozygous for a paracentric inversion. However, obtaining flies that were homozygous for the inversion was impossible even after many attempts. What is the most likely explanation for this inability to produce a homozygous inversion?

Answer: The data suggest that one or both breakpoints of the inversion are located within an essential gene, causing a recessive lethal mutation.

48. Orangutans are an endangered species in their natural environment (the islands of Borneo and Sumatra), and so a captive-breeding program has been established using orangutans currently held in zoos throughout the world. One component of this program is research into orangutan cytogenetics. This research has shown that all orangutans from Borneo carry one form of chromosome 2, as shown in the accompanying diagram, and all orangutans from

Sumatra carry the other form. Before this cytogenetic difference became known, some matings were carried out between animals from different islands, and 14 hybrid progeny are now being raised in captivity.

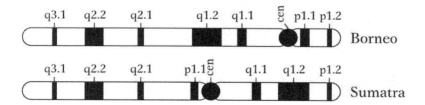

a. What term or terms describe the differences between these chromosomes?

b. Draw the chromosomes 2, paired in the first meiotic prophase, of such a hybrid orangutan. Be sure to show all the landmarks indicated in the accompanying diagram, and label all parts of your drawing.

c. In 30 percent of meioses, there will be a crossover somewhere in the region between bands p1.1 and q1.2. Draw the gamete chromosomes 2 that would result from a meiosis in which a single crossover occurred within band q1.1.

d. What fraction of the gametes produced by a hybrid orangutan will give rise to viable progeny, if these chromosomes are the only ones that differ between the parents? (Problem 48 is from Rosemary Redfield.)

Answer:

a. The Sumatra chromosome contains a pericentric inversion when compared with the Borneo chromosome.

b.

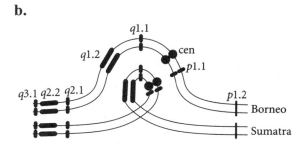

c.

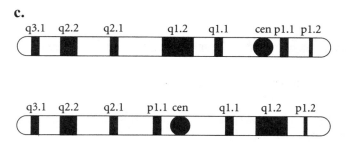

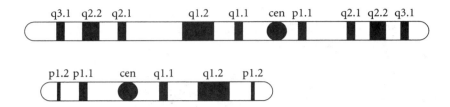

d. Recall that all single crossovers within the inverted region will lead to four meiotic products: two that will be viable, nonrecombinant (parental) types and two that will be extremely unbalanced, (most likely nonviable), recombinant types. In other words, if 30 percent of the meioses have a crossover in this region, 15 percent of the gametes will not lead to viable progeny. That means that 85 percent of the gametes should produce viable progeny.

49. In corn, the genes for tassel length (alleles T and t) and rust resistance (alleles R and r) are known to be on separate chromosomes. In the course of making routine crosses, a breeder noticed that one T/t ; R/r plant gave unusual results in a testcross with the double-recessive pollen parent t/t ; r/r. The results were

Progeny:
T/t ; R/r 98
t/t ; r/r 104
T/t ; r/r 3
t/t ; R/r 5

Corncobs: Only about half as many seeds as usual

a. What key features of the data are different from the expected results?
b. State a concise hypothesis that explains the results.
c. Show genotypes of parents and progeny.
d. Draw a diagram showing the arrangement of alleles on the chromosomes.
e. Explain the origin of the two classes of progeny having three and five members.

Unpacking Problem 49

1. What do a "gene for tassel length" and a "gene for rust resistance" mean?

Answer: A "gene for tassel length" means that there is a gene with at least two alleles (T and t) that controls the length of the tassel. A "gene for rust resistance" means that there is a gene that determines whether the corn plant is resistant to a rust infection or not (R and r).

2.   Does it matter that the precise meaning of the allelic symbols *T, t, R*, and *r* is not given? Why or why not?

Answer: The precise meaning of the allelic symbols for the two genes is irrelevant to solving the problem because what is being investigated is the distance between the two genes.

3.   How do the terms *gene* and *allele*, as used here, relate to the concepts of locus and gene pair?

Answer: A locus is the specific position occupied by a gene on a chromosome. It is implied that gene loci are the same on both homologous chromosomes. The gene pair can consist of identical or different alleles.

4.   What prior experimental evidence would give the corn geneticist the idea that the two genes are on separate chromosomes?

Answer: Evidence that the two genes are normally on separate chromosomes would have come from previous experiments showing that the two genes independently assort during meiosis.

5.   What do you imagine "routine crosses" are to a corn breeder?

Answer: Routine crosses could consist of F_1 crosses, F_2 crosses, backcrosses, and testcrosses.

6.   What term is used to describe genotypes of the type *T/t* ; *R/r*?

Answer: The genotype *T/t* ; *R/r* is a double heterozygote, or dihybrid, or F_1 genotype.

7.   What is a "pollen parent"?

Answer: The pollen parent is the "male" parent that contributes to the pollen tube nucleus, the endosperm nucleus, and the progeny.

8.   What are testcrosses, and why do geneticists find them so useful?

Answer: Testcrosses are crosses that involve a genotypically unknown and a homozygous recessive organism. They are used to reveal the complete genotype of the unknown organism and to study recombination during meiosis.

9. What progeny types and frequencies might the breeder have been expecting from the testcross?

Answer: The breeder was expecting to observe 1 *T/t* ; *R/r*:1 *T/t* ; *r/r*:1 *t/t* ; *R/r*:1 *t/t* ; *r/r*.

10. Describe how the observed progeny differ from expectations.

Answer: Instead of a 1:1:1:1 ratio indicating independent assortment, the testcross indicated that the two genes were linked, with a genetic distance of 100% (3 + 5)/210 = 3.8 map units.

11. What does the approximate equality of the first two progeny classes tell you?

Answer: The equality and predominance of the first two classes indicate that the parentals were *T R/t r*.

12. What does the approximate equality of the second two progeny classes tell you?

Answer: The equality and lack of predominance of the second two classes indicate that they represent recombinants.

13. What were the gametes from the unusual plant, and what were their proportions?

Answer: The gametes leading to this observation were:
    46.7% *T R*   1.4% *T r*
    49.5% *t r*   2.4% *t R*

14. Which gametes were in the majority?

Answer:
    46.7% *T R*
    49.5% *t r*

15. Which gametes were in the minority?

Answer:

.4% *T r*
2.4% *t R*

16. Which of the progeny types seem to be recombinant?

   Answer: *Tr* and *t R*

17. Which allelic combinations appear to be linked in some way?

   Answer: *T* and *R* are linked, as are *t* and *r*.

18. How can there be linkage of genes supposedly on separate chromosomes?

   Answer: Two genes on separate chromosomes can become linked through a translocation.

19. What do these majority and minority classes tell us about the genotypes of the parents of the unusual plant?

   Answer: One parent of the hybrid plant contained a translocation that linked the *T* and *R* alleles and the *t* and *r* alleles.

20. What is a corncob?

   Answer: A corncob is a structure that holds on its surface the seeds that will become the next generation of corn.

21. What does a normal corncob look like? (Sketch one and label it.)

   Answer:

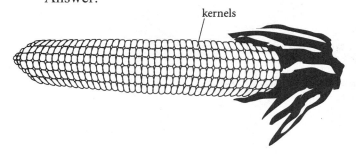
kernels

22. What do the corncobs from this cross look like? (Sketch one.)

Answer:

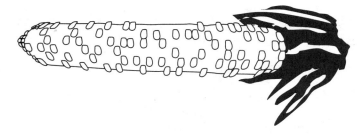

23. What exactly is a kernel?

Answer: A kernel is one progeny on a corn cob.

24. What effect could lead to the absence of half the kernels?

Answer: Absence of half the kernels, or 50 percent aborted progeny (semisterility), could result from the random segregation of one normal with one translocated chromosome (T1 + N2 and T2 + N1) during meioses in a parent that is heterozygous for a reciprocal translocation.

25. Did half the kernels die? If so, was the female or the male parent the reason for the deaths?

Answer: Approximately 50 percent of the progeny died. It was the "female" that was heterozygous for the translocation.

Now try to solve the problem.

Solution to the Problem

Answer:

a. The progeny are not in the 1:1:1:1 ratio expected for independent assortment; instead, the data indicate close linkage. Also, half the progeny did not develop, indicating semisterility.

b. These observations are best explained by a translocation of material between the two chromosomes.

c. Parents: $T R/t r \quad \times \quad t/t \; ; \; r/r$

Progeny:
98	$T R/t \; ; \; r$
104	$t r/t \; ; \; r$
3	$T r/t \; ; \; r$
5	$t R/t \; ; \; r$

d. Assume a translocation heterozygote in coupling. If pairing is as diagrammed below, then you would observe the following:

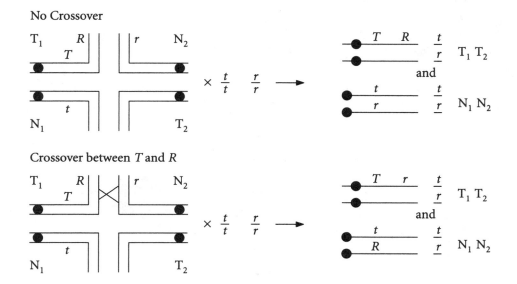

The locations of the genes on the chromosomes is not critical, as long as the cumulative distance between the genes and the breakpoints allows approximately 3.8 percent recombination.

e. The two recombinant classes result from a recombination event followed by proper segregation of chromosomes, as diagrammed above.

50. A yellow body in *Drosophila* is caused by a mutant allele *y* of a gene located at the tip of the X chromosome (the wild-type allele causes a gray body). In a radiation experiment, a wild-type male was irradiated with X rays and then crossed with a yellow-bodied female. Most of the male progeny were yellow, as expected, but the scanning of thousands of flies revealed two gray-bodied (phenotypically wild-type) males. These gray-bodied males were crossed with yellow-bodied females, with the following results:

	Progeny
gray male 1 × yellow female	females all yellow
	males all gray
gray male 2 × yellow female	1/2 females yellow
	1/2 females gray
	1/2 males yellow
	1/2 males gray

a. Explain the origin and crossing behavior of gray male 1.
b. Explain the origin and crossing behavior of gray male 2.

Answer: The cross was

P X^{e^+}/Y (irradiated) $\times$ X^e/X^e
F_1 Most X^e/Y yellow males
      Two ? gray males

a. Gray male 1 was crossed with a yellow female, yielding yellow females and gray males, which is reversed sex linkage. If the e^+ allele was translocated to the Y chromosome, the gray male would be X^e/Y^{e^+} or gray. When crossed with yellow females, the results would be

X^e/Y^{e^+} gray males
X^e/X^e yellow females

b. Gray male 2 was crossed with a yellow female, yielding gray and yellow males and females in equal proportions. If the e^+ allele was translocated to an autosome, the progeny would be as below, where "A" indicates autosome:

P A^{e^+}/A ; X^e/Y $\times$ A/A X^e/X^e

F_1 A^{e^+}/A ; X^e/X^e gray female
      A^{e^+}/A ; X^e/Y gray male
      A/A ; X^e/X^e yellow female
      A/A ; X^e/Y yellow male

Unpacking the Problem

51. In corn, the allele *Pr* stands for green stems, *pr* for purple stems. A corn plant of genotype *pr/pr* that has standard chromosomes is crossed with a *Pr/Pr* plant that is homozygous for a reciprocal translocation between chromosomes 2 and 5. The F_1 is semisterile and phenotypically *Pr*. A backcross with the parent with standard chromosomes gives 764 semi-sterile *Pr*, 145 semisterile *pr*, 186 normal *Pr*, and 727 normal *pr*. What is the map distance between the *Pr* locus and the translocation point?

Answer: The break point can be treated as a gene with two "alleles," one for normal fertility and one for semisterility. The problem thus becomes a two-point cross.

Parentals	764	Semisterile *Pr*
	727	Normal *pr*
Recombinants	145	Semisterile *pr*
	186	Normal *Pr*
	1822	

$100\%(145 + 186)/1822 = 18.17$ m.u.

52. Distinguish among Klinefelter, Down, and Turner syndromes. Which syndromes are found in both sexes?

Answer:
Klinefelter syndrome	XXY male
Down syndrome	Trisomy 21
Turner syndrome	XO female

53. Show how you could make an allotetraploid between two related diploid plant species, both of which are $2n = 28$.

Answer: Create a hybrid by crossing the two plants and then double the chromosomes with a treatment that disrupts mitosis, such as colchicine treatment. Alternatively, diploid somatic cells from the two plants could be fused and then grown into plants through various culture techniques.

54. In *Drosophila*, trisomics and monosomics for the tiny chromosome 4 are viable, but nullisomics and tetrasomics are not. The *b* locus is on this chromosome. Deduce the phenotypic proportions in the progeny of the following crosses of trisomics.

   a. $b^+/b/b \times b/b$
   b. $b^+/b^+/b \times b/b$
   c. $b^+/b^+/b \times b^+/b$

Answer:

a. $b^+/b/ \quad \times \quad b/b$
$$\downarrow \qquad\qquad \downarrow$$
Gametes: 1/6 $b^+\ b$
1/3 b
1/3 b^+/b
1/6 b/b

Among the progeny of this cross, the phenotypic ratio will be 1 wild-type (b^+) : 1 b.

b. $b^+/b^+/b \times b/b$
$$\downarrow \qquad \downarrow$$
Gametes: 1/6 b $\qquad b$
1/3 b^+
1/3 b^+/b
1/6 b^+/b^+

Among the progeny of this cross, the phenotypic ratio will be 5 wild-type (b^+) : 1 b.

c.
$$b^+/b^+/b \qquad \times b^+/b$$
$$\downarrow \qquad\qquad \downarrow$$

Gametes: 1/6 b 1/2 b
             1/3 b^+ 1/2 b^+
             1/3 b^+/b
             1/6 b^+/b^+

Among the progeny of this cross, the phenotypic ratio will be 11 wild-type (b^+) : 1 b.

55. A woman with Turner syndrome is found to be color-blind (an X-linked recessive phenotype). Both her mother and her father have normal vision.

a. Explain the simultaneous origin of Turner syndrome and color blindness by the abnormal behavior of chromosomes at meiosis.

b. Can your explanation distinguish whether the abnormal chromosome behavior occurred in the father or the mother?

c. Can your explanation distinguish whether the abnormal chromosome behavior occurred at the first or second division of meiosis?

d. Now assume that a color-blind Klinefelter man has parents with normal vision, and answer parts *a, b,* and *c*.

Answer:

a., b., and c. One of the parents of the woman with Turner syndrome (XO) must have been a carrier for color blindness, an X-linked recessive disorder. Because her father has normal vision, she could not have obtained her only X from him. Therefore, nondisjunction occurred in her father. A sperm lacking a sex chromosome fertilized an egg with the X chromosome carrying the color blindness allele. The nondisjunctive event could have occurred during either meiotic division.

d. If the color-blind patient had Klinefelter syndrome (XXY), then both X's must carry the allele for color blindness. Therefore, nondisjunction had to occur in the mother. Remember that during meiosis I, given no crossover between the gene and the centromere, allelic alternatives separate from each other. During meiosis II, identical alleles on sister chromatids separate. Therefore, assuming there have been no crossovers between the color-blind allele and the centromere, the nondisjunctive event had to occur during meiosis II because both alleles are identical. If the gene is far from the centromere, it would be difficult to determine if nondisjunction happened at M_I or M_{II} without molecular studies on haplotypes near the centromere.

56. **a.** How would you synthesize a pentaploid?

    b. How would you synthesize a triploid of genotype *A/a/a*?

    c. You have just obtained a rare recessive mutation a* in a diploid plant, which Mendelian analysis tells you is *A/a**. From this plant, how would you synthesize a tetraploid (4*n*) of genotype *A/A/a*/a**?

    d. How would you synthesize a tetraploid of genotype *A/a/a/a*?

Answer:
a. If a 6x were crossed with a 4x, the result would be 5x.

    b. Cross *A/A* with *a/a/a/a* to obtain *A/a/a*.

    c. The easiest way is to expose the *A/a** plant cells to colchicine for one cell division. This will result in a doubling of chromosomes to yield *A/A/a*/a**.

    d. Cross 6x (*a/a/a/a/a/a*) with 2x (*A/A*) to obtain *A/a/a/a*.

57. Suppose you have a line of mice that has cytologically distinct forms of chromosome 4. The tip of the chromosome can have a knob (called 4^K) or a satellite (4^S) or neither (4). Here are sketches of the three types:

You cross a $4^K/4^S$ female with a 4/4 male and find that most of the progeny are $4^K/4$ or $4^S/4$, as expected. However, you occasionally find some rare types as follows (all other chromosomes are normal):

    a. $4^K/4^K/4$
    b. $4^K/4^S/4$
    c. 4^K

Explain the rare types that you have found. Give, as precisely as possible, the stages at which they originate, and state whether they originate in the male parent, the female parent, or the zygote. (Give reasons briefly.)

Answer: The following answers make the simplifying assumption that there are no crossovers. Without this assumption, it would be hard to tell which stage of meiosis led to the nondisjunction.

Type a: the extra chromosome must be from the mother. Because the chromosomes are identical, nondisjunction had to have occurred at M_{II}.

Type b: the extra chromosome must be from the mother. Because the chromosomes are not identical, nondisjunction had to have occurred at M_I.

Type c: the mother correctly contributed one chromosome, but the father did not contribute any chromosome 4. Therefore, nondisjunction occurred in the male during either meiotic division.

58. A cross is made in tomatoes between a female plant that is trisomic for chromosome 6 and a normal diploid male plant that is homozygous for the recessive allele for potato leaf (*p/p*). A trisomic F_1 plant is backcrossed to the potato-leaved male.

    a. What is the ratio of normal-leaved plants to potato-leaved plants when you assume that *p* is located on chromosome 6?

    b. What is the ratio of normal-leaved to potato-leaved plants when you assume that *p* is not located on chromosome 6?

Answer:
a. The cross is *P/P/p* × *p/p*.

The gametes from the trisomic parent will occur in the following proportions:

1/6	*p*
2/6	*P*
1/6	*P/P*
2/6	*P/p*

Only gametes that are *p* can give rise to potato leaves, because potato is recessive. Therefore, the ratio of normal to potato will be 5:1.

b. If the gene is not on chromosome 6, there should be a 1:1 ratio of normal to potato.

59. A tomato geneticist attempts to assign five recessive mutations to specific chromosomes by using trisomics. She crosses each homozygous mutant (*2n*) with each of three trisomics, in which chromosomes 1, 7, and 10 take part. From these crosses, the geneticist selects trisomic progeny (which are less vigorous) and backcrosses them to the appropriate homozygous recessive. The *diploid* progeny from these crosses are examined. Her results, in which the ratios are wild type:mutant, are as follows:

Trisomic chromosome	*d*	*y*	*c*	*h*	*cot*
1	48:55	72:29	56:50	53:54	32:28
7	52:56	52:48	52:51	58:56	81:40
10	45:42	36:33	28:32	96:50	20:17

Which of the mutations can the geneticist assign to which chromosomes? (Explain your answer fully.)

Answer: The generalized cross is *A/A/A* × *a/a*, from which *A/A/a* progeny were selected. These progeny were crossed with *a/a* individuals, yielding the results given. Assume for a moment that each allele can be distinguished from the other, and let 1 = *A*, 2 = *A* and 3 = *a*. The gametic combinations possible are

1-2 (*A/A*) and 3 (*a*)
1-3 (*A/a*) and 2 (*A*)
2-3 (*A/a*) and 1 (*A*)

Because only diploid progeny were examined in the cross with *a/a*, the progeny ratio should be 2 wild type:1 mutant if the gene is on the trisomic chromosome. With this in mind, the table indicates that *y* is on chromosome 1, *cot* is on chromosome 7, and *h* is on chromosome 10. Genes *d* and *c* do not map to any of these chromosomes.

60. A petunia is heterozygous for the following autosomal homologs:

A	*B*	*C*	*D*	*E*	*F*	*G*	*H*	*I*
a	*b*	*c*	*d*	*h*	*g*	*f*	*e*	*i*

a. Draw the pairing configuration that you would see at metaphase I, and identify all parts of your diagram. Number the chromatids sequentially from top to bottom of the page.

b. A three-strand double crossover occurs, with one crossover between the *C* and *D* loci on chromatids 1 and 3, and the second crossover between the *G* and *H* loci on chromatids 2 and 3. Diagram the results of these recombination events as you would see them at anaphase I, and identify all parts of your diagram.

c. Draw the chromosome pattern that you would see at anaphase II after the crossovers described in part *b*.

d. Give the genotypes of the gametes from this meiosis that will lead to the formation of viable progeny. Assume that all gametes are fertilized by pollen that has the gene order *A B C D E F G H I*.

Answer:

a.

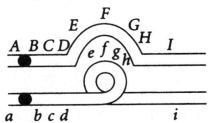

b.

1   *A B C d h G F E D C B A*   Dicentric

2   *I H g f e i*   Acentric

3   *a b c D E F G H I*   Viable

4   *a b c d h g f e i*

c.

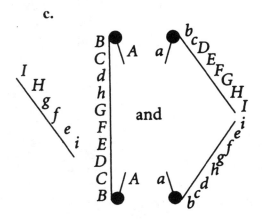

and

d. The chromosomes numbered 3 and 4 will give rise to viable progeny. The genotypes of those progeny will be *A B C D E F G H I/a b c D E F G H I* and *A B C D E F G H I/a b c d h g f e i*.

61. Two groups of geneticists, in California and in Chile, begin work to develop a linkage map of the medfly. They both independently find that the loci for body color (B = black, b = gray) and eye shape (R = round, r = star) are linked 28 m.u. apart. They send strains to each other and perform crosses; a summary of all their findings is shown here:

Cross	F_1	Progeny of F_1 × any $b\ r/b\ r$	
$B\ R/B\ R$ (Calif.) × $b\ r/b\ r$ (Calif.)	$B\ R/b\ r$	$B\ R/b\ r$	36%
		$b\ r/b\ r$	36
		$B\ r/b\ r$	14
		$b\ R/b\ r$	14
$B\ R/B\ R$ (Chile) × $b\ r/b\ r$ (Chile)	$B\ R/b\ r$	$B\ R/b\ r$	36
		$b\ r/b\ r$	36
		$B\ r/b\ r$	14
		$b\ R/b\ r$	14
$B\ R/B\ R$ (Calif.) × $b\ r/b\ r$ (Chile) or $b\ r/b\ r$ (Calif.) × $B\ R/B\ R$ (Chile)	$B\ R/b\ r$	$B\ R/b\ r$	48
		$b\ r/b\ r$	48
		$B\ r/b\ r$	2
		$b\ R/b\ r$	2

a. Provide a genetic hypothesis that explains the three sets of testcross results.

b. Draw the key chromosomal features of meiosis in the F_1 from a cross of the Californian and Chilean lines.

Answer:

a. Two crosses show 28 map units between the loci for body color and eye shape in a testcross of the F_1: California × California and Chile × Chile. The third type of cross, California × Chile, leads to only four map units between the two genes when the hybrid is testcrossed. This indicates that the genetic distance has decreased by 24 map units, or 100% (24/28) = 85.7%. A deletion cannot be used to explain this finding, nor can a translocation. Most likely the two lines are inverted with respect to each other for 85.7 percent of the distance between the two genes.

b.

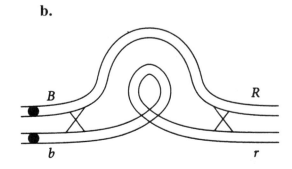

A single crossover in either region would result in 4 percent crossing-over between *B* and *R*. The products are

62. An aberrant corn plant gives the following RF values when testcrossed:

	Interval				
	d–f	*f–b*	*b–x*	*x–y*	*y–p*
Control	5	18	23	12	6
Aberrant plant	5	2	2	0	6

(The locus order is centromere-*d–f–b–x–y–p*.) The aberrant plant is a healthy plant, but it produces far fewer normal ovules and pollen than does the control plant.

a. Propose a hypothesis to account for the abnormal recombination values and the reduced fertility in the aberrant plant.

b. Use diagrams to explain the origin of the recombinants according to your hypothesis.

Answer:
a. The aberrant plant is semisterile, which suggests an inversion. Because the *d–f* and *y–p* frequencies of recombination in the aberrant plant are normal, the inversion must involve *b* through *x*.

b. To obtain recombinant progeny when an inversion is involved, either a double crossover occurred within the inverted region or single crossovers occurred between *f* and the inversion, which occurred someplace between *f* and *b*.

63. The following corn loci are on one arm of chromosome 9 in the order indicated (the distances between them are shown in map units):

c–bz–wx–sh–d–centromere
12 8   10   20 10

C gives colored aleurone; *c*, white aleurone.
Bz gives green leaves; *bz*, bronze leaves.
Wx gives starchy seeds; *wx*, waxy seeds.

Sh gives smooth seeds; *sh*, shrunken seeds.
D gives tall plants; *d*, dwarf.

A plant from a standard stock that is homozygous for all five recessive alleles is crossed with a wild-type plant from Mexico that is homozygous for all five dominant alleles. The F_1 plants express all the dominant alleles and, when backcrossed to the recessive parent, give the following progeny phenotypes:

colored, green, starchy, smooth, tall	360
white, bronze, waxy, shrunk, dwarf	355
colored, bronze, waxy, shrunk, dwarf	40
white, green, starchy, smooth, tall	46
colored, green, starchy, smooth, dwarf	85
white, bronze, waxy, shrunk, tall	84
colored, bronze, waxy, shrunk, tall	8
white, green, starchy, smooth, dwarf	9
colored, green, waxy, smooth, tall	7
white, bronze, starchy, shrunk, dwarf	6

Propose a hypothesis to explain these results. Include

a. a general statement of your hypothesis, with diagrams if necessary;
b. why there are 10 classes;
c. an account of the origin of each class, including its frequency; and
d. at least one test of your hypothesis.

Answer: The cross is

P   *c bz wx sh d/c bz wx sh d* × *C Bz Wx Sh D/C Bz Wx Sh D*

F_1 *C Bz Wx Sh D/c bz wx sh d*

Backcross   *C Bz Wx Sh D/c bz wx sh d* × *c bz wx sh d/c bz wx sh d*

a. The total number of progeny is 1000. Classify the progeny as to where a crossover occurred for each type. Then, total the number of crossovers between each pair of genes. Calculate the observed map units.

Region	#COs	M.U. observed	M.U. expected
C–Bz	103	10.3	12
Bz–Wx	13	1.3	8
Wx–Sh	13	1.3	10
A–D	186	18.6	20

Notice that a reduction of map units, or crossing-over, is seen in two intervals. Results like this are suggestive of an inversion. The inversion most likely involves the *Bz*, *Wx*, and *Sh* genes.

Further, notice that all those instances in which crossing-over occurred in the proposed inverted region involved a double crossover. This is the expected pattern.

b. A number of possible classes are missing: four single-crossover classes resulting from crossing-over in the inverted region, eight double-crossover classes involving the inverted region and the noninverted region, and triple crossovers and higher. The 10 classes detected were the only classes that were viable. They involved a single crossover outside the inverted region or a double crossover within the inverted region.

c. Class 1: parental; increased due to nonviability of some crossovers

Class 2: parental; increased due to nonviability of some crossovers

Class 3: crossing-over between *C* and *Bz*; approximately expected frequency

Class 4: crossing-over between *C* and *Bz*, approximately expected frequency

Class 5: crossing-over between *Sh* and *D*; approximately expected frequency

Class 6: crossing-over between *Sh* and *D*; approximately expected frequency

Class 7: double crossover between *C* and *Bz* and between *Sh* and *D*; approximately expected frequency

Class 8: double crossover between *C* and *Bz* and between *A* and *D*; approximately expected frequency

Class 9: double crossover between *Bz* and *Wx* and between *Wx* and *Sh*; approximately expected frequency

Class 10: double crossover between *Bz* and *Wx* and between *Wx* and *Sh*; approximately expected frequency

d. Cytological verification could be obtained by looking at chromosomes during meiotic pairing. Genetic verification could be achieved by mapping these genes in the wild-type strain and observing their altered relationships.

64. Chromosomally normal corn plants have a *p* locus on chromosome 1 and an *s* locus on chromosome 5.

> *P* gives dark green leaves; *p,* pale green leaves.
> *S* gives large ears; *s,* shrunken ears.

An original plant of genotype *P/p* ; *S/s* has the expected phenotype (dark green, large ears) but gives unexpected results in crosses as follows:

- On selfing, fertility is normal, but the frequency of *p/p* ; *s/s* types is 1/4 (not 1/16 as expected).
- When crossed with a normal tester of genotype *p/p* ; *s/s*, the F₁ progeny are 1/2 ; *P/p* ; *S/s* and 12; *p/p* ; *s/s*; fertility is normal.
- When an F₁ *P/p* ; *S/s* plant is crossed with a normal *p/p* ; *s/s* tester, it proves to be semisterile, but, again, the progeny are 12; *P/p* ; *S/s* and 12; *p/p* ; *s/s*.

Explain these results, showing the full genotypes of the original plant, the tester, and the F₁ plants. How would you test your hypothesis?

Answer: The original plant was homozygous for a translocation between chromosomes 1 and 5, with breakpoints very close to genes *P* and *S*. Because of the close linkage, a ratio suggesting a monohybrid cross, instead of a dihybrid cross, was observed, both with selfing and with a testcross. All gametes are fertile because of homozygosity.

> original plant: *P S/p s*
> tester: *p s/p s*

F₁ progeny: heterozygous for the translocation:

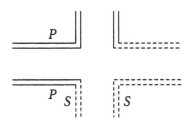

This figure is an example of one configuration that fits the data. One way to test for the presence of a translocation is to look at the chromosomes of heterozygotes during meiosis I.

65. A male rat that is phenotypically normal shows reproductive anomalies when compared with normal male rats, as shown in the following table. Propose a genetic explanation of these unusual results, and indicate how your idea could be tested.

	Embryos (mean number)			
Mating	Implanted in the uterine wall	Degeneration after implantation	Normal	Degeneration (%)
exceptional ♂ × normal ♀	8.7	5.0	3.7	37.5
normal ♂ × normal ♀	9.5	0.6	8.9	6.5

Answer: The percent degeneration seen in the progeny of the exceptional rat is roughly 50 percent larger than that seen in the progeny from the normal male. Fifty percent emisterility is an important diagnostic for translocation heterozygotes. This could be verified by cytological observation of the meiotic cells from the exceptional male.

66. A tomato geneticist working on *Fr*, a dominant mutant allele that causes rapid fruit ripening, decides to find out which chromosome contains this gene by using a set of lines of which each is trisomic for one chromosome. To do so, she crosses a homozygous diploid mutant with each of the wild-type trisomic lines.

   a. A trisomic F_1 plant is crossed with a diploid wild-type plant. What is the ratio of fast- to slow-ripening plants in the diploid progeny of this second cross if *Fr* is on the trisomic chromosome? Use diagrams to explain.

   b. What is the ratio of fast- to slow-ripening plants in the diploid progeny of this second cross if *Fr* is not located on the trisomic chromosome? Use diagrams to explain.

   c. Here are the results of the crosses. On which chromosome is *Fr*, and why?

Trisomic chromosome	Fast ripening:slow ripening in diploid progeny
1	45:47
2	33:34
3	55:52
4	26:30
5	31:32
6	37:41
7	44:79
8	49:53
9	34:34
10	37:39

(Problem 66 is from Tamara Western.)

Answer:

   a. The cross is *Fr/Fr* × *fr/fr/fr*.

Trisomic progeny are then crossed to a diploid wild-type plant.

$$Fr/fr/fr \times fr/fr$$

Because only diploid progeny of this cross are evaluated, the ratio of fast- to slow-ripening plants will be 1:2.

b. If *Fr* is not located on the trisomic chromosome, the crosses are

$$Fr/Fr \times fr/fr$$

and

$$Fr/fr \times fr/fr$$

Therefore, the ratio of fast- to slow-ripening plants will be 1:1.

c. The 1:2 ratio of fast- to slow ripening plants indicates that the *Fr* gene is on chromosome 7.

CHALLENGING PROBLEMS

67. The *Neurospora un-3* locus is near the centromere on chromosome 1, and crossovers between *un-3* and the centromere are very rare. The *ad-3* locus is on the other side of the centromere of the same chromosome, and crossovers occur between *ad-3* and the centromere in about 20 percent of meioses (no multiple crossovers occur).

a. What types of linear asci (see Chapter 4) do you predict, and in what frequencies, in a normal cross of *un-3 ad-3* × wild type? (Specify genotypes of spores in the asci.)

b. Most of the time such crosses behave predictably, but, in one case, a standard *un-3 ad-3* strain was crossed with a wild type isolated from a field of sugarcane in Hawaii. The results follow:

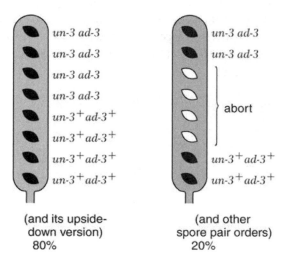

Explain these results, and state how you could test your idea. (**Note:** In *Neurospora*, ascospores with extra chromosomal material survive and are the normal black color, whereas ascospores lacking any chromosome region are white and inviable.)

Answer:

a. Single crossovers between a gene and its centromere lead to a tetratype (second-division segregation.) Thus, a total of 20 percent of the asci should show second division segregation, and 80 percent will show first-division segregation. The following are representative asci

$un3^+\ ad3^+$	$un3^+\ ad3^+$	$un3^+\ ad3^+$	$un3^+\ ad3$	$un3^+\ ad3$
$un3^+\ ad3^+$	$un3^+\ ad3^+$	$un3^+\ ad3^+$	$un3^+\ ad3$	$un3^+\ ad3$
$un3^+\ ad3^+$	$un3^+\ ad3$	$un3^+\ ad3$	$un3^+\ ad3^+$	$un3^+\ ad3^+$
$un3^+\ ad3^+$	$un3^+\ ad3$	$un3^+\ ad3$	$un3^+\ ad3^+$	$un3^+\ ad3^+$
$un3\ ad3$	$un3\ ad3^+$	$un3\ ad3$	$un3\ ad3^+$	$un3\ ad3$
$un3\ ad3$	$un3\ ad3^+$	$un3\ ad3$	$un3\ ad3^+$	$un3\ ad3$
$un3\ ad3$	$un3\ ad3$	$un3\ ad3^+$	$un3\ ad3$	$un3\ ad3^+$
$un3\ ad3$	$un3\ ad3$	$un3\ ad3^+$	$un3\ ad3$	$un3\ ad3^+$
80%	5%	5%	5%	5%

In all cases, the "upside-down" version would be equally likely.

b. The aborted spores could result from a crossing-over event within an inversion of the wild type, compared with the standard strain. Crossing-over within heterozygous inversions leads to unbalanced chromosomes and nonviable spores. This could be tested by using the wild type from Hawaii in mapping experiments of other markers on chromosome 1 in crosses with the standard strain and looking for altered map distances.

68. Two mutations in *Neurospora*, *ad-3* and *pan-2*, are located on chromosomes 1 and 6, respectively. An unusual *ad-3* line arises in the laboratory, giving the results shown in the table below. Explain all three results with the aid of clearly labeled diagrams. (**Note:** In *Neurospora*, ascospores with extra chromosomal material survive and are the normal black color, whereas ascospores lacking any chromosome region are white and inviable.)

	Ascospore appearance	RF between *ad-3* and *pan-2*
1. Normal *ad-3* × normal *pan-2*	All black	50%
2. Abnormal *ad-3* × normal *pan-2*	About 1/2 black and 1/2 white (inviable)	1%
3. Of the black spores from cross 2, about half were completely normal and half repeated the same behavior as the original abnormal *ad-3* strain		

Answer:
Cross 1:   independent assortment of the 2 genes (expected for genes on separate chromosomes).

Cross 2:   the 2 genes now appear to be linked (the observed RF is 1 percent); also, half of the progeny are inviable. These data suggest a reciprocal translocation occurred and both genes are very close to the breakpoints.

Cross 3:   the viable spores are of 2 types: half contain the normal (nontranslocated chromosomes) and half contain the translocated chromosomes.

69.   Deduce the phenotypic proportions in the progeny of the following crosses of autotetraploids in which the a^+/a locus is very close to the centromere. (Assume that the four homologous chromosomes of any one type pair randomly two by two and that only one copy of the a^+ allele is necessary for the wild-type phenotype.)

a. $a^+/a^+/a/a \times a/a/a/a$
b. $a^+/a/a/a \times a/a/a/a$
c. $a^+/a/a/a \times a^+/a/a/a$
d. $a^+/a^+/a/a \times a^+/a/a/a$

Answer:

a. $a^+/a^+/a/a \;\times\; a/a/a/a$
                              $\downarrow$ $\downarrow$
Gametes: 1/6 a^+/a^+ a/a
                2/3 a^+/a
                1/6 a/a

Among the progeny of this cross, the phenotypic ratio will be 5 wild-type (a^+) : 1 a.

b. $a^+/a/a/a \;\times\; a/a/a/a$
                              $\downarrow$ $\downarrow$
Gametes: 1/2 a^+/a a/a
                1/2 a/a

Among the progeny of this cross, the phenotypic ratio will be 1 wild-type (a^+) : 1 a.

c. $a^+/a/a/a \;\times\; a^+/a/a/a$
                              $\downarrow$ $\downarrow$
Gametes: 1/2 a^+/a 1/2 a^+/a
                1/2 a/a 1/2 a/a

Among the progeny of this cross, the phenotypic ratio will be 3 wild-type (a^+) : 1 a.

d.

$$a^+/a^+/a/a \times a^+/a/a/a$$
$$\downarrow \qquad\qquad \downarrow$$

Gametes: 1/6 a^+/a^+ 1/2 a^+/a
          2/3 a^+/a 1/2 a/a
          1/6 a/a

Among the progeny of this cross, the phenotypic ratio will be 11 wild-type (a^+) : 1 a.

70. The New World cotton species *Gossypium hirsutum* has a $2n$ chromosome number of 52. The Old World species *G. thurberi* and *G. herbaceum* each have a $2n$ number of 26. Hybrids between these species show the following chromosome pairing arrangements at meiosis:

Hybrid	Pairing arrangement
G. hirsutum × *G. thurberi*	13 small bivalents + 13 large univalents
G. hirsutum × *G. herbaceum*	13 large bivalents + 13 small univalents
G. thurberi × *G. herbaceum*	13 large univalents + 13 small univalents

Draw diagrams to interpret these observations phylogenetically, clearly indicating the relationships between the species. How would you go about proving that your interpretation is correct? (Problem 70 is adapted from A. M. Srb, R. D. Owen, and R. S. Edgar, *General Genetics,* 2nd ed. W. H. Freeman and Company,1965.)

Answer: Consider the following table, in which "L" and "S" stand for 13 large and 13 small chromosomes, respectively:

Hybrid	Chromosomes
G. hirsutum × *G. thurberi*	S, S, L
G. hirsutum × *G. herbaceum*	S, S, L
G. thurberi × *G. herbaceum*	S, L

Each parent in the cross must contribute half its chromosomes to the hybrid offspring. It is known that *G. hirsutum* has twice as many chromosomes as the other two species. Furthermore, because the *G. hirsutum* chromosomes form bivalent pairs in the hybrids, the *G. hirsutum* karyotype must consist of chromosomes donated by the other two species. Therefore, the genome of *G. hirsutum* must consist of one large and one small set of chromosomes. Once this is realized, the rest of the problem essentially solves itself. In the first hybrid, the genome of *G. thurberi* must consist of one set of small chromosomes. In the second hybrid, the genome of *G. herbaceum* must consist of one set of large chromosomes. The third hybrid confirms the conclusions reached from the first two hybrids.

The original parents must have had the following chromosome constitution:

G. hirsutum   26 large, 26 small
G. thurberi   26 small
G. herbaceum   26 large

G. hirsutum is a polyploid derivative of a cross between the two Old World species. This could easily be checked by looking at the chromosomes.

71. There are six main species in the *Brassica* genus: *B. carinata, B. campestris, B. nigra, B. oleracea, B. juncea,* and *B. napus*. You can deduce the interrelationships among these six species from the following table:

Species or F_1 hybrid	Chromosome number	Number of bivalents	Number of univalents
B. juncea	36	18	0
B. carinata	34	17	0
B. napus	38	19	0
B. juncea × B. nigra	26	8	10
B. napus × B. campestris	29	10	9
B. carinata × B. oleracea	26	9	8
B. juncea × B. oleracea	27	0	27
B. carinata × B. campestris	27	0	27
B. napus × B. nigra	27	0	27

   a. Deduce the chromosome number of *B. campestris, B. nigra*, and *B. oleracea*.
   b. Show clearly any evolutionary relationships between the six species that you can deduce at the chromosomal level.

Answer:
   a. *B. campestris* was crossed with *B. napus*, and the hybrid had 29 chromosomes consisting of 10 bivalents; and 9 univalents. *B. napus* had to have contributed a total of 19 chromosomes to the hybrid. Therefore, *B. campestris* had to have contributed 10 chromosomes. The $2n$ number of *B. campestris* is 20.

   When *B. nigra* was crossed with *B. napus*, *B. nigra* had to have contributed 8 chromosomes to the hybrid. The $2n$ number of *B. nigra* is 16.

   B. oleracea had to have contributed 9 chromosomes to the hybrid formed with *B. juncea*. The $2n$ number in *B. oleracea* is 18.

b. First, list the haploid and diploid number for each species:

Species	Haploid	Diploid
B. nigra	8	16
B. oleracea	9	18
B. campestris	10	20
B. carinata	17	34
B. juncea	18	36
B. napus	19	38

Now, recall that a bivalent in a hybrid indicates that the chromosomes are essentially identical. Therefore, the more bivalents formed in a hybrid, the closer the two parent species. Three crosses result in no bivalents, suggesting that the parents of each set of hybrids are not closely related:

Cross	Haploid number
B. juncea × *B. oleracea*	18 vs. 9
B. carinata × *B. campestris*	17 vs. 10
B. napus × *B. nigra*	19 vs. 8

Three additional crosses resulted in bivalents, suggesting a closer relationship among the parents:

Cross	Haploid #	Bivalents	Univalents
B. juncea × *B. nigra*	18 vs. 8	8	10
B. napus × *B. campestris*	19 vs. 10	10	9
B. carinata × *B. oleracea*	17 vs. 9	9	8

Note that in each cross the number of bivalents is equal to the haploid number of one species. This suggests that the species with the larger haploid number is a hybrid composed of the second species and some other species. In each case, the haploid number of the unknown species is the number of univalents. Therefore, the following relationships can be deduced:

B. juncea is an amphidiploid formed by the cross of *B. nigra* and *B. campestris*.

B. napus is an amphidiploid formed by the cross of *B. campestris* and *B. oleracea*.

B. carinata is an amphidiploid formed by the cross of *B. nigra* and *B. oleracea*.

These conclusions are in accord with the three crosses that did not yield bivalents:

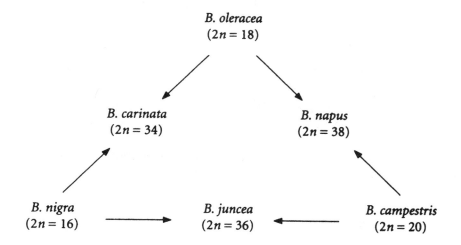

72. Several kinds of sexual mosaicism are well documented in humans. Suggest how each of the following examples may have arisen by nondisjunction at mitosis:

   a. XX/XO (that is, there are two cell types in the body, XX and XO)
   b. XX/XXYY
   c. XO/XXX
   d. XX/XY
   e. XO/XX/XXX

   Answer:
   a. Loss of one X in the developing fetus after the two-cell stage

   b. Nondisjunction leading to Klinefelter syndrome (XXY), followed by a nondisjunctive event in one cell for the Y chromosome after the two-cell stage, resulting in XX and XXYY

   c. Nondisjunction of the X at the one-cell stage

   d. Fused XX and XY zygotes (from the separate fertilizations either of two eggs or of an egg and a polar body by one X-bearing and one Y-bearing sperm)

   e. Nondisjunction of the X at the two-cell stage or later

73. In *Drosophila*, a cross (cross 1) was made between two mutant flies, one homozygous for the recessive mutation bent wing (*b*) and the other homozygous for the recessive mutation eyeless (*e*). The mutations *e* and *b* are alleles of two different genes that are known to be very closely linked on the tiny autosomal chromosome 4. All the progeny had a wild-type phenotype. One of the female progeny was crossed with a male of genotype *b e/b e*; we will call this cross 2.

Most of the progeny of cross 2 were of the expected types, but there was also one rare female of wild-type phenotype.

a. Explain what the common progeny are expected to be from cross 2.

b. Could the rare wild-type female have arisen by (1) crossing over or (2) nondisjunction? Explain.

c. The rare wild-type female was testcrossed to a male of genotype *b e/b e* (cross 3). The progeny were

1/6 wild type
1/6 bent, eyeless
1/3 bent
1/3 eyeless

Which of the explanations in part *b* is compatible with this result? Explain the genotypes and phenotypes of the progeny of cross 3 and their proportions.

Unpacking Problem 73

1. Define *homozygous, mutation, allele, closely linked, recessive, wild type, crossing over, nondisjunction, testcross, phenotype,* and *genotype.*

Answer:
Homozygous means that an organism has two identical alleles.
A *mutation* is any deviation from wild type.
An *allele* is one particular form of a gene.
Closely linked means two genes are almost always transmitted together through meiosis.
Recessive refers to a type of allele that is expressed only when it is the sole type of allele for that gene found in an individual.
Wild type is the most frequent type found in a laboratory population or in a population in the "wild."
Crossing over refers to the physical exchange of alleles between homologous chromosomes.
Nondisjunction is the failure of separation of either homologous chromosomes or sister chromatids in the two meiotic divisions.
A *testcross* is a cross to a homozygous recessive organism for the trait or traits being studied.
Phenotype is the appearance of an organism.
Genotype is the genetic constitution of an organism.

2. Does this problem concern sex linkage? Explain.

Answer: No. The genes in question are on an autosome, specifically, number 4.

3.   How many chromosomes does *Drosophila* have?

Answer: The most common lab species, *Drosophila melanogaster*, has eight chromosomes.

4.   Draw a clear pedigree summarizing the results of crosses 1, 2, and 3.

Answer:
P $b\,e^+/b\,e^+ \times b^+\,e/b^+\,e$ (cross 1)

F1 $b\,e^+/b^+\,e \times b\,e/b\,e$ (cross 2)

Progeny $b\,e^+/b\,e$ expected parental
          $b^+\,e/b\,e$ expected parental
          $b^+\,e^+/b\,e$ unexpected recombinant ("very closely linked" so rare)
          $b\,e/b\,e$ unexpected recombinant ("very closely linked" so rare)

          rare wild type $\times\ b\,e/b\,e$ (cross 3)

Progeny   1/6   wild type
          1/6   bent, eyeless
          1/3   bent
          1/3   eyeless

5.   Draw the gametes produced by both parents in cross 1.

Answer: $b\,e^+$ and $b^+\,e$

6.   Draw the chromosome 4 constitution of the progeny of cross 1.

Answer: $b\,e^+/b^+\,e$

7.   Is it surprising that the progeny of cross 1 are wild-type phenotype? What does this outcome tell you?

Answer: It is not at all surprising that the F_1 are wild type. This means that both mutations are recessive and complement (are in different genes).

8. Draw the chromosome 4 constitution of the male tester used in cross 2 and the gametes that he can produce.

Answer: $b\ e/b\ e\ \rightarrow$ gametes: $b\ e$

9. With respect to chromosome 4, what gametes can the female parent in cross 2 produce in the absence of nondisjunction? Which would be common and which rare?

Answer: The two common gametes are $b\ e^+$ and $b^+\ e$. The two rare gametes are $b^+\ e^+$ and $b\ e$.

10. Draw first- and second-division meiotic nondisjunction in the female parent of cross 2, as well as in the resulting gametes.

Answer:
Normal

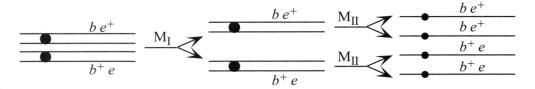

First-Division Nondisjunction: all gametes are aneuploid

Second-Division Nondisjunction: half the gametes are aneuploid

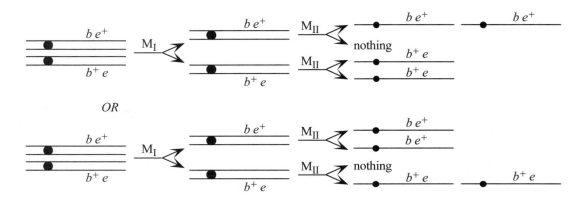

11. Are any of the gametes from part 10 aneuploid?

Answer: This is answered in 10, above.

12. Would you expect aneuploid gametes to give rise to viable progeny? Would these progeny be nullisomic, monosomic, disomic, or trisomic?

Answer: Viable progeny may be able to arise from aneuploidic gametes because chromosome 4 is very small and, percentage-wise, contributes little to the genome. The progeny would be monosomic and trisomic.

13. What progeny phenotypes would be produced by the various gametes considered in parts 9 and 10?

Answer: Listed below are the gametes from 9 and 10 above, the contribution of the male parent, and the phenotype of the progeny.

Female gamete	Male gamete	Phenotype
$b^+\ e$	$b\ e$	eyeless
$b\ e^+$	$b\ e$	bent
$b^+\ e^+$	$b\ e$	wild type
$b\ e$	$b\ e$	bent, eyeless
$b^+\ e/b\ e^+$	$b\ e$	wild type
—	$b\ e$	bent, eyeless
$b\ e^+/b\ e^+$	$b\ e$	bent
$b^+\ e/b^+\ e$	$b\ e$	eyeless

14. Consider the phenotypic ratio in the progeny of cross 3. Many genetic ratios are based on halves and quarters, but this ratio is based on thirds and sixths. To what might this ratio point?

Answer: The ratio points to meiosis of a trisomic.

15. Could there be any significance to the fact that the crosses concern genes on a very small chromosome? When is chromosome size relevant in genetics?

Answer: Research with artificial chromosomes has indicated that extremely small chromosomes segregate improperly at higher rates than longer chromosomes. It is suspected that the chromatids from homologous chromosomes need to intertwine in order to remain together until the onset of anaphase. Very short chromosomes are thought to have some difficulty in doing this and therefore have a higher rate of nondisjunction. In this instance, which deals with natural chromosomes as opposed to artificial chromosomes, very small chromosomes would be expected to have very little genetic material in them, and therefore their loss or gain may not be of too much importance during development.

16. Draw the progeny expected from cross 3 under the two hypotheses, and give some idea of relative proportions.

Answer:
rare wild type $\times$ e b/e b (cross 3)

If the rare wild type is from recombination, then the cross becomes

$$b^+ e^+/b\ e \times b\ e/b\ e$$

Progeny $b^+ e^+/b\ e$ parental: wild type
           $b\ e/b\ e$ parental: bent, eyeless
           $b^+ e/b\ e$ rare recombinant: eyeless
           $b\ e^+/b\ e$ rare recombinant: bent

If the rare wild type is from nondisjunction, then the cross becomes

$$b\ e^+/b^+ e/b\ e \times b\ e/b\ e$$

Progeny $b\ e^+/b^+ e/b\ e$ wild type
           $b\ e^+/b\ e/b\ e$ bent
           $b^+ e/b\ e/b\ e$ eyeless
           $b\ e^+/b\ e$ bent
           $b^+ e/b\ e$ eyeless
           $b\ e/b\ e$ bent, eyeless

Solution to the Problem

Cross 1: P $b\ e^+/b\ e^+ \times b^+ e/b^+ e$
          F_1 $b^+ e/b\ e^+$

Cross 2: P X/X ; $b^+ e/b\ e^+ \times X/Y$; $b\ e/b\ e$
          F_1 expect 1 $b\ e^+/b\ e$: 1 $b^+ e/b\ e$, X/X and X/Y one rare observed X/X ; $b^+ e^+$

a. The common progeny are $b^+ e/b\ e$ and $b\ e^+/b\ e$.

b. The rare female could have come from crossing-over, which would have resulted in a gamete that was $b^+ e^+$. The rare female could also have come from nondisjunction that gave a gamete that was $b\ e^+/b^+ e$. Such a gamete might give rise to viable progeny.

c. If the female had been wild type ($b^+ e^+/b\ e$) as a result of crossing-over, her progeny would have been as follows:

Parental:	$b^+ e^+/b\ e$	wild type (common)
	$b\ e/b\ e$	bent, eyeless (common)
Recombinant:	$b\ e^+/b\ e$	bent (rare)
	$b^+ e/b\ e$	eyeless (rare)

These expected results are very far from what was observed, so the rare female was not the result of recombination.

If the female had been the product of nondisjunction ($b\ e^+/b^+ e/b\ e$), her progeny when crossed to $b\ e/b\ e$ would be as follows:

1/6	$b^+ e/b\ e$	eyeless
1/6	$b\ e^+/b\ e/b\ e$	bent
1/6	$b^+ e/b\ e/b\ e$	eyeless
1/6	$b\ e^+/b\ e$	bent
1/6	$b\ e/b\ e$	bent, eyeless
1/6	$b\ e^+/b^+ e/b\ e$	wild type

Overall, 2 bent:2 eyeless:1 bent eyeless:1 wild type

These results are in accord with the observed results, indicating that the female was a product of nondisjunction.

74. In the fungus *Ascobolus* (similar to *Neurospora*), ascospores are normally black. The mutation *f*, producing fawn-colored ascospores, is in a gene just to the right of the centromere on chromosome 6, whereas mutation *b*, producing beige ascospores, is in a gene just to the left of the same centromere. In a cross of fawn and beige parents ($+f \times b+$), most octads showed four fawn and four beige ascospores, but three rare exceptional octads were found, as shown in the accompanying illustration. In the sketch, black is the wild-type phenotype, a vertical line is fawn, a horizontal line is beige, and an empty circle represents an aborted (dead) ascospore.

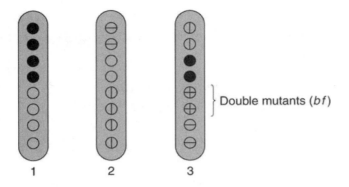

1   2   3

Double mutants (*bf*)

a. Provide reasonable explanations for these three exceptional octads.
b. Diagram the meiosis that gave rise to octad 2.

Answer: Recall that ascospores are haploid. The normal genotype associated with the phenotype of each spore is given below.

1	2	3
$b^+ f^+$	$b f^+$	$b^+ f$
$b^+ f^+$	$b f^+$	$b^+ f$
$b^+ f^+$	abort	$b^+ f^+$
$b^+ f^+$	abort	$b^+ f^+$
abort	$b^+ f$	$b f$
abort	$b^+ f$	$b f$
abort	$b^+ f$	$b f^+$
abort	$b^+ f$	$b f^+$

a. For the first ascus, the most reasonable explanation is that nondisjunction occurred at the first meiotic division. Second-division nondisjunction or chromosome loss are two explanations of the second ascus. Crossing-over best explains the third ascus.

b.

Chromosome Loss

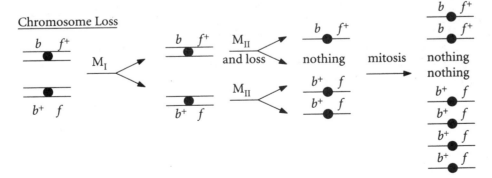

Second-Division Nondisjunction

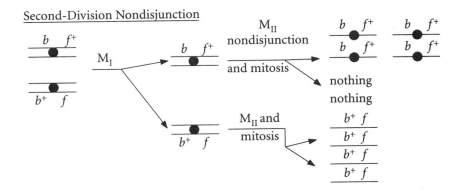

75. The life cycle of the haploid fungus *Ascobolus* is similar to that of *Neurospora*. A mutational treatment produced two mutant strains, 1 and 2, both of which when crossed with wild type gave unordered tetrads, all of the following type (fawn is a light brown color; normally, crosses produce all black ascospores):

    spore pair 1 black   spore pair 3 fawn
    spore pair 2 black   spore pair 4 fawn

a. What does this result show? Explain.

The two mutant strains were crossed. Most of the unordered tetrads were of the following type:

    spore pair 1 fawn   spore pair 3 fawn
    spore pair 2 fawn   spore pair 4 fawn

b. What does this result suggest? Explain.

When large numbers of unordered tetrads were screened under the microscope, some rare ones that contained black spores were found. Four cases are shown here:

	Case A	Case B	Case C	Case D
spore pair 1	black	black	black	black
spore pair 2	black	fawn	black	abort
spore pair 3	fawn	fawn	abort	fawn
spore pair 4	fawn	fawn	abort	fawn

(**Note:** Ascospores with extra genetic material survive, but those with less than a haploid genome abort.)

c. Propose reasonable genetic explanations for each of these four rare cases.
d. Do you think the mutations in the two original mutant strains were in one single gene? Explain.

Answer:

a. Each mutant is crossed with wild type, or

$$m \times m^+$$

The resulting tetrads (octads) show 1:1 segregation, indicating that each mutant is the result of a mutation in a single gene.

b. The results from crossing the two mutant strains indicate either both strains are mutant for the same gene,

$$m_1 \times m_2$$

or, that they are mutant in different but closely linked genes

$$m_1 \, m_2{}^+ \times m_1{+}m_2$$

c. and d. Because phenotypically black offspring can result from nondisjunction (notice that in Case C and Case D, black appears in conjunction with aborted spores), it is likely that mutant 1 and mutant 2 are mutant in different but closely linked genes. The cross is therefore

$$m_1 \, m_2{}^+ \times m_1{+}m_2$$

Case A is an NPD tetrad and would be the result of a four-strand double crossover.

$m_1{}^+ \, m_2{}^+$	black
$m_1{}^+ \, m_2{}^+$	black
$m_1 \, m_2$	fawn
$m_1 \, m_2$	fawn

Case B is a tetratype and would be the result of a single crossover between one of the genes and the centromere.

$m_1{}^+ \, m_2{}^+$	black
$m_1{}^+ \, m_2$	fawn
$m_1 \, m_2{}^+$	fawn
$m_1 \, m_2$	fawn

Case C is a the result of nondisjunction during meiosis I.

$m_1{}^+ \, m_2 \, ; \, m_1 \, m_2{}^+$	black
$m_1{}^+ \, m_2 \, ; \, m_1 \, m_2{}^+$	black
no chromosome	abort
no chromosome	abort

Case D is a the result of recombination between one of the genes and the centromere followed by nondisjunction during meiosis II. For example:

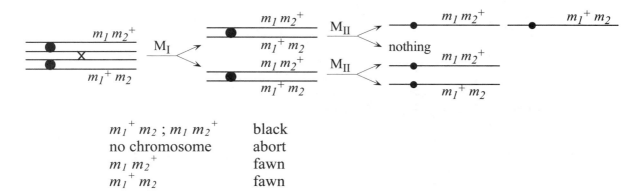

$m_1{}^+ m_2$; $m_1 m_2{}^+$ black
no chromosome   abort
$m_1 m_2{}^+$ fawn
$m_1{}^+ m_2$ fawn

18

Population Genetics

1. Which individual in Figure 18-3 has the most heterozygous loci, and which individual has the fewest?

   Answer: Individual 3 has three heterozygous loci (microsatellites 1, 3 and 4), while individual 1 has none.

2. Suppose that the seven haplotypes in Figures 18-1 and 18-4a represent a random sample of haplotypes from a population.

   a. Calculate gene diversity (*GD*) separately for the indel, the microsatellite locus, and the SNP at position 3.

   b. If the sequence was shortened so that you had data only for positions 1 through 24, how many haplotypes would there be?

   c. Calculate the linkage disequilibrium parameter (*D*) between the SNPs at positions 29 and 33.

   Answer:
   a. *GD* (gene diversity) is calculated as:
      1- $(p1^2 + p2^2 + ...pn^2)$ where p is allele frequency.
        2/7 chromosomes are different at the indel site, so $GD = 1 - 0.59 = 0.41$
        4/7 at the microsatellite locus, so $GD = 0.49$
        1/7 differences based on SNP in position 3 and $GD = 0.26$

   b. There would be five haplotypes, instead of six: A, B, C, D (IIa and IIc) and E (both IIb).

   c. D stands for the difference between the observed frequency of a haplotype and the expected frequency of a haplotype (with random association among alleles at the two loci). $D = P_{AB} - p_A p_B$. At position 29, G is present in 6 out of 7 haplotypes for a frequency of 0.857. A is present at position 33 in

3 out   of 7 haplotypes for a frequency of 0.429. A random combination of G and A   would be expected to occur with a frequency of $0.857 \times 0.429 = 0.368$, so $p_A p_B$ would equal 0.368. The observed frequency of   the   GA   haplotype, P_{AB},   is 3 out of 7, which equals 0.429. $P_{AB} - p_A p_B$ equals $0.429 - 0.368 = 0.061 = $ D. D is different than 0, so population is in linkage disequilibrium for the given haplotypes at the positions 29 and 33.

3. Looking at Figure 18-6, can you count how many mitochondrial haplotypes were carried from Asia into the Americas? How many of these made it all the way to South America?

Answer: Five mtDNA haplotypes carried over from Asia to North and South America: A ×, A, B, C and D. Four of these made it to South America (A × did not), with haplotype B having a somewhat different path than A, C, and D.

4. In Figure 18-13, the "general" column for Japan is higher than the "general" column for Europe. What does this tell you?

Answer: This column tells us that the frequency of first-cousin marriages is higher in Japan (0.05) than in Europe (0.02) and that consequently, frequency of genetic disorders caused by inbreeding depression is higher in Japan.

5. In Figure 18-14, some individuals have unique SNP alleles—for example, the T allele at SNP4 occurs only in individual 12. Can you identify two individuals each of whom have unique alleles at two SNPs?

Answer: These individuals are: 21 (Native American) and 14 (South African). Individual 21 has AT in SNP positions 6 and 7, while individual 14 has a T in SNP 11 and A in SNP 13. Both with two unique alleles compared to the rest of the sample.

6. Looking at Figure 18-20, do people of the Middle East tend to have higher or lower levels of heterozygosity compared to the people of East Asia? Why might this be the case?

Answer: Yes, people in Middle East appear to have higher heterozygosity (haplotype and microsatellite), when compared to people in East Asia. This is most likely due to the anthropological circumstances, considering that Middle Eastern populations had strong migratory history and the present populations have genetic signatures from a number of ancestral populations. East Asia had been somewhat isolated, with less migration from Middle East, Europe or Central Asia. It is also likely that genetic drift played a role.   Human populations migrated out of Africa and Middle Eastern populations were established before East Asian populations.   East Asian populations were established from later migrations out of the Middle East. With each migration, a reduction in population size would lead to bottlenecks and a greater influence from drift, which would lower heterozygosity. Those populations established after two successive migrations would be expected to have less heterozygosity than those established after one migration.

BASIC PROBLEMS

7. What are the forces that can change the frequency of an allele in a population?

Answer: The frequency of an allele in a population can be altered by natural selection, mutation, migration, and genetic drift (sampling errors).

8. In a population of mice, there are two alleles of the A locus (A_1 and A_2). Tests showed that, in this population, there are 384 mice of genotype A_1/A_1, 210 of A_1/A_2, and 260 of A_2/A_2. What are the frequencies of the two alleles in the population?

Answer: There are a total of $(2)(384) + (2)(210) + (2)(260) = 1708$ alleles in the population. Of those, $(2)(384) + 210 = 978$ are $A1$ and $210 + (2)(260) = 730$ are A_2. The frequency of A_1 is $978/1708 = 0.57$, and the frequency of A_2 is $730/1708 = 0.43$.

9. In a randomly mating laboratory population of *Drosophila*, 4 percent of the flies have black bodies (encoded by the autosomal recessive b), and 96 percent have brown bodies (the wild type, encoded by B). If this population is assumed to be in Hardy–Weinberg equilibrium, what are the allele frequencies of B and b and the genotypic frequencies of B/B and B/b?

Answer: The given data are $q^2 = 0.04$ and $p2 + 2pq = 0.96$. Assuming Hardy–Weinberg equilibrium, if $q^2 = 0.04$, $q = 0.2$, and $p = 0.8$. The frequency of B/B is $p^2 = 0.64$, and the frequency of B/b is $2pq = 0.32$.

10. In a population of a beetle species, you notice that there is a 3 : 1 ratio of shiny to dull wing covers. Does this ratio prove that the *shiny* allele is dominant? (Assume that the two states are caused by two alleles of one gene.) If not, what does it prove? How would you elucidate the situation?

Answer: No, a 3:1 ratio stands for F2 offspring from heterozygous F1 genotypes (with two alleles of one gene and dominant-recessive interaction). This observation only proves that the shiny wing phenotype is more common among the beetles observed.

11. The relative fitnesses of three genotypes are $w_{A/A} = 1.0$, $w_{A/a} = 1.0$, and $w_{a/a} = 0.7$

    a. If the population starts at the allele frequency $p = 0.5$, what is the value of p in the next generation?

    b. What is the predicted equilibrium allele frequency if the rate of mutation from $A \rightarrow a$ is 2×10^{-5}?

Answer:
a. This is the case of selection against a recessive allele. The needed equations for the frequency of dominant allele in the next generation (p') are

$$p' = p\frac{pW_{AA} + qW_{Aa}}{\overline{W}}$$

    and

$$\overline{W} = p^2W_{AA} + 2pqW_{Aa} + q^2W_{aa}$$

$$p' = 0.5\,[0.5(1.0) + 0.5(1.0)]/[(0.25)(1.0) + (0.5)(1.0) + (0.25)(0.7)] = 0.54$$

b. The needed equation for the equilibrium frequency is

$\hat{q} = \sqrt{\mu/s}$, where q is the equilibrium frequency, μ is the mutation rate, and s is the selection coefficient, $1-w_{a/a}$. The mutation rate of A to a is 2×10^{-5}, or 0.00002 and the selection coefficient is $1-.7 = 0.3$. The square root of $0.00002/0.30 = 0.008$.

12. *A/A* and *A/a* individuals are equally fertile. If 0.1 percent of the population is *a/a*, what selection pressure exists against *a/a* if the $A \rightarrow a$ mutation rate is 10^{-5}? Assume that the frequencies of the alleles are at their equilibrium values.

Answer:
The needed equation is
$$q^2 = \mu/s$$
or
$$s = \mu/q^2 = 10^{-5}/10^{-3} = 0.01$$

13. If the recessive allele for an X-linked recessive disease in humans has a frequency of 0.02 in the population, what proportion of individuals in the population will have the disease? Assume that the population is 50 : 50 male : female.

Answer: Since males need only one copy of the affected X to have the disease, the frequency of affected males will be the frequency of the disease allele, 0.02. Females must inherit two copies and be homozygotes, so by the Hardy-Weinberg law their chance of having the disease is q^2, which equals 0.02^2, or 0.0004. Since half of the population is male and half female, the overall frequency of the disease will be $(0.5) (0.02) + (0.5) (0.0004) = 0.01 + 0.0002$. Thus, the expected frequency of affected individuals in this population is 0.0102.

14. It seems clear that most new mutations are deleterious. Can you explain why?

Answer: Proper function results from the right gene products in the proper ratio to all other gene products. A mutation will change the gene product, eliminate the gene product, or change the ratio of it to all other gene products. All three outcomes upset a previously adapted system. While a new or even "better" balance may be achieved with a random mutation, this outcome is expected to be rarer and less likely than a new mutation being deleterious or neutral.

15. In a population of 50,000 diploid individuals, what is the probability that a new neutral mutation will ultimately reach fixation? What is the probability that it will ultimately be lost from the population?

Answer: The probability that any allele will reach fixation through genetic drift is $1/2N$, where N is the population size. Therefore, the probability that a new neutral allele would drift to fixation in this case is $1/2(50,000)$, or $1/100,000$. The probability that it will be lost is $1-1/2N$, or 99,999 out of 100,000

CHALLENGING PROBLEMS

16. Figure 18-14 presents haplotype data for the *G6PD* gene in a worldwide sample of people.

   a. Draw a haplotype network for these haplotypes. Label the branches on which each SNP occurs.

   b. Which of the haplotypes has the most connections to other haplotypes?

   c. On what continents is this haplotype found?

   d. Counting the number of SNPs along the branches of your network, how many differences are there between haplotypes 1 and 12?

Answer:

a.

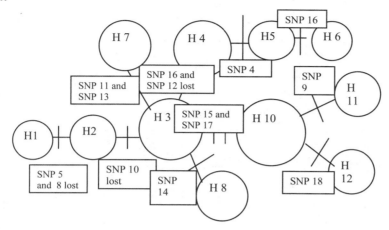

b. There are 12 haplotypes and Haplotype 3 has the most connections with other haplotypes.

c. Haplotype 3 was found in Central Africa and in East Asia.

d. There are 6 differences between Haplotypes 1 and 12: SNP's 5 and 8 lost; SNP 10 lost; SNP's 15 and 17 lost; and SNP 18 gained.

17. Figure 18-12 shows a pedigree for the offspring of a half-sib mating.

    a. If the inbreeding coefficient for the common ancestor (*A*) in Figure 18-12 is 1/2, what is the inbreeding coefficient of I?

    b. If the inbreeding coefficient of individual I in Figure 18-12 is 1/8, what is the inbreeding coefficient of the common ancestor, A?

Answer:

a.

$$F_I = (1/2)^3 (1 + F_A) = 1/8 \times (1 + 1/2) = 1/8 \times 3/2 = 3/16$$

b.

$$F_I = 1/8 \quad F_A = ?$$

$$F_I = (1/2)^3 (1 + F_A)$$
$$1/8 = (1/2)^3 (1 + F_A)$$
$$1/8 = 1/8 \times 1 + F_A$$
$$F_A = 0$$

18. Consider 10 populations that have the genotypes shown in the following table:

Population	A/A	A/a	a/a
1	1.0	0.0	0.0
2	0.0	1.0	0.0
3	0.0	0.0	1.0
4	0.50	0.25	0.25
5	0.25	0.25	0.50
6	0.25	0.50	0.25
7	0.33	0.33	0.33
8	0.04	0.32	0.64
9	0.64	0.32	0.04
10	0.986049	0.013902	0.000049

a. Which of the populations are in Hardy–Weinberg equilibrium?

b. What are p and q in each population?

c. In population 10, the $A \rightarrow a$ mutation rate is discovered to be 5×10^{-6}. What must be the fitness of the a/a phenotype if the population is at equilibrium?

d. In population 6, the a allele is deleterious; furthermore, the A allele is incompletely dominant; so A/A is perfectly fit, A/a has a fitness of 0.8, and a/a has a fitness of 0.6. If there is no mutation, what will p and q be in the next generation?

Answer:

a. and b. For each, p and q must be calculated and then compared with the predicted genotypic frequencies of $p^2 + 2pq + q^2 = 1$.

Population	p	q	Equilibrium?
1	1.0	0.0	Yes
2	0.5	0.5	No
3	0.0	1.0	Yes
4	0.625	0.375	No
5	0.375	0.625	No
6	0.5	0.5	Yes
7	0.5	0.5	No
8	0.2	0.8	Yes
9	0.8	0.2	Yes
10	0.993	0.007	Yes

c. The formulas to use are $q^2 = \mu/s$ and $s = 1-W$.
$4.9 \times 10^{-5} = 5 \times 10^{-6}/s$; $s = 0.102$, so $W = 0.898$

d. For simplicity, assume that the differences in survivorship occur prior to reproduction. Thus, each genotype's fitness can be used to determine the relative percentage each contributes to the next generation.

Genotype	Frequency	Fitness	Contribution	A	a
A/A	0.25	1.0	0.25	0.25	0.0
A/a	0.50	0.8	0.40	0.20	0.20
a/a	0.25	0.6	0.15	0.0	0.15
				0.45	0.35

$$p' = 0.45/(0.45 + 0.35) = 0.56$$
$$q' = 0.35/(0.45 + 0.35) = 0.44$$

Alternatively, the formulas to use are

$$p' = p\frac{pW_{AA} + qW_{Aa}}{\overline{W}}$$

and

$$\overline{W} = p^2 W_{AA} + 2pq W_{Aa} + q^2 W_{aa}$$

$$p' = (0.5)[(0.5)(1.0) + (0.5)(0.8)]/[(0.25)(1.0) + (0.5)(0.8) + (0.25)(0.6)]$$
$$= (0.5)(0.9)/(0.8) = 0.56$$

19. The hemoglobin B gene (Hb) has a common allele (A) of an SNP ($rs334$) that encodes the Hb^A form of (adult) hemoglobin and a rare allele (T) that encodes the sickling form of hemoglobin, Hb^S. Among 57 members of the Yoruba tribe of Nigeria, 44 were A/A and 13 were A/T. No T/T individuals were observed. Use the χ^2 test to determine whether these observed genotypic frequencies fit Hardy–Weinberg expectations.

Answer:

	A/A	A/T	T/T	Total
Observed	44	13	0	57

p (f A) = ((44 × 2) + 13) / 2 × 57 = 101/ 114 = 0.8859

q (f T) = 1- p = 0.1141

Expected frequencies, if population is in the Hardy- Weinberg equilibrium would be

f(A/A) = p^2 ; f (A/T) = 2pq and f (T/T) = q^2

Expected frequency: p^2= 0.785; 2pq = 0.202 and q^2 = 0.013

Expected values:	A/A	A/T	T/T	Total 57
f (genotype) × 57	44.73	11.52	0.74	

χ^2 calculation:	A/A	A/T	T/T	Total
O Observed	44	13	0	57

E	Expected	44.73	11.52	0.74	57

$\chi^2 = (44\text{-}44.73)^2/44.73 + (13\text{-}11.52)^2/11.52 + (0\text{-}0.74)^2/0.74 = 0.012 + 0.19 + 0.74 = 0.942$.

df = 3-2 = 1

Null hypothesis that this population is in Hardy–Weinberg equilibrium cannot be rejected, because the χ^2 of 0.942 is less than the critical value of 3.81 at $p = 0.05$, 1 df. A difference this great or greater between the expected and observed is most likely due to by random chance. The Yoruba tribe can be considered to be in Hardy–Weinberg equilibrium for the two alleles of hemoglobin gene.

20. A population has the following gametic frequencies at two loci: $AB = 0.40$ $Ab = 0.1$, $aB = 0.1$, and $ab = 0.4$. If the population is allowed to mate at random until linkage equilibrium is achieved, what will be the expected frequency of individuals that are heterozygous at both loci?

Answer: If the population comes to linkage equilibrium, there will be an equal proportion of allele B on chromosome A and allele B on chromosome a, etc. In the beginning, frequency of B on chromosome A is 0.4 and B on chromosome a is 0.1. Equilibrium would be at 0.25. Therefore, frequency of Ab would be = 0.25 and frequency of aB would be 0.25. Expected frequency of both "heterozygous" individuals is 0.25, if the population reaches linkage equilibrium for this pair of genes.

For each locus the frequency of each allele is 0.5, so that p = q = 0.5, For the individual loci A = a = 0.5 and B = b = 0.5. The expected frequency of heterozygotes for each allele is 2pq or 2 × 0.5 × 0.5 = 0.5. At linkage equilibrium the frequency of heterozygotes for both alleles will be the product of their individual frequencies, or 0.5 × 0.5 = 0.25.

21. Two species of palm trees differ by 50 bp in a 5000-bp stretch of DNA that is thought to be neutral. The mutation rate for these species is 2×10^{-8} substitutions per site per generation. The generation time for these species is five years. Estimate the time since these species had a common ancestor.

Answer:
The two species of palm trees differ by: 50/5000 bp = 0.01
We can estimate the time of their divergence using the formula:
t= d/ 2k (where d is the genetic difference between the species/ neutral sequence, since the divergence and k is the rate of substitutions).
Thus: t = 0.01/2 × (2 × 10^{-8}) = 250,000 × 5 = 1.25 million years ago

22. Color blindness in humans is caused by an X-linked recessive allele. Ten percent of the males of a large and randomly mating population are color-blind.

A representative group of 1000 people from this population migrates to a South Pacific island, where there are already 1000 inhabitants and where 30 percent of the males are color-blind. Assuming that Hardy–Weinberg equilibrium applies throughout (in the two original populations before the migration and in the mixed population immediately after the migration), what fraction of males and females can be expected to be color-blind in the generation immediately after the arrival of the migrants?

Answer: Prior to migration, $q^A = 0.1$ and $q^B = 0.3$ in the two populations. Because the two populations are equal in number, immediately after migration, $q^{A+B} = 1/2(q^A + q^B) = 1/2(0.1 + 0.3) = 0.2$. At the new equilibrium, the frequency of affected males is $q = 0.2$, and the frequency of affected females is $q^2 = (0.2)^2 = 0.04$. (Color blindness is an X-linked trait.)

23. Using pedigree diagrams, calculate the inbreeding coefficient (F) for the offspring of (a) parent–offspring matings; (b) first-cousin matings; (c) aunt–nephew or uncle–niece matings; (d) self-fertilization of a hermaphrodite.

Answer:
For all four $F_A = 0$

a.

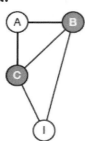

$p = (1/2)^2 = 1/4$

b.

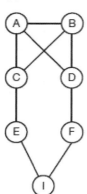

Individual I is the offspring of E and F (first cousins). There are two paths to I and we will calculate those separately:

$$F_I = (1/2)^5 + (1/2)^5 = 1/32 + 1/32 = 1/16$$

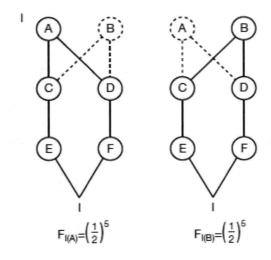

c.

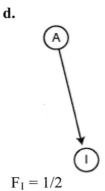

There are 3 paths $F_I = (1/2)^3 = 1/8$

d.

$$F_I = 1/2$$

24. A group of 50 men and 50 women establish a colony on a remote island. After 50 generations of random mating, how frequent would a recessive trait be if it were at a frequency of 1/500 back on the mainland? The population remains the same size over the 50 generations, and the trait has no effect on fitness.

Answer: Since the frequency of the recessive phenotype is 0.002, which equals q^2, the frequency of the recessive allele is $q = 0.045$. Inbreeding will increase over time as a function of population size and increase the frequency of homozygotes. The inbreeding coefficient after 50 generations can be calculated as:

$$F_{50} = (1 - (1 - 1/2N)^{50} (1 - F_0).$$

Taking F_0 in the initial population of 100 founders to be 0.0, $F_{50} = 1 - (1 - 0.005)^{50}$, which equals 0.222. Inbreeding increases the frequency of homozygotes at the expense of heterozygotes by the inbreeding coefficient F, so if the allele frequency remains constant, the new frequency of a/a homozygotes after 50 generations would be $q^2 + pqF_{50}$ or $0.045^2 + (0.045)(0.995)(0.222) = 0.012$. One or two members of the new population would be expected to express the recessive trait after 50 generations.

This calculation would only apply where allele frequencies did not change. The recessive allele has no effect on fitness, so changes in frequency in a randomly mating population will be due to genetic drift, which is by nature random and thus unpredictable. This population is relatively small, so the effects of drift will be greater than in a larger population and the allele could either increase or decrease in frequency in any given generation. The frequency of the recessive allele after 50 generations of random mating cannot be determined, but could be substantially different than in the original population of founders. An estimate of its ultimate fate in the population can be made. The likelihood of an allele drifting to fixation is equal to its frequency. So there is a 0.045 chance that the allele will ultimately be fixed and a 1 - 0.045 or 0.955 chance that it will be eliminated.

25. Figure 18-22 shows 10 haplotypes from a population before a selective sweep and another 10 haplotypes many generations later after a selective sweep has occurred for this chromosomal region. There are 11 loci defining each haplotype, including one with a red allele that was the target of selection. In the figure, two loci are designated as *A* and *B*. These loci each have two alleles: one black and the other gray. Calculate the linkage disequilibrium parameter (*D*) between A and B, both before and after the selective sweep. What effect has the selective sweep had on the level of linkage disequilibrium?

Answer: There are 10 haplotypes in the population before and after selective sweep, caused by the advantage of the red allele over gray, on the locus in position 6.

Let us assume that at each locus A and B have a pair of alleles: Black (b) and Gray (g). Before selection, we can calculate the linkage disequilibrium parameter (D) as follows:

$$D = (b\,A \times b\,B) \times (g\,A \times g\,B) - (b\,A \times g\,B) \times (g\,A \times b\,B) =$$

$$(2/10 \times 3/10) - (3/10 \times 2/10) = 0.06 - 0.06 = 0$$

D is 0, and this population was in the linkage equilibrium, before the sweep.

After selection, the number of haplotypes has changed favoring chromosomes with red allele:

$$D = (0/10 \times 1/10) - (3/10 \times 6/10) = -0.18$$

D is different than 0 (- 0.18), and this population is in linkage disequilibrium

Selective sweep allows physically linked alleles at nearby loci to "hitchhike" to high frequency or simply cause LD between the haplotypes.

26. The recombination fraction (r) between linked loci A and B is 0.25. In a population, we observe the following haplotypic frequencies:

AB	0.49
aB	0.49
Ab	0.00
ab	0.02

a. What is the level of linkage disequilibrium as measured by D in the present generation?

b. What will D be in the next generation?

c. What is the expected frequency of the Ab haplotype in the next generation?

Answer:
a. In the present generation we can estimate LD value as D_0 (before recombination at the rate "r"). From the data above:

$$D_0 = (g\,AB \times g\,ab) - (g\,aB \times g\,aB) = (0.49 \times 0.02) - (0.49 \times 0) = 0.0098$$

b. D_1 will be a value of LD in the next generation with 0.25 recombination rate:

$$D_1 = D_0 (1-r) = 0.0098 (1- 0.25) = 0.00735$$

Thus, value of D declines with recombination.

c.   The expected frequency of the Ab haplotype, based on the frequency of A (0.49) and b (0.02) in the population is $0.49 \times 0.02 = 0.0098$. D_0 is equal to the difference between the observed and expected frequencies of the Ab haplotype, which also equals 0.0098 because there are no Ab haplotypes. D_1 (equal to 0.00735), represents D after one generation of recombination.   Since the expected frequency of Ab is 0.0098 and D_1 is 0.00735, the actual frequency of the Ab haplotype after recombination should be the difference between them or $0.0098 - 0.000735 = 0.00245$.

27.   Allele B is a deleterious autosomal dominant. The frequency of affected individuals is 4.0×10^{-6}. The reproductive capacity of these individuals is about 30 percent that of normal individuals. Estimate μ, the rate at which b mutates to its deleterious allele B. Assume that the frequencies of the alleles are at their equilibrium values.

Answer: Affected individuals $= B/b = 2pq = 4 \times 10^{-6}$. Because q is almost equal to 1.0, $2p = 4 \times 10^{-6}$. Therefore, $p = 2 \times 10^{-6}$.

$$\mu = hsp = (1.0)(0.7)(2 \times 10^{-6}) = 1.4 \times 10^{-6}$$

where h = degree of dominance of the deleterious allele.

28.   What is the equilibrium heterozygosity for an SNP in a population of 50,000 when the mutation rate is 3×10^{-8}?

Answer: We can use the formula based on heterozygosity (H), as a function of population size (could be lost by drift) and new mutations (M or Greek "mu"), as the change in $H = 2M(1-H)$

$1-H$ stands for the frequency of homozygotes in the population.
When derived for equilibrium between the two factors:

$^{\wedge}H = 4NM/ 4NM + 1 = 4 \times 50,000 \times (3 \times 10^{-8)} / 4 \times 50,000 \times (3 \times 10^{-8)} + 1$

Or:
Equilibrium heterozygosity for such an SNP will be 0.0059.

29.   Of 31 children born of father–daughter matings, 6 died in infancy, 12 were very abnormal and died in childhood, and 13 were normal. From this information, calculate roughly how many recessive lethal genes we have, on average, in our human genomes. (**Hint:** If the answer were 1, then a daughter would stand a 50 percent chance of carrying the lethal allele, and the probability of the union's producing a lethal combination would be $1/2 \times 1/4 = 1/8$. So 1 is not the

answer.) Consider also the possibility of undetected fatalities in utero in such matings. How would they affect your result?

Answer: The probability of not getting a recessive lethal genotype for one gene is $1-1/8 = 7/8$. If there are n lethal genes, the probability of not being homozygous for any of them is $(7/8)^n = 13/31$. Solving for n, an average of 6.5 recessive lethals are predicted. If the actual percentage of "normal" children is less owing to missed in utero fatalities, the average number of recessive lethals would be higher.

30. The *sd* gene causes a lethal disease of infancy in humans when homozygous. One in 100,000 newborns die each year of this disease. The mutation rate from *Sd* to *sd* is 2×10^{-4}. What must the fitness of the heterozygote be to explain the observed gene frequency in view of the mutation rate? Assign a relative fitness of 1.0 to *Sd/Sd* homozygotes. Assume that the population is at equilibrium with respect to the frequency of *sd*.

Answer:

This is the case of recessive lethal (S =1 for sd/sd)

Based on H-W prediction: if p^2 is f (Sd/ Sd); 2pq is f(Sd/sd) and q^2 is f(sd/sd)

Frequency of a recessive lethal form a H–W q^2 value (1/100,000 children who die) should be: $q = 0.003$ and $p = 1- q = 0.997$

Genotype:	*Sd/Sd*	*Sd/sd*	*sd/sd*
Frequency:	p^2 W $_{Sd/Sd}$	2pq W $_{Sd/sd}$	q^2 W $_{sd/sd}$
	1	1-hs	1-s

We substitute Wsd/Sd with 1 and W sd/sd with 0.

The rate of mutation from Sd to sd M or 2×10^{-4}

At equilibrium q 2 = M/S and since this allele has some effect on Sd/sd fitness, that is measure as "h":

$\hat{q} = M/ h \times S$ or

$2 \times 10^{-4} / h \times 1 = 0.003$

h = 0.067

The relative fitness of a heterozygous Sd/sd genotype will be lower than 1 for value h:

$$W_{Sd/sd} = 1 - 0.067 = 0.933$$

31. If we define the *total selection cost* to a population of deleterious recessive genes as the loss of fitness per individual affected (s) multiplied by the frequency of affected individuals (q^2), then selection cost $= sq^2$.

  a. Suppose that a population is at equilibrium between mutation and selection for a deleterious recessive allele, where $s = 0.5$ and $\mu = 10^{-5}$. What is the equilibrium frequency of the allele? What is the selection cost?

  b. Suppose that we start irradiating individual members of the population so that the mutation rate doubles. What is the new equilibrium frequency of the allele? What is the new selection cost?

  c. If we do not change the mutation rate but we lower the selection coefficient to 0.3 instead, what happens to the equilibrium frequency and the selection cost?

  Answer:
  a. The formula needed is

  $$\hat{q} = \sqrt{\mu/s}$$
  $$= 4.47 \times 10^{-3}$$

  so,   Genetic cost $= sq^2 = 0.5(4.47 \times 10{-}3)^2 = 10^{-5}$

  b. Using the same formulas as part *a*,

  $$\hat{q} = 6.32 \times 10^{-3}$$

  Genetic cost $= sq^2 = 0.5(6.32 \times 10{-}3)^2 = 2 \times 10^{-5}$

  c. $\hat{q} = 5.77 \times 10^{-3}$

  Genetic cost $= sq^2 = 0.3(5.77 \times 10{-}3)^2 = 10^{-5}$

32. Balancing selection acts to maintain genetic diversity at a locus since the heterozygous class has a greater fitness than the homozygous classes. Under this form of selection, the allele frequencies in the population approach an equilibrium point somewhere between 0 and 1. Consider a locus with two alleles A and a with frequencies p and q, respectively. The relative genotypic fitnesses are shown below, where s and g are the selective disadvantages of the two homozygous classes.

Genotype	A/A	A/a	a/a
Relative fitness	$1 - s$	1	$1 - g$

a. At equilibrium, the mean fitness of the A alleles (w_A) will be equal to the mean fitness of the a alleles (w_a) (see Box 18-7). Set the mean fitness of the A alleles (w_A) equal to the mean fitness of the a alleles (w_a). Solve the resulting equation for the frequency of the A allele. This is the expression for the equilibrium frequency of A ($\hat{p}$).

b. Using the expression that you just derived, find $\hat{p}$ when $s = 0.2$ and $g = 0.8$.

Answer: This is the case of overdominance, where heterozygous have the highest fitness (W = 1). Selection acting against the A/A is s and against a/a is g

a. If the mean fitness of allele A (W $_A$) is equal to W $_a$ (s = g); and we could substitute (1-p) for q, equilibrium frequency of the allele is:

$$\hat{}$$
$$f(A) = p = \text{Waa} - \text{W Aa} / \text{ W AA} - 2 \text{ (Waa } + \text{W aa)}$$

Finally, substituting fitness into the pervious equation:

$$\hat{}$$
$$f(A) = p = s/ s + g$$

since (s=g) our equilibrium value for f (A) is 0.5

b. If s = 0.2 and g = 0.8

$$\hat{}$$
$$p = \frac{0.8}{0.2 + 0.8} = 0.8$$

19

The Inheritance of Complex Traits

WORKING WITH THE FIGURES

1. Figure 19-9 shows the trait distributions before and after a cycle of artificial selection. Does the variance of the trait appear to have changed as a result of selection? Explain.

   Answer: The phenotypic variance of the trait does not appear to have changed as a result of selection. Neither the range nor the shape of the phenotypic distribution have changed, so the variance would not have changed because it is a function of these two parameters. Although it might be expected that phenotypic variation would decrease after selection, there are reasons why it might not. First, selection would act on additive genetic effects, which in this case are calculated to account for 50 percent of the phenotypic variance. Phenotypic variance due to dominance and environmental effects would be unchanged and could help to sustain phenotypic variation even after a reduction in additive genetic variation.   Second, and probably more important, most populations have a large store of unexpressed additive genetic variation, as shown by the long term responses to most artificial selection efforts.   One generation of selection would likely favor the expression of some of this untapped variation, extending the range of phenotypic variation beyond that present in the original population. Together these two factors could lead to a shift in the population mean without a corresponding reduction in phenotypic variation.

2. Figure 19-11 shows the expected distributions for the three genotypic classes if the B locus is a QTL affecting the trait value.

   a. As drawn, what is the dominance/additive D/A ratio?
   b. How would you redraw this figure if the B locus had no effect on the trait value?
   c. How would the positions along the x-axis of the curves for the different genotypic classes of the B locus change if A/D = 1.0?

Answer:

a. As drawn, the *Bb* distribution is intermediate between *bb* and *BB*. This indicates additive gene action and the D/A ratio will be 0.

b. If the *B* locus had no effect on the trait value, all three distributions would coincide and there would be three completely overlapping distributions that looked like the one depicted for *bb*.

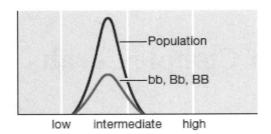

c. If the A/D ratio = 1.0, this would mean total dominance so that the *BB* and *Bb* phenotypes would be the same. There would be two distributions – a small one for bb and a large one for *BB* and *Bb* combined.

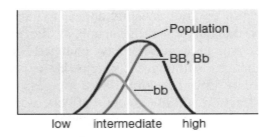

3. Figure 19-16 shows the results of a QTL fine-mapping experiment. Which gene would be implicated as controlling fruit weight if the mean fruit weight for each line was as follows?

Line	Fruit weight (g)
1	181.4
2	169.3
3	170.7
4	171.2
5	171.4
6	182.2
7	180.6
8	180.7
9	181.8
10	169.3

Answer: Lines 1, 6, 7, 8, and 9 all have high fruit weight and Beefmaster chromosomal segments that include the *ald2* gene. Lines 2, 3, 4, 5 and 10 all have low fruit weight and Sungold chromosomal segments that include *ald2*. *ald 2* is the only gene that is Beefmaster in high fruit weight lines and Sungold in low fruit weight lines. This indicates that *ald2* controls the difference in fruit weight.

4. Figure 19-17 shows a set of haplotypes. Suppose these are haplotypes for a chromosomal segment from 18 haploid yeast strains. On the right edge of the figure, the S and D indicate whether the strain survives (S) or dies (D) at high temperature (40°C). Using the χ^2 test (see Chapter 3) and Table 3-1, does either SNP1 or SNP6 show evidence for an association with the growth phenotype? Explain.

Answer: These samples may be too small for a meaningful chi-square test, because with samples of this size the differences between the expected and observed outcomes would have to be relatively very large to reach significance. However, a chi-square can be calculated. Overall, there were 10 strains that survived high temperatures. If SNP 1 has no effect, these survivors should be distributed randomly between the A and G variants. Out of 18 strains, 12 were A strains and 6 were G strains, so 2/3 of the survivors, or 6.67 should be A and 1/3 or 3.33 should be G. This constitutes the expected outcome or null hypothesis. The observed outcome is that 7 survivors were A strain and 3 were G. Using the chi-square formula $\sum (O-E)^2/E$, the chi-square test for this data set is 0.049 with 1df, $P > 0.05$. The null hypothesis cannot be rejected. This data set falls well within the range of normal variation for a sample this size. For SNP 6 half the strains were A and half were G, so the expected number of survivors out of 10 would be 5 each. The observed number of survivors was G = 8 and A = 2. Using the same chi-square formula, a chi-square of 3.6 is obtained. The critical value for a chi-square with 1df is 3.81, so this value is close to significance but also falls within the range of normal variation. Even though G appears to be strongly associated with survival at high temperatures, the null hypothesis cannot be rejected because the sample size is too small to resolve a difference of this magnitude, even if it is "real." A larger sample would be necessary.

5. Figure 19-18a shows a plot of P values (represented by the dots) along the chromosomes of the dog genome. Each P value is the result of a statistical test of association between an SNP and body size. Other than the cluster of small P values near *IGF1*, do you see any chromosomal regions with evidence for a significant association between an SNP and body size? Explain.

Answer: The only interval with values that exceeds the threshold for significance is that near *IGF1*, so it is the only region with evidence of significant association with body size. Other regions seem to exceed the usual significance level of 0.05 but are not considered significant in this case. These are not considered significant because in any analysis where many comparisons are made the threshold for significance must be raised to account for the increased chance of a "significant" outcome that was actually due to random chance and would not be repeatable.

6. Figure 19-19 shows plots of P values (represented by the dots) along the chromosomes of the human genome. Each P value is the result of a statistical test of association between an SNP and a disease condition. There is a cluster, or spike, of statistically significant P values (green dots) at the gene *HLA-DRB1* for two diseases. Why might this particular gene contribute to susceptibility for the autoimmune diseases rheumatoid arthritis and type 1 diabetes?

Answer: HLA is one of the major histocompatibility complex (MHC) genes that regulates immune response. Rheumatoid arthritis and type 1 diabetes are both autoimmune diseases. Mutations in immune response genes are thus associated with diseases that result from abnormal immune responses.

BASIC PROBLEMS

7. Distinguish between continuous and discontinuous variation in a population, and give some examples of each.

Answer: There are many traits that vary more or less continuously over a wide range. For example, height, weight, shape, color, reproductive rate, metabolic activity, etc., vary quantitatively rather than qualitatively. Continuous variation can often be represented by a bell-shaped curve, where the "average" phenotype is more common than the extremes. Discontinuous variation describes the easily classifiable, discrete phenotypes of simple Mendelian genetics: seed shape, auxotrophic mutants, sickle-cell anemia, etc. These traits show a simple relationship between genotype and phenotype.

8. The table below shows a distribution of bristle number in a *Drosophila* population. Calculate the mean, variance, and standard deviation for these data.

Bristle number	Number of individuals
1	1
2	4
3	7
4	31
5	56
6	17
7	4

Answer: The mean (or average) is calculated by dividing the sum of all measurements by the total number of measurements, or in this case, the total number of bristles divided by the number of individuals.

$$\text{mean} = \overline{x} = \frac{[1 + 4(2) + 7(3) + 31(4) + 56(5) + 17(6) + 4(7)]}{(1 + 4 + 7 + 31 + 56 + 17 + 4)}$$

$$= {}^{564}/_{120} = 4.7 \text{ average number of bristle/individual}$$

The variance is useful for studying the distribution of measurements around the mean and is defined in this example as

variance $= s^2 =$ average of the (actual bristle count $-$ mean)^2

$s^2 = {}^{1}/_{N} \Sigma (x_i - \overline{x})^2$
$ = 1/120\Sigma[(1 - 4.7)^2 + (2 - 4.7)^2 + (3 - 4.7)^2 + (4 - 4.7)^2 + (5 - 4.7)^2 + (6 - 4.7)^2 + (7 - 4.7)^2]$
$ = 0.26$

The standard deviation, another measurement of the distribution, is simply calculated as the square root of the variance.

standard deviation $= \quad s = \sqrt{0.26} = 0.51$

9. Suppose that the mean IQ in the United States is roughly 100 and the standard deviation is 15 points. People with IQs of 145 or higher are considered "geniuses" on some scales of measurement. What percentage of the population is expected to have an IQ of 145 or higher? In a country with 300 million people, how many geniuses are there expected to be?

Answer: Fifteen IQ points is equal to three standard deviations. Values that fall within $\pm$ three standard deviations comprise 99.7 percent of a normally distributed population. In this example that would include all people with IQs of 145 or higher and all those with IQs of 55 and lower. Half of those, or 0.15 percent, will have IQs or 145 or higher. In a country with 300 million people, this would equal 450,000 people.

10. A bean breeder is working with a population in which the mean number of pods per plant is 50 and the variance is 10 pods^2. The broad-sense heritability is known to be 0.8. Given this information, can the breeder be assured that the population will respond to selection for an increase in the number of pods per plant in the next generation?

Answer: Although a response to selection in the next generation is likely, the breeder cannot be assured of it. The response will depend on the type of genetic variation present. Broad-sense heritability consists of both additive and dominant genetic effects. If all genetic variation were due to additive effects, a strong response to selection would occur because the phenotype of selected individuals would correlate strongly with their genotypes and would be transmissable to the next generation. If there were no additive genetic variance (all genetic effects were due to dominant gene action) selected phenotypes would not be correlated with specific genotypes and would not be heritable. Variation due to dominant gene effects is not transmissible to the next generation. The actual response to selection would depend on the relative contributions of additive and dominant genetic effects, the D/A ratio.

CHALLENGING PROBLEMS

11. In a large herd of cattle, three different characters showing continuous distribution are measured, and the variances in the following table are calculated:

	Characters		
Variance	Shank length	Neck length	Fat content
Phenotypic	310.2	730.4	106.0
Environmental	248.1	292.2	53.0
Additive genetic	46.5	73.0	42.4
Dominance genetic	15.6	365.2	10.6

a. Calculate the broad- and narrow-sense heritabilities for each character.
b. In the population of animals studied, which character would respond best to selection? Why?
c. A project is undertaken to decrease mean fat content in the herd. The mean fat content is currently 10.5 percent. Animals with a mean of 6.5 percent fat content are interbred as parents of the next generation. What mean fat content can be expected in the descendants of these animals?

Answer:
a. Broad heritability measures that portion of the total variance that is due to genetic variance. The equation to use is

H^2 = the genetic variance/phenotypic variance

where genetic variance = phenotypic variance – environmental variance

$$H^2 = \frac{s_p^2 - s_e^2}{s_p^2}$$

Narrow heritability measures that portion of the total variance that is due to the additive genetic variation. The equation to use is

$$h^2 = \frac{\text{additive genetic variance}}{\text{additive genetic variance} + \text{dominance variance} + \text{environmental variance}}$$

$$h^2 = \frac{s_a^2}{s_a^2 + s_d^2 + s_e^2}$$

Shank length:
$$H^2 = (310.2 - 248.1)/(310.2) = 0.200$$
$$h^2 = 46.5/(46.5 + 15.6 + 248.1) = 0.150$$

Neck length:
$$H^2 = (730.4 - 292.2/(730.4) = 0.600$$
$$h^2 = 73.0/(73.0 + 365.2 + 292.2) = 0.010$$

Fat content:
$$H^2 = (106.0 - 53.0)/(106.0) = 0.500$$
$$h^2 = 42.4/(42.4 + 10.6 + 53.0) = 0.400$$

b. The larger the value of h^2, the greater the difference between selected parents and the population as a whole, and the more that characteristic will respond to selection. Therefore, fat content would respond best to selection.

c. The formula needed is

$$\text{selection response} = h^2 \times \text{selection differential}$$

Therefore, selection response = $(0.400)(10.5\% - 6.5\%) = 1.6\%$ decrease in fat content, or 8.9% fat content.

12. Suppose that two triple heterozygotes A/a ; B/b ; C/c are crossed. Assume that the three loci are on different chromosomes.

   a. What proportions of the offspring are homozygous at one, two, and three loci, respectively?
   b. What proportions of the offspring carry 0, 1, 2, 3, 4, 5, and 6 alleles (represented by capital letters), respectively?

Answer:
   a. The probability of any gene being homozygous is 1/2 (e.g., for A: A/A or a/a), and the probability of being heterozygous (or not homozygous) is also 1/2. Thus, the probability for any one gene being homozygous while the other two are heterozygous is $(1/2)^3$. Because there are three ways for this to happen (homozygosity at A or at B or at C), the total probability is

$$p(\text{homozygous at 1 locus}) = 3(1/2)^3 = 3/8$$

The same logic can be applied to any two genes being homozygous

$$p(\text{homozygous at 2 loci}) = 3(1/2)^3 = 3/8$$

There are two ways for all three genes to be homozygous, so

$$p(\text{homozygous at 3 loci}) = (1/2)^3 = 1/8$$

b. $p(\text{0 capital letters}) = p(\text{all homozygous recessive}) = (1/4)^3 = 1/64$

$p(\text{1 capital letter}) = p(\text{1 heterozygote and 2 homozygous recessive})$
$$= 3(1/2)(1/4)(1/4) = 3/32$$

$p(\text{2 capital letters}) = p(\text{1 homozygous dominant and 2 homozygous recessive})$

or

$p(\text{2 heterozygotes and 1 homozygous recessive})$

$$= 3(1/4)^3 + 3(1/4)(1/2)^2 = 15/64$$

$p(\text{3 capital letters}) = p(\text{all heterozygous})$

or

$p(\text{1 homozygous dominant, 1 heterozygous, and 1 homozygous recessive})$

$$= (1/2)^3 + 6(1/4)(1/2)(1/4) = 10/32$$

$p(\text{4 capital letters}) = p(\text{2 homozygous dominant and 1 homozygous recessive})$

or

$p(\text{1 homozygous dominant and 2 heterozygous})$

$$= 3(1/4)^3 + 3(1/4)(1/2)^2 = 15/64$$

$p(\text{5 capital letters}) = p(\text{2 homozygous dominant and 1 heterozygote})$

$$= 3(1/4)^2(1/2) = 3/32$$

$p(\text{6 capital letters})\ \ = p(\text{all homozygous dominant}) = (1/4)^3 = 1/64$

13. In Problem 12, suppose that the average phenotypic effect of the three genotypes at the *A* locus is *A/A* = 3, *A/a* = 2, and *a/a* = 1 and that similar effects exist for the *B* and *C* loci. Moreover, suppose that the effects of loci add to one another. Calculate and graph the distribution of phenotypes in the population (assuming no environmental variance).

Answer: For three genes, there are a total of 27 genotypes that will occur in predictable proportions. For example, there are three genotypes that have two genes that are heterozygous and one gene that is homozygous recessive (*A/a* ; *B/b* ; *c/c, A/a* ; *b/b* ; *C/c, a/a* ; *B/b* ; *C/c*). The frequency of this combination is 3(1/2)(1/2)(1/4) = 3/16, and the phenotypic score is 3 + 3 + 1 = 7. For all the genotypes possible, the total distribution of phenotypic scores is as follows:

Score	Proportion
3	1/64
5	3/32
6	3/64
7	3/16
8	3/16
9	11/64
10	3/16
11	3/32
12	1/64

And the plot of these data will be

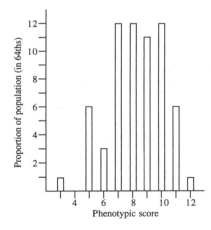

14. In a species of the Darwin's finches (*Geospiza fortis*), the narrow-sense heritability of bill depth has been estimated to be 0.79. Bill depth is correlated with the ability of the finches to eat large seeds. The mean bill depth for the population is 9.6 mm. A male with a bill depth of 10.8 mm is mated with a female with a bill depth of 9.8 mm. What is the expected value for bill depth for the offspring of this mating pair?

Answer: If the narrow-sense heritability (h^2) were 1.0, the expected value for the offspring would be midway between the two parent means. Since h^2 is less than one, it will be correspondingly closer to the population mean. The calculation for the expected bill depth is then:

$$\text{Female deviation} = \quad 10.8 - 9.6 = 1.2$$
$$\text{Male deviation} = \quad 9.8 - 9.6 = 0.2$$
$$\text{Mean deviation} = \quad 1.2 + 0.2 \, / \, 2 = 0.7$$

The expected deviation from the population mean with $h^2 = 1.0$ would be 0.7. With $h^2 = 0.79$ it would be $0.79 \times 0.7 = 0.55$. The expected mean for the offspring would be $9.6 + 0.55 = 10.15$.

15. The table below contains measurements of total serum cholesterol (mg/dl) for 10 sets of monozygotic twins who were reared apart. Calculate the following: overall mean, overall variance, covariance between the twins, broad-sense heritability (H^2).

X′	X″
228	222
186	152
204	220
142	185
226	210
217	190
207	226
185	213
179	159
170	129

Answer: The overall mean for serum cholesterol is the sum of all individual measures divided by the sample size of 20, which equals 192.5. The overall variance is the sum of the differences between the individual measures and the overall mean, squared, divided by the sample size. For example, in this data set: $(228 - 192.5)^2 + (186 - 192.5)^2 + \ldots + (159 - 192.5)^2 + (129 - 192.5)^2$. The variance for this data set equals 849.8. To calculate the covariance, the mean of the two sets of twins is calculated. In this case the mean for X′ is 194.4 and for X″ it is 190.6. For each pair of twins, the individual value for each is taken from the mean for that set of twins, then the two values are multiplied together. For the first set of twins this calculation would be: $(228 - 194.5)(222 - 190.6) = 1055.0$

The values for all sets of twins are:

X′	X″	
228	222	1055.0
186	152	324.2
204	220	282.2
142	185	293.4
226	210	613.0

217	190	−13.6
207	226	446.0
185	213	−210.6
179	159	486.6
170	129	1503.0

Mean/Total

194.4	190.6	4779.2

When this has been done for all 10 sets of twins, the values are totaled (4779.2) and divided by the total sample size, 20. This gives the covariance, which equals 239.0 for this data set. The broad-sense heritability is the covariance divided by the overall variance, which is 239.0 / 849.8 = 0.28.

16. Population A consists of 100 hens that are fully isogenic and that are reared in a uniform environment. The average weight of the eggs they lay is 52 g, and the variance is 3.5 g^2. Population B consists of 100 genetically variable hens that produce eggs with a mean weight of 52 g and a variance of 21.0 g^2. Population B is raised in an environment that is equivalent to that of Population A. What is the environmental variance (Ve) for egg weight? What is the genetic variance in Population B? What is the broad-sense heritability in Population B?

Answer: Population A is fully isogenic, so variation in this population is due to environmental factors. Therefore, environmental variance (V_e) is equal to 3.5 g^2. Population B has a total variance (V_x) of 21.0 g^2 under the same environmental conditions as population A. Since V_x is equal to $V_e + V_g$ (genetic variance) then $V_g = V_x - V_e$ or 21.0 − 3.5 = 17.5 g^2 for population B. Broad-sense heritability (H^2) for population B is V_g / V_x or 17.5 / 21.0 = 0.83.

17. A population of maize plants are on average 180 cm tall. Narrow-sense heritability for plant height in this population is 0.5. A breeder selects plants that are 10 cm taller on average than the population mean to produce the next generation, and the breeder continues applying this level of selection for eight generations. What will the average height of the plants be after eight generations of selection? Assume that h^2 remains 0.5 over the course of the experiment.

Answer: If narrow sense heritability were 1.0, the plants would be 10 cm taller each generation because the selected parents would produce offspring as tall as they were. Since it is only 0.5, only half the height difference between generations is heritable and the plants will be 5 cm taller each generation. After eight generations the plants will be 40 cm taller than in the initial generation, or 220 cm tall.

18. GWA studies reveal statistical correlations between the genotypes at marker loci in genes and complex traits. Do GWA studies prove that allelic variation in a gene actually causes the variation in the trait? If not, what experiments could prove that allelic variants in a gene in a population are responsible for variation in a trait?

Answer: Association studies do not prove that allelic variation in a gene causes variation in a trait. These studies assess association between phenotypic variation and particular SNPs. However, SNPs for several genes may be in strong linkage disequilibrium. Since the SNPs in linkage disequilibrium are inherited as a unit (as a haplotype block) any of the genes in the unit could be the actual cause of the phenotypic variation. The best way to establish that allelic variants in a specific gene are responsible for variation in a trait would be a molecular analysis in which the function of each alternate allele in a haplotype block was established. One could determine what the cellular function of each gene was and whether mRNA or functional proteins were produced by the two alternate alleles. This would help establish whether the gene function and variation in the function associated with an SNP could be linked to the trait in question. Another way would be to seek recombinants in which linkage disequilibrium was disrupted and find out whether the association between an SNP and the trait persisted.

19. The *ocular albinism-2* (*OCA2*) gene and the *melanocortin1-receptor* (*MC1R*) gene are both involved in melanin metabolism in skin cells in humans. To test whether variation at these genes contributes to sun sensitivity and the associated risk of being afflicted with skin cancer, you perform association analyses. A sample of 1000 people from Iceland were asked to classify themselves as having tanning or burning (nontanning) skin when exposed to the sun. The individuals were also genotyped for an SNP in each gene (rs7495174 and rs1805007). The table shows the number of individuals in each class.

	OCA2 (rs7495174)			*MC1R* (rs1805007)		
	A/A	A/G	G/G	C/C	C/T	T/T
Burning	245	56	1	192	89	21
Tanning	555	134	9	448	231	19

    a. What are the frequencies of tanning and burning phenotypes in Iceland?
    b. What are the allelic frequencies at each locus (SNP)?
    c. Using the χ^2 test (see Chapter 3) and Table 3-1, test the null hypothesis that there is no association between these SNPs and sun-sensitive skin. Does either SNP show evidence for an association?
    d. If you find evidence for an association between the gene and the trait, what is the mode of gene action?
    e. If the P value is greater than 0.05, does that prove that the gene does not contribute to variation for sun sensitivity? Why?

Answer: The frequencies of tanning and burning phenotypes are the total for each phenotype divided by the sample size. The frequency for burning would be (245 + 56 + 1) / 1000, which equals 302 / 1000 = 0.302. The frequency for tanning is 1 – 0.302 = 0.698. There are 2000 alleles in a population of 1000, with two copies in each homozygote and one copy in each heterozygote. The allelic frequencies for *OCA2* are A = (245 × 2) + (555 × 2) + 56 + 134 / 2000 = 0.895 and G = (9 × 2) + (1 × 2) + 56 + 134 / 2000 = 0.105. Following the same procedure for *MC1R*, the allele frequencies are C = 0.80 and T = 0.20.

A contingency chi-square would be used to test for non-random associations between the alleles and tendency to burn. First, the row and column totals would be calculated as shown below (the row totals are the same for both SNPs and are only included once).

	OCA2 (rs7495174)			*MC1R* (rs1805007)			
	A/A	A/G	G/G	C/C	C/T	T/T	Totals
Burning	245	56	1	192	89	21	302
Tanning	555	134	9	448	231	19	698
Total	800	190	10	640	320	40	1000

This table gives the observed values for each genotype/train combination. To calculate the expected values for each, the overall frequency of the genotype is multiplied by the overall frequency of the trait then by the population size, 1000. This gives the expected number of individuals with that combination if the two combine at random. For Burning A/A, the calculation would be .302 × .800 × 1000, or 241.6. Tanning A/A would be .698 × .800 × 1000 or 558.4. Calculation of the remaining values produces a table of expected outcomes shown below (note that the row and column totals have not changed).

	OCA2 (rs7495174)			*MC1R* (rs1805007)			
	A/A	A/G	G/G	C/C	C/T	T/T	Totals
Burning	241.6	57.4	3.0	193.3	96.6	12.1	302
Tanning	558.4	132.6	7.0	446.7	223.4	27.9	698
Total	800	190	10	640	320	40	1000

The next step is to calculate the deviation from expected for each cell by the formula $(O – E)^2 / E$. For the Burning A/A cell, this would be $(245 – 241.6)^2 / 241.6$, which equals 0.05. The next table contains the deviations for each cell, rounded to the nearest hundredth.

	OCA2 (rs7495174)			*MC1R* (rs1805007)		
	A/A	A/G	G/G	C/C	C/T	T/T
Burning	0.05	0.03	1.33	0.01	0.60	6.54
Tanning	0.02	0.02	0.57	0.00	0.26	2.84

Now the deviations from expected for each cell are added to obtain a chi-square for each SNP. The total chi-square for *OCA2* is 2.02. The df for this test are (columns – 1)(rows – 1) or 2. The critical value for a chi-square with 2df is

5.99, so the null hypothesis cannot be rejected with this data set. There is no evidence of nonrandom association between tendency to burn and *OCA2*. For *MC1R*, the total chi-square is 10.25, greater than the critical value of 5.99. In fact, a chi-square this large indicates a probability of less than 0.01: a difference at least this great would be obtained by random chance alone. The null hypothesis can be rejected. *MC1R* is significantly associated with variation in the tendency to burn. The data indicate that individuals with a tendency to burn are homozygous for the T allele more frequently than expected. Heterozygosity for T has no apparent effect, so the trait is recessive.

A P value of greater than 0.05 does not prove that the gene does not contribute. It means that a difference between expected and observed outcomes at least this great would be likely due to random chance alone. It does not mean that it was certainly due to by random chance alone.

20

Evolution of Genes and Traits

WORKING WITH THE FIGURES

1.   Examining Figure 20-4, explain why the rate of evolution at nonsynonymous sites is lower. Do you expect this to be true only of globin genes or of most genes?

   Answer: Both types of mutations are expected to arise at the same rate. However, synonymous substitutions do not change the amino acid sequence of a protein and generally have no phenotypic effect. They are therefore not subject to natural selection and are expected to accumulate freely. On the other hand, nonsynonymous substitutions do change the amino acid sequence and potentially have a wide range of affects on phenotype. These effects are subject to natural selection. Most nonsynonymous substitutions will be at least slightly harmful and selected against, reducing their frequency, while a few will be nearly neutral or advantageous; so nonsynonymous mutations will tend to accumulate much more slowly and be fixed at a lower rate than synonymous ones. This general tendency would not change from gene to gene, so the effect would be expected to occur for most genes.

2.   In Figure 20-7, the overall survival rates of three genotypes are plotted. Explain the reasons for the differences between the three survival curves.

   Answer: The decrease in survivorship common to all three genotypes could be due to any of many factors affecting early childhood mortality.   The differences between the genotypes can be explained as differences in resistance to infection by malaria parasites (*P. falciparum*), an important source of early childhood mortality in areas where malaria is common. The graph indicates that *SS* genotypes suffer the greatest mortality, likely due to the consequences of red blood cell sickling in homozygotes. Untreated, this is frequently a lethal condition. *AA* has the second highest mortality. *AA* homozygotes have no sickling but are subject to high childhood mortality from infection by malaria parasites. *AS* has the lowest mortality because heterozygotes are partially resistant to infection by the malarial parasite and do not exhibit harmful sickling of red blood cells. There are at least two possible explanations for a decrease in the difference between *AS* and *AA* as the children age. First, malaria mortality is highest among young children, who are more susceptible because their natural immunity has not yet developed. Second, *AA* genotypes that survive may have acquired immunity to infection and are less likely to be infected as adults. In areas where malaria is common, mortality from the disease drops dramatically as children age and is uncommon among adults.

3. From Table 20-4, would you expect the noncoding mutation g4205a to be fixed before or after the coding mutation G238S in a population of bacteria evolving resistance to the antibiotic cefotaxime? Give at least two reasons for your answer.

   Answer: The noncoding mutation g4205a would be fixed after the coding mutation G238S. First, the noncoding mutation had a positive effect on only 8 alleles, whereas the coding mutation had a positive effect on all 16. Also, the coding mutation had a positive effect independently of the state of other genes. This gives it a strong advantage. Second, the noncoding mutation had a negative effect on 2 other alleles, increasing its disadvantage relative to the coding gene. The strength of the advantage of the coding gene is seen in the mean proportional increase column—the increase in the coding mutation is almost 3 orders of magnitude greater than the noncoding gene.

4. Examining Table 20-5, what do you think would be the order of mutations fixed during selection in a third evolving virus line? Would the mutations become fixed in the same order as the TX or ID virus?

   Answer: The differences in the order of mutations fixed between the two different lines indicates that the order is not predictable—even though several of the fixed mutations were the same in the two lines, the mutations were not fixed in the same order. It is likely that the mutations in a third virus line would be fixed in a different order but involve at least some of the same mutations, for example those that were the same between TX and ID. It also seems likely that the mutations in a third line would involve the A and F proteins, since the fixed mutations commonly occur in those two.

5. Using Figure 20-17, explain how the mutation in the GATA sequence of the *Duffy* gene imparts resistance to *P. vivax* infection.

   Answer: The GATA sequence activates Duffy expression that is specific to red blood cells, which results in Duffy protein on the surface of red blood cells. The malarial parasite, *P.vivax,* invades red blood cells by using Duffy protein as a recognition site. The mutation GACA prevents binding of the GATA1 protein to the enhancer and prevents expression of Duffy protein on the red blood cells, but not on other cells. The lack of Duffy protein means that *P. vivax* are no longer able to recognize and infect the red blood cells, and resistance to malaria is conferred.

6.   In Figure 20-18, what is the evidence that polyploid formation has been important in plant evolution?

Answer: Polyploidy doubles the diploid chromosome number, which automatically produces an even haploid number.   In the absence of polyploidy, the haploid number would be expected to be randomly distributed between even and odd. The high frequency of even haploid chromosome numbers compared to odd numbers indicates that polyploidy is common and likely important in plant evolution.

BASIC PROBLEMS

7.   Compare Darwin's description of natural selection as quoted on page 730 with Wallace's description of the tendency of varieties to depart from the original type quoted on page 731. What ideas do they have in common?

Answer: The ideas of Darwin and Wallace were remarkably similar. Both authors recognized a struggle for existence and that survival and reproduction were not assured. Both also recognized the existence of variations that could confer a reproductive advantage. Finally, both realized that if beneficial variations were passed to the offspring they would increase in frequency over time.

8.   What are the three principles of the theory of evolution by natural selection?

Answer: The three principles are: (1) organisms within a species vary from one another, (2) the variation is heritable, and (3) different types leave different numbers of offspring in future generations.

9.   Why was the neutral theory of molecular evolution a revolutionary idea?

Answer: Before the neutral theory was developed, evolutionary biologists considered all change to be due to natural selection. In the absence of molecular data, mutations were thought of as beneficial or harmful, even if only slightly so, and so subject to natural selection. It was difficult to conceive that modification of a highly specific enzyme, for example, could be neutral. With a more complete understanding of the genetic code and the molecular basis for mutation and protein function came the surprising realization that neutral or nearly neutral mutations could arise. Neutral theory established the even more revolutionary idea that evolution of neutral genes by genetic drift was not only possible but common, and that most amino acid differences between species had not arisen through natural selection of adaptive variation but through genetic drift of neutral alleles.

10. What would you predict to be the relative rate of synonymous and nonsynonymous substitutions in a globin pseudogene?

Answer: A pseudogene is a nonfunctional duplication of an active gene. Because it is nonfunctional, changes in the pseudogene are not subject to natural selection—all amino acid substitutions are neutral. As a result, there should be no difference in the rate of synonymous and nonsynonymous substitutions in a globin pseudogene or any other pseudogene.

11. Are *AS* heterozygotes completely resistant to malarial infection? Explain the evidence for your answer.

Answer: *AS* heterozygotes are not completely resistant to malarial infection. The primary evidence for this is that *AS* heterozygotes have lower rates of infection than *AA* homozygotes, but still may be infected by the parasite. One study showed a 27.9 percent rate of infection for *AS* and a 45.7 percent rate for *AA*. Other indirect evidence indicates mortality among *AS* heterozygotes associated with the infection. Malaria is a primary source of early child mortality where it is present, so complete resistance would be expected to greatly increase survivorship for *AS* heterozygotes. However, survivorship curves between *AA* and *AS* indicate a relatively moderate difference. Additionally, the selective advantage for the *AS* genotype calculated using the Hardy-Weinberg equilibrium suggests a moderate advantage, inconsistent with complete resistance to a disease that is often lethal.

CHALLENGING PROBLEMS

Unpacking the Problem

12. If the mutation rate to a new allele is 10^{-5}, how large must isolated populations be to prevent chance differentiation among them in the frequency of this allele?

Answer: A population will not differentiate from other populations by local inbreeding if

$$\mu \geq 1/N$$

so

$$N \geq 1/\mu$$
$$N \geq 10^5$$

13.   Glucose-6-phosphate dehydrogenase (G6PD) is a critical enzyme involved in the metabolism of glucose, especially in red blood cells. Deficiencies in the enzyme are the most common human enzyme defect and occur at a high frequency in Tanzanian children.

    a.   Offer one hypothesis for the high incidence of G6PD mutations in Tanzanian children.
    b.   How would you test your hypothesis further?
    c.   Scores of different G6PD mutations affecting enzyme function have been found in human populations. Offer one explanation for the abundance of different G6PD mutations.

Answer:

a. One hypothesis is that G6PD mutations affect the metabolism of glucose in red blood cells in such a way that they reduce infection by malarial parasites, which are common in Tanzania and a significant source of childhood mortality. It is possible, for example, that malaria parasites cannot reproduce in a red blood cell deficient for glucose metabolism. It is also possible that G6PD deficiency modifies the cell in some way that prevents infection.

b. This hypothesis could be tested by comparing G6PD mutants with wild types and measuring the rate of infection by parasites. If the mutations confer resistance, infection rates should be lower.

c. The abundance of many G6PD mutations could be explained in at least two different ways. One is that any of a wide variety of mutations is capable of conferring resistance to parasitic infection and that different mutations have arisen and spread in different populations (or lineages). In this case, the variety of different mutations would be due to the randomness of the initial mutation, but their spread would be due to a selective advantage. It would be predicted from this hypothesis that G6PD variation would be greatest in populations with higher rates of parasitic infection. Another explanation is that existing G6PD mutations have a weak effect on fitness, or are neutral— G6PD metabolic function may not be particularly sensitive to amino acid changes. This seems unlikely given the importance of G6PD in glucose metabolism, but some G6PD mutations may have minimal effects on enzyme activity, or there may be other enzymes or biochemical pathways that can compensate for reduction of G6PD activity. In this case, the various mutants would have spread by drift, according to the neutral theory of evolution, and G6PD variation would be widespread, rather than limited to certain populations. Of course, both of these means could contribute to the existing variety of G6PD mutations.

14.   Large differences in *HbS* frequencies among Kenyan and Ugandan tribes had been noted in surveys conducted by researchers other than Tony Allison. These researchers offered alternative explanations different from the malarial

linkage proposed by Allison. Offer one counterargument to, or experimental test for, the following alternative hypotheses:

a. The mutation rate is higher in certain tribes.

b. There is a low degree of genetic mixing among tribes, so the allele rose to high frequency through inbreeding in certain tribes.

Answer:

a. That mutation rates are higher in some tribes than others is very unlikely, but this could be tested by comparing the synonymous substitutions in the hemoglobin gene among the different tribes. This has the advantage of targeting the hemoglobin molecule specifically, but the number of substitutions in hemoglobin might be too low to be informative. A better alternative would be to measure mutation rates in noncoding DNA sequences such as intergenic spacers or introns. The most useful sequences would be those near or within the hemoglobin gene.

b. The S allele is common in geographical areas that have a high incidence of malaria and much less common in areas without malaria, which argues against the inbreeding hypothesis. However, the hypothesis could be tested directly. The inbreeding hypothesis predicts that areas of high frequencies of the S allele (and areas of high malaria) should be areas of high inbreeding. A molecular analysis of neutral variation (at several loci) for different tribes would allow an estimate of the extent of inbreeding within the different populations and provide a test of the inbreeding hypothesis.

15. How many potential evolutionary paths are there for an allele to evolve six different mutations? Seven different mutations? Ten different mutations?

Answer: The number of potential evolutionary pathways is n!, where n is the number of different mutations. For six mutations it would be 6! or $6 \times 5 \times 4 \times 3 \times 2 \times 1 = 720$. For seven mutations, it would be 5,040, and for ten mutations it would be 3,628,800.

16. The *MC1R* gene affects skin and hair color in humans. There are at least 13 polymorphisms of the gene in European and Asian populations, 10 of which are nonsynonymous. In Africans, there are at least 5 polymorphisms of the gene, none of which are nonsynonymous. What might be one explanation for the differences in *MC1R* variation between Africans and non-Africans?

Answer: When amino acid changes have been driven by positive adaptive selection, there should be an excess of nonsynonymous changes. The *MC1R* gene (melanocortin 1 receptor) encodes a key protein controlling the amount of melanin in skin and hair. Asian and European populations appear to have experienced positive adaptive selection for more lightly pigmented skin relative to their African counterparts. By contrast, since high levels of melanin protect from damaging UV rays, purifying selection is likely to have maintained the sequence of *MC1R* in Africans.

17.   Opsin proteins detect light in photoreceptor cells of the eye and are required for color vision. The nocturnal owl monkey, the nocturnal bush baby, and the subterranean blind mole rat have different mutations in an opsin gene that render it nonfunctional. Explain why all three species can tolerate mutations in this gene that operates in most other mammals.

Answer: Both the owl monkey and bush baby are nocturnal while the subterranean mole rat is blind. Genes necessary for color vision are therefore not necessary for these three species. When species shift their habitats or life styles, mutations that cause inactivation of genes whose functions are no longer necessary is expected.

18.   Full or partial limblessness has evolved many times in vertebrates (snakes, lizards, manatees, whales). Do you expect the mutations that occurred in the evolution of limblessness to be in the coding or noncoding sequences of toolkit genes? Why?

Answer: Noncoding sequences. A major constraint on gene evolution is the potential pleiotropic effects of mutations in coding regions. These can be circumvented by mutations in regulatory sequences which play a major role in the evolution of body form. Changes in noncoding sequences provide a mechanism for altering one aspect of gene expression while preserving the role of pleiotropic proteins in other essential developmental processes.

19.   Several *Drosophila* species with unspotted wings are descended from a spotted ancestor. Would you predict the loss of spot formation to entail coding or noncoding changes in pigmentation genes? How would you test which is the case?

Answer: Noncoding changes. The difference in pigment expression between spotted and unspotted species of *Drosophila* is more likely due to the differences in cis-acting regulatory sequences that regulate the pigment gene. Mutations in coding regions might be expected to be highly pleiotropic. To test this, the activity of the cis-acting regulatory sequences from the different species could be tested by placing them upstream of a reporter gene and then placing the constructs into *D. melanogaster*. The regulatory sequences from a spotted species should drive high levels of reporter gene expression in just a spot while those from unspotted species should drive low-level expression of the reporter gene across the whole wing blade.

20. It has been claimed here that "evolution repeats itself." What is the evidence for this claim from

  a. the analysis of *HbS* alleles?
  b. the analysis of antibiotic resistance in bacteria?
  c. the analysis of experimentally selected bacteriophage φX174?
  d. the analysis of *Oca2* mutations in cave fish?
  e. the analysis of stickleback *Pitx1* loci?

Answer:

a. The analysis of *HbS* alleles shows that this allele has arisen independently five times in areas where malaria is common. In this case, evolution has repeated itself.

b. The analysis of antibiotic resistance in bacteria indicates that only a small proportion of the potential evolutionary pathways to resistance will actually be used. This reduces the number of available pathways and increases the chances of the same pathway occurring twice, but it also shows that many different pathways still remain.

c. The experimentally selected bacteriophage had several of the same substitutions, but they occurred in different orders. Also several substitutions were different. In cases like this, the pathway differs even if the outcome is the same, and evolution cannot be considered to have repeated itself.

d. Lack of pigmentation in different populations of cave fish was found to be due to mutations (deletions) in the *Oca2* gene, but the deletions were in different places. Again, the outcome was the same but the process differed so evolution did not repeat.

e. Loss of spines due to deletions in the *Pitx1* locus, a regulatory sequence, has evolved independently in many populations of stickleback fish. However, the size and location of the deletions differed in different populations.

The overall pattern is that evolution can repeat itself in a general way, but rarely follows exactly the same process. For example an evolutionary result is unlikely to occur in the same way when multiple mutations are involved—when the evolutionary pathway is complex. When the evolutionary pathway is simpler, involving only one gene, it is more likely that mutations in that gene will arise repeatedly and spread when they confer an advantage.

21.   What is the molecular evidence that natural selection includes the "rejection of injurious change"?

Answer: Purifying selection, or the "rejection of injurious variations," as Darwin termed it, explains why many protein sequences are unchanged, or nearly so, over vast spans of evolutionary time. A lower-than-expected ratio of nonsynonymous to synonymous changes is a signature of purifying selection. For example, a set of about 500 genes that exist is all Domains-of-life encode proteins whose sequences have been largely conserved over three billion years of evolution. To preserve such sequences, variants that have arisen at random in individuals in tens of millions of species have been rejected by selection over and over again.

22.   What are three alternative fates of a new gene duplicate?

Answer: First, a new gene duplication can become a pseudogene through complete inactivation. Pseudogenes have no function and are invisible to natural selection, so these genes will accumulate mutations relatively rapidly. Second, the duplicate can remain active and a new or modified function for the protein can evolve (neofunctionalization). An example of this is seen in human hemoglobin coding genes. Finally, the duplicate can become subfunctionalized, in which the original gene function is split between the original gene and the duplicate. The evolution of regulatory sequences may facilitate this path.

23.   What is the evidence that gene duplication has been the source of the α and ß gene families for human hemoglobin?

Answer: The α and ß gene families show remarkable amino acid sequence similarities (see table in the companion text). Within each gene family, sequence similarities are greater and, in some cases, member genes have identical intron-exon structure.

24.   DNA-sequencing studies for a gene in two closely related species produce the following numbers of sites that vary:

Synonymous polymorphisms	50
Nonsynonymous species differences	2
Synonymous species differences	18
Nonsynonymous polymorphisms	20

Does this result support neutral evolution of the gene? Does it support an adaptive replacement of amino acids? What explanation would you offer for the observations?

Answer: For polymorphic sites within a species, let nonsynonymous = a and synonymous = b. For polymorphic sites between the species, let nonsynonymous = c and synonymous = d. If divergence is due to neutral evolution, then

$a/b = c/d$

If divergence is due to selection, then

$a/b < c/d$

However, in this example, a/b = $20/50$ > c/d = $2/18$, which fits neither expectation.

Because the ratio of nonsynonymous to synonymous polymorphisms (a/b) is relatively high, the gene being studied may encode a protein tolerant of substitution (like fibrinopeptides, discussed in the companion text). The relatively fewer species differences may suggest that speciation was a recent event, so few polymorphisms have been fixed in one species that are not variants in the other.

25. In humans, two genes encoding the opsin visual pigments that are sensitive to green and red wavelengths of light are found adjacent to one another on the X chromosome. They encode proteins that are 96 percent identical. Nonprimate mammals possess just one gene encoding an opsin sensitive to the red/green wavelength.

a. Offer one explanation for the presence of the two opsin genes on the human X chromosome.

b. How would you test your explanation further and pinpoint when in evolutionary history the second gene arose?

Answer:
a. One way the two genes could arise is a duplication of the opsin gene sometime after the divergence of primates from other mammals. Since both of these genes are functional, this would be an example of neofunctionalization. Another way would be transposition, through the movement of a transposable element. A third way would involve retrotransposition.

b. A molecular analysis of the genes could differentiate between these three alternatives. The three mechanisms for existence of the two genes would have very different molecular characteristics. A direct duplication would produce an identical copy of the gene. Transposable elements have characteristic sequences that could be detected, either by DNA sequencing or analysis by polymerase chain reaction. Retrotransposition produces a duplicate that lacks introns. The timing of this event could be roughly estimated by looking for the two duplicate genes in other primates. The order of divergence of the primates is known, so the timing of the duplication could be inferred from its first appearance in other primates. A comparison of synonymous substitutions in the two genes would also provide information about the timing of the duplication—most would be the same, but the different ones would have occurred since the duplication. A greater number would indicate a more distant divergence.

26.   About 9 percent of Caucasian males are color-blind and cannot distinguish red-colored from green-colored objects.

   a.   Offer one genetic model for color blindness.
   b.   Explain why and how color blindness has reached a frequency of 9 percent in this population.

Answer:
a. Color blindness could result from a mutation in any of several steps involved in the perception of color. One would be a mutation in a gene controlling light absorbing pigments in photoreceptors, so that both "red" cones and "green" cones respond to the same wavelength of light. Another could be a mutation in the way the information is processed by the nervous system, so that "red" and "green" are processed the same way and become indistinguishable.

b. Color blindness may be evolutionarily neutral. In that case its frequency of 9 percent would have been achieved by drift. However, it is also possible that inability to distinguish certain colors leads to greater ability to distinguish among others, or to greater visual acuity. Effectively neutralizing some cones, for example, may place a greater emphasis on rods, which are capable of greater acuity.